Paper-Based Functional Materials

Advanced Chemical Products and Materials

Paper-Based Functional Materials: Preparation and Applications, First Edition, 2026. Xianping Yao, Genrong Chen, Zhen Yao, and Liping Zheng

Paper-Based Functional Materials

Preparation and Applications

Xianping Yao
Genrong Chen
Zhen Yao
Liping Zheng

WILEY VCH

CIP

Authors

Prof. Xianping Yao
Hangzhou Res. Inst. of Chem. Industry
Hangzhou, 310014
China

Prof. Genrong Chen
Hangzhou Res. Inst. of Chem. Industry
Hangzhou, 310014
China

Prof. Zhen Yao
Hangzhou Res. Inst. of Chem. Industry
Hangzhou, 310014
China

Prof. Liping Zheng
Hangzhou Res. Inst. of Chem. Industry
Hangzhou, 310014
China

Cover Image: © Roland Birke/Getty Images

Typesetting: Straive, Chennai, India
Printing and Binding: CPI Group (UK) Ltd, Croydon, CR0 4YY

Library of Congress Card No.: applied for

British Library Cataloguing-in-Publication Data A catalogue record for this book is available from the British Library.

Bibliographic information published by the Deutsche Nationalbibliothek
The Deutsche Nationalbibliothek lists this publication in the Deutsche Nationalbibliografie; detailed bibliographic data are available on the Internet at <http://dnb.d-nb.de>.

Print ISBN: 978-3-527-35352-1
ePDF ISBN: 978-3-527-84609-2
ePub ISBN: 978-3-527-84610-8
oBook ISBN: 978-3-527-84611-5

Published by WILEY-VCH GmbH, Boschstr. 12, 69469 Weinheim, Germany, under exclusive license granted by Chemical Industry Press Co., Ltd.

C9783527353521_050326

Contents

About the Author *xi*
Preface *xiii*

1 Introduction *1*
1.1 Definition of Paper-based Functional Materials *1*
1.2 Raw Materials of Paper-based Functional Materials *2*
1.2.1 Natural Polymers *3*
1.2.2 Inorganic Minerals *4*
1.2.3 Petroleum-based Organic Polymers *4*
1.3 Properties and Applications of Paper-based Functional Materials *5*
1.4 Economic and Social Value of Paper-based Functional Materials *6*
1.5 Development Trends of Paper-based Functional Materials *8*
1.5.1 Development of New Technologies *8*
1.5.2 Development of New Products *8*
References *10*

2 High-performance Starch Paper-based Functional Materials *11*
2.1 Overview *11*
2.1.1 Starch and Starch-based Functional Products *11*
2.1.2 Types and Functions of Starch-based Functional Products for Papermaking *13*
2.1.3 Application of Starch-based Functional Products in Papermaking *13*
2.1.4 The Significance of Starch-based Functional Products in Maintaining Forest Resources and Energy Conservation and Emissions Reduction *15*
2.2 Preparation Methods of Starch-based Functional Products *18*
2.2.1 Wet-processing Technology *18*
2.2.2 Dry Process *19*
2.2.3 Semi-dry Process *20*
2.2.4 Continuous Flow and Pipeline-based Green Manufacturing New Process *20*

2.3 High-performance Starch-based Functional Products for Papermaking *24*
2.3.1 Special Modified Starch for Non-wood Fiber *24*
2.3.2 High-retention Surface-sizing Starch *27*
2.3.3 Interference-resistant Specialized Enhancer for Recycled Paper *32*
2.3.4 Environmentally Friendly Paper Coating Adhesive *33*
2.3.5 Specialized Functional Starch for Heat Transfer Printing Paper *39*
2.3.6 High-substitution High-molecular-weight Cationic Starch *43*
2.4 Application Technologies of Starch-based Functional Products in Papermaking *46*
2.4.1 Wet-end Application Technology *47*
2.4.2 Surface-sizing Technology *49*
2.4.3 Interlayer Application Technology *53*
2.4.4 Coating Application Technology *57*
2.4.5 Relevant Green Technology Applications and Equipment *59*
2.4.6 Comprehensive Economic and Technical Evaluation *64*
2.5 Outlook and Recommendations *65*
References *67*

3 Functional Materials Based on Nanocellulose Paper *69*
3.1 Overview *69*
3.1.1 Classification of Nanocellulose *71*
3.1.2 Cellulose Nanofibrils - CMF (or MFC) *73*
3.2 Preparation and Characterization of Nanocellulose *74*
3.2.1 Preparation *74*
3.2.2 Modification Techniques *79*
3.2.3 Main Characterization Methods *80*
3.3 Functional Application of Nanocellulose in PaperMaking *82*
3.3.1 Paper Reinforcement *83*
3.3.2 Paper Filling *86*
3.3.3 Paper Barrier *87*
3.3.4 Other Applications *91*
3.4 Existing Problems and Prospects *93*
3.4.1 Existing Problem *93*
3.4.2 Prospect *94*
References *95*

4 High-performance Fiber Paper-based Functional Materials *97*
4.1 Introduction *97*
4.2 Typical High-performance Fiber *97*
4.2.1 High-performance Organic Fibers *98*
4.2.2 High-performance Inorganic Fiber *104*
4.3 Wet-laid Forming Technology of High-performance Fiber Paper *111*
4.3.1 Challenges in Preparing High-performance Fiber Paper *112*
4.3.2 Key Technologies *112*
4.4 Typical High-performance Fiber-based Functional Paper Materials *122*

4.4.1 Aramid Fiber-based Electrical Insulation Paper *123*
4.4.2 Aramid/ Mica Insulation Paper *125*
4.4.3 Lightweight and High-strength Para-aramid Honeycomb Core Material *127*
4.4.4 High-temperature-resistant Polyimide Fiber Paper-based Functional Materials *128*
References *129*

5 Long-fiber Filter Paper-based Functional Materials *133*
5.1 Long Fibers for Filter Paper-based Functional Materials *133*
5.1.1 Overview *133*
5.1.2 Classification and Characteristics of Long Fibers *133*
5.2 Mechanisms of Filtration *140*
5.2.1 Air Filtration *140*
5.2.2 Liquid Filtration *143*
5.3 Applications of Filter Paper-based Functional Materials *145*
5.3.1 Paper-based Functional Materials for Engine Filtration *145*
5.3.2 Filter Paper-based Functional Materials for Analysis and Testing *153*
5.3.3 Paper-based Functional Materials for Tea Filtration *159*
References *163*

6 Hydrophobic/Oleophobic/Hydrophilic/Lipophilic Paper-based Functional Materials *165*
6.1 Introduction to the Mechanism of Surface Wettability of Paper-based Materials *165*
6.1.1 Wetting Model of the Surface of the Object *165*
6.1.2 Definition of Different Wetting Interfaces *167*
6.2 Hydrophobic and Oleophobic Paper-based Functional Materials *169*
6.2.1 Preparation *169*
6.2.2 Application *175*
6.3 Hydrophobic and Lipophilic Paper-based Functional Materials *177*
6.3.1 Preparation *177*
6.3.2 Application *178*
6.4 Hydrophilic and Lipophilic Paper-based Functional Materials *180*
6.4.1 Preparation *180*
6.4.2 Application *185*
6.5 Hydrophilic and Oleophobic Paper-based Functional Materials *188*
6.5.1 Preparation *188*
6.5.2 Application *188*
6.6 Prospect *192*
References *193*

7 Flame-retardant and Heat Insulation Paper-based Functional Materials *199*
7.1 Flame-retardant Paper-based Functional Materials *199*
7.1.1 Classification of Flame Retardants for Paper-based Functional Materials *199*

7.1.2 Flame-retardant Mechanism of Paper-based Functional Materials *203*
7.1.3 Preparation of Flame-retardant Paper-based Functional Materials *205*
7.1.4 Application of Flame-retardant Paper-based Functional Materials *208*
7.1.5 Prospect *211*
7.2 Thermal Insulation Paper-based Functional Materials *212*
7.2.1 Classification of Thermal Insulation Paper-based Functional Materials *212*
7.2.2 Thermal Transfer Principle of Paper-based Functional Materials *213*
7.2.3 Preparation of Thermal Insulation Paper-based Functional Materials *214*
7.2.4 Applications of Thermal Insulation Paper-based Functional Materials *216*
References *218*

8 Paper-based Functional Materials for Filtration and Separation *221*
8.1 Overview *221*
8.2 Fundamentals of Filtration and Separation *224*
8.2.1 Gas Filtration and Separation *225*
8.2.2 Liquid Filtration and Separation *226*
8.3 Preparation of Paper-based Functional Materials for Filtration and Separation *229*
8.3.1 Preparation of Wet-laid Nonwoven Fabric *230*
8.3.2 Nonwoven Fabric Composite Technology *234*
8.4 Applications of Paper-based Functional Materials for Filtration and Separation *238*
8.4.1 Automotive Filtration *238*
8.4.2 High-efficiency Air Filtration *239*
8.4.3 Water Filtration *240*
8.4.4 Battery Separator Paper *241*
8.4.5 Face Masks *242*
References *243*

9 Paper-based Functional Materials for Analysis and Detection *245*
9.1 Development of Paper-based Functional Materials for Analysis and Detection *245*
9.2 Fabrication Methods for Paper-based Analytical Detection Chips *249*
9.2.1 Physical Barrier/Deposition Method *251*
9.2.2 Chemical Modification Method *253*
9.2.3 Paper Modification Method *255*
9.3 Detection Methods of Paper-based Functional Materials for Analysis and Detection *255*
9.3.1 Electrochemical Method *256*
9.3.2 Chemiluminescence Method *256*

9.3.3 Electrochemiluminescence Method *257*
9.3.4 Fluorescence Method *257*
9.3.5 Colorimetric Method *258*
9.4 Applications of Paper-based Functional Materials for Analysis and Detection *258*
9.4.1 Application of Paper-based Detection Chips in Medical Diagnosis *258*
9.4.2 Application of Paper-based Detection Chips in Environmental Monitoring *260*
9.4.3 Application of Paper-based Detection Chips in Food Safety Control *261*
9.5 Application Cases of Paper-based Functional Materials for Analysis and Detection *262*
9.5.1 Optimization and Regulation of Blood Typing Paper-chip and Paper-based Structure *263*
9.5.2 Paper-based Chip for Measuring Heavy Metal Ions in Water Based on Optical Transmittance Density *266*
9.5.3 Paper-based Chip for High-throughput Detection of Antibiotics in Milk *266*
9.5.4 Construction of Composite Substrate for Paper-based Chips and Optimization of Physical and Detection Performance *270*
9.6 Future Prospects of Paper-based Functional Materials for Analysis and Detection *271*
References *272*

10 Other Paper-based Functional Materials *281*
10.1 Cream Paper-based Functional Materials *281*
10.1.1 Overview *281*
10.1.2 Preparation Technology and Principles of Cream *287*
10.1.3 Preparation of Cream Paper *293*
10.1.4 Outlook *294*
10.2 Photocatalytic Paper-based Functional Materials *294*
10.2.1 Overview of Photocatalytic Materials *294*
10.2.2 Nano TiO_2 Photocatalytic Material *297*
10.2.3 Photocatalytic Paper-based Functional Materials Application Research *300*
10.2.4 Prospect *307*
10.3 Basic Functional Materials for Fruit and Vegetable Preservation Paper *307*
10.3.1 Overview *307*
10.3.2 Types and Characteristics *312*
10.3.3 Preparation Technology *324*
10.3.4 Outlook *328*
10.4 Electromagnetic Shielding Paper-based Functional Materials *328*
10.4.1 Overview *328*
10.4.2 Shielding Mechanism of Electromagnetic Shielding Materials *329*

10.4.3 Classification and Development of Electromagnetic Shielding Materials *330*
10.4.4 Research Status of Electromagnetic Shielding Paper-based Functional Materials at Home and Abroad *332*
10.4.5 Preparation Methods and Functional Characteristics of Electromagnetic Shielding Paper-based Functional Materials *333*
10.4.6 Applications of Electromagnetic Shielding Paper-based Functional Materials *336*
10.4.7 Outlook *338*
10.5 Paper-based Functional Materials for Decorative Base Paper *339*
10.5.1 Overview *339*
10.5.2 The Quality Requirements of Decorative Base Paper *341*
10.5.3 Functional Materials and Characteristics of Decorative Base Paper *342*
10.5.4 The Application of Functional Materials in Decorative Base Paper *347*
10.5.5 Outlook *350*
10.6 Thermal-sensitive Paper *351*
10.6.1 Introduction *351*
10.6.2 Quality and Requirements *353*
10.6.3 Thermal Functional Materials *354*
10.6.4 Production Process *358*
10.6.5 Prospect of Industry *359*
References *360*

Index *367*

About the Author

Yao Xianping is a professor-level senior engineer and an academician of the Russian Academy of Natural Sciences. He is the director of the Hangzhou Research Institute of Chemical Industry and the director of the National Engineering & Technology Research Center for Paper Chemicals. He has been engaged in the research, development and industrialization of paper chemicals, starch derivatives, and biomass-based functional materials for a long time.

He has led the development of over 80 new products, including 10 national key new products. He has won one Second Prize of the National Science and Technology Progress Awards, three First Prizes of Provincial and Ministerial-level Prize for Progress in Science and Technology, one First Prize of Provincial and Ministerial-level Award for Technological Invention, and over 20 other awards. He holds 22 invention patents and has won two China Patent Excellence Awards. He has published over 130 theses and authored six books. All his achievements have been industrialized, generating a total of over 10 billion yuan in revenue from technology transfer, 1.5 billion yuan in profits and taxes, and 800 million yuan in national tax revenue. The products have been used in over 100 million tons of paper and paperboard, achieving significant economic and social benefits. He is one of the top 10 leaders in China's paper industry.

Preface

In an era of rapid technological advancement, paper-based functional materials have emerged as a field of interdisciplinary research, bridging materials science, chemistry, biology, and engineering. Traditionally known primarily for its writing and packaging purposes, paper has undergone a remarkable transformation. Thanks to innovative manufacturing techniques and the incorporation of functional additives, modern paper-based functional materials exhibit a diverse array of unique properties, such as exceptional mechanical strength, superior conductivity, remarkable optical characteristics, and outstanding biocompatibility. These attributes have propelled paper-based functional materials into various applications, ranging from flexible electronics and energy storage devices to sensors, drug delivery systems, and environmental remediation, heralding a new chapter in materials science.

This book aims to provide a comprehensive and in-depth exploration of paper-based functional materials. As this field rapidly expands, there is an urgent need for a coherent resource that integrates the latest research findings, technological advancements, and application prospects. By compiling the collective expertise and cutting-edge research in this domain, we seek to offer a valuable reference for researchers, engineers, students, and industry professionals alike. This book not only facilitates an understanding of the fundamental principles and synthesis methods of paper-based functional materials but also inspires further innovation and development, driving the translation of laboratory research into practical applications.

The writing of this book follows a logical and progressive structure. It begins by introducing the historical development and classification of paper-based functional materials, laying a solid foundation for subsequent discussions. The book then delves into the fundamental principles of various functionalization approaches, covering high-performance starch-based paper materials, nanocellulose-based paper materials, high-performance fiber-based paper materials, long-fiber filtration paper-based materials, hydrophobic/oleophobic/hydrophilic/oleophilic paper-based materials, flame-retardant and thermal-insulating paper-based materials, filtration and separation paper-based materials, and paper-based analytical and detection materials, among others.

This book is authored by Xianping Yao and others, with Chapter 1 written by Xianping Yao; Chapter 2 by Xianping Yao, Zhen Yao and Liping Zheng; Chapter 3 by Xianping Yao, Jianmei Jiang, and Huang Huang; Chapter 4 by Meiyun Zhang and Shunxi Song; Chapter 5 by Xiangrong Yao and Lan Shao; Chapter 6 by Haisong Wang; Chapter 7 by Lizheng Sha, Daliang Guo and Yanjun Tang; Chapter 8 by Ziyang Chang; Chapter 9 by Lizi Li and Junfei Tian; Chapter 10 by Juan Li, Chaoyu Zhu, Hui Chen, You Zhou, Lifeng Wang, Xi Chen, Shasha Zhu and Peiwen Bian, and the entire book is coordinated by Xianping Yao and Shasha Zhu.

Due to the limited expertise of the authors, there may be omissions or errors in the manuscript, and we welcome readers to provide suggestions.

Xianping Yao
Hangzhou, February 2026

1 Introduction

As modern technology advances, the demand for advanced functional materials continues to grow. However, most functional materials have significant drawbacks, such as high costs and complex manufacturing processes, which make it challenging to meet the increasing demands of the application market. In this context, paper-based functional materials have emerged as a key focus of cutting-edge technology development. Compared to traditional functional materials, paper-based functional materials offer several advantages, including strong functional properties, a wide range of applications, ease of production, low cost, biodegradability, and recyclability. These benefits make paper-based functional materials a preferred choice for future technological development.

1.1 Definition of Paper-based Functional Materials

Paper-based functional materials are advanced composite materials made primarily from plant fibers. By incorporating various functional materials—such as inorganic mineral fibers and high-performance fibers (e.g., aramid fibers, carbon fibers, and polyimide fibers)—and using modern papermaking techniques, these materials form a three-dimensional network structure with specific properties and high added value [1]. As a key component in the production of specialty paper and paperboard, the functional characteristics of these materials depend significantly on the high-performance fibers or inorganic mineral fibers added during their manufacture [2].

In paper-based functional materials, the paper serves as the substrate, with various functional materials loaded onto it [3]. These functional materials mainly include modified natural polymers like starch and fibers, synthetic organic polymers, and inorganic mineral materials, collectively known as papermaking chemicals. For many years, papermaking chemicals have been recognized as new functional materials and are included in the national catalog of industries, products, and technologies encouraged for development. Papermaking chemicals have two primary applications [4]. First, the chemicals include high-performance starch derivatives, water-soluble polymers, and newly developed nanocellulose that are used for the enhancement of ordinary paper (or paperboard). They are used to improve paper

Paper-Based Functional Materials: Preparation and Applications, First Edition.
Xianping Yao, Genrong Chen, Zhen Yao, and Liping Zheng.

strength (dry strength, wet strength, and surface strength), retention, and filtration; reduce fiber and filler loss; and mitigate wastewater pollution. They also enhance paper properties like whiteness, smoothness, printability, softness, stiffness, tear resistance, water resistance, and breathability. Second, papermaking chemicals are applied in the production of specialty paper, which is designed for specific functions such as hydrophobicity/oleophobicity, hydrophilicity/oleophilicity, filtration/separation, flame retardancy, insulation, conductivity, luminescence, high barrier properties, detection and separation, construction and home decoration, fruit and vegetable preservation, electromagnetic shielding, dust and static resistance, corrosion and rust resistance, and antimicrobial and insect resistance. These papers have unique properties and are used in various specialized applications.

1.2 Raw Materials of Paper-based Functional Materials

The various functional materials incorporated into or composited with paper-based substrates can be categorized based on their raw material types into natural polymer materials, inorganic mineral materials, and organic polymer materials (Table 1.1).

Table 1.1 Major raw material composition of functional materials incorporated into (or composited with) paper-based substrates.

Classification	Raw materials	Functional products	Main raw material sources
Natural polymers	Starch	Starch-based biological products	Corn, potatoes, cassava, etc.
	Plant fibers	Nanofibers, cellulose-based biological products	Wood, bamboo, rattan, cotton, etc.
Inorganic minerals	Inorganic minerals	Functional materials such as conductive, magnetic, flame-retardant, reinforced, thermal insulation, wear-resistant, insulating, air purification, and barrier materials	Inorganic minerals, carbon-based materials, magnetic materials, rare earth materials, metallic materials, etc.
Petroleum-based organic polymers		Functional polymer materials, such as strengthening agents, retention aids, filtration aids, adhesives, defoamers, corrosion inhibitors, oil-resistant agents, water-resistant agents, softening agents, dispersants, and deinking agents	Petrochemical raw materials
		High-performance synthetic fibers	Aramid fibers, polyacrylonitrile fibers, polyester fibers, nylon fibers, vinylon, polypropylene fibers, etc.

1.2.1 Natural Polymers

Key raw materials include starch and cellulose, which are abundant renewable resources with strong compatibility with papermaking fibers and full biodegradability. Modified starches, produced through physical, chemical, or biological modifications, are essential in papermaking for enhancing strength, retention, filtration, surface sizing, coating adhesion, and interlayer bonding.

Starch-based functional materials are derived from natural starch that has been modified through physical, chemical, or biological methods to impart various functional properties. This kind of starch is commonly referred to as "modified starch." In the papermaking industry, modified starches include enzyme-converted starch, anionic starch (such as oxidized starch, phosphate starch, and carboxymethyl starch), cationic starch, amphoteric and multifunctional modified starch, cross-linked starch, hydroxyalkyl starch, composite modified starch, and grafted starch.

The benefits of these modified starches include low gelatinization temperature, good transparency and stability of the starch paste, strong affinity with fibers, excellent film-forming properties, and resistance to acids and alkalis. Modified starches are suitable for almost all types of paper and paperboard, typically used at a dosage of around 1%. They serve various functions such as strengthening, retention aid, filtration aid, surface sizing, coating adhesion, and interlayer bonding. These properties make modified starches indispensable in the papermaking industry [5], earning them the nickname "industrial MSG."

Nanocellulose consists of cellulose fibers with a one-dimensional size less than 100 nm, exhibiting unique properties characteristic of nanomaterials. These properties include size effects, quantum effects, surface effects, and macroscopic quantum tunneling effects, making their physical, chemical, and mechanical characteristics significantly different from those of macroscopic materials. As a promising new material for the "post-carbon fiber era," nanocellulose has the potential to replace metals and plastics in the future.

Nanocellulose is produced through chemical, physical, and biological methods, and its development and industrialization are still in the early stages globally. Nanocellulose and its modified products offer exceptional properties such as high strength, oil resistance, water resistance, and barrier functions, making them highly promising as biomass-based functional materials in the papermaking industry. Additionally, cellulose derivatives like carboxymethyl cellulose and hydroxypropyl cellulose are also important in papermaking coatings, contributing to improved paper properties [6].

Plant-based specialty long fibers primarily include bast fibers (such as hemp, mulberry bark, paper mulberry, and jute) and cotton fibers. Specialty papers made from these plant-based long fibers through papermaking processes exhibit high breathability and high strength, making them widely used as filtration materials for gases and liquids.

1.2.2 Inorganic Minerals

Common materials include talc, kaolin, and bentonite, which are used to enhance paper whiteness, smoothness, retention, and filtration. Other inorganic materials like mineral wool, expanded perlite, aerogel blankets, and foam glass serve as thermal insulators. Metal-based materials are often used for heat insulation, while flame retardants typically include inorganic phosphates, borates, metal hydroxides, oxides, and salts. Carbon-based materials, magnetic materials, rare earth materials, and metals also play significant roles in various functional papers.

1.2.3 Petroleum-based Organic Polymers

Derived from petrochemicals, these materials include superhydrophobic and superoleophobic functional papers used for self-cleaning, corrosion resistance, and oil transport. Common thermal insulation materials include expanded polystyrene, polyurethane sponge, cork, phenolic foam, and cellulose. These materials are typically processed through papermaking technologies to meet high-temperature and insulation requirements. Flame-retardant materials often involve halogen, phosphorus, and nitrogen compounds, and noncombustible fibers.

Paper-based functional materials derived from petrochemical raw materials come in various forms. For example, durable super-oleophobic and hydrophobic paper-based functional materials are primarily created by combining perfluoroalkanes and block copolymers with functional end groups (fluorinated monomers) through coating or similar methods. These super-repellent materials, which are both oleophobic and hydrophobic, are widely used in applications such as self-cleaning, corrosion resistance, oil transportation, antibiofouling devices, oil collection, antifouling, microdroplet transfer, and oil–water separation.

Common organic polymer thermal insulation materials include expanded polystyrene, polyurethane sponge, cork, phenolic foam, and cellulose. These materials are typically made using fibers as the main raw material and formed through papermaking techniques. They differ significantly in structure and properties from traditional paper-based materials and are often used in high-temperature fields such as aerospace and as outer protective layers for various insulation materials.

Organic polymer flame-retardant materials mainly include halogenated, phosphorus-based, and nitrogen-based compounds. Additionally, flame-resistant or noncombustible fibers can be used as raw materials in paper making processes to produce flame-retardant papers. These materials are widely used in construction, packaging, interior decoration, automotive filtration, cable wrapping, etc.

High-performance fiber paper-based functional materials are made from natural plant fibers mixed with specialty fibers (including inorganic and synthetic fibers). Through dispersion, formation, and enhancement techniques, these materials are processed to meet specific performance requirements, resulting in a range of paper-based functional materials with both structural and functional properties.

Typical high-performance synthetic fibers used include aromatic polyamide fibers (poly (p-phenylene terephthalamide)), ultra-high-molecular-weight polyethylene fibers, and polyimide fibers. These high-performance fiber paper-based materials are widely used in personal protection, bulletproof armor, rubber products, asbestos substitutes, automotive friction materials, high-grade insulation paper, and paper-based honeycomb structures and are essential in advanced fields such as aerospace, national defense, electronic communication, petrochemicals, and marine development.

1.3 Properties and Applications of Paper-based Functional Materials

Paper-based functional materials have a structure and properties that are completely different from traditional paper. They primarily exhibit mechanical properties; thermal characteristics; electrical, electronic, and magnetic properties; optical characteristics; other physical properties (such as adhesion, separation and filtration, water, oil, and gas adsorption properties); chemical properties; and biochemical properties. These materials overcome the limitations of single natural plant fiber materials, which cannot withstand harsh conditions such as high impact, high temperature, high humidity, and high corrosion. Paper-based functional materials are widely used in important fields such as industry, agriculture, aerospace, electronic information, healthcare, food safety, and transportation.

Starch-based functional materials are the most widely used biomass-based products in the papermaking industry, with the broadest range of applications and the most comprehensive functional roles. They have been widely adopted in China, with an annual consumption of over a million tons in the papermaking industry. These materials play an irreplaceable role in improving quality and performance, saving energy and reducing emissions, conserving wood resources, and protecting the ecological environment [7].

High-performance fiber paper-based functional materials possess excellent properties such as low density, high specific strength, high specific stiffness, and high-temperature resistance. They are strategic functional materials with significant applications in fields such as rail transportation, aerospace, and national defense technology. The preparation technology for these materials is internationally recognized as a challenging issue and has been tightly restricted by developed countries. There are several scientific problems and technological bottlenecks in key areas such as fiber preparation and dispersion, fluid delivery and forming, and hot pressing and reinforcement, which severely restrict the development of advanced domestic functional materials, such as insulation and structural weight reduction, required for high-speed trains and aircraft manufacturing.

Nanocellulose is lightweight, biodegradable, and derived from a wide range of sources. It has excellent properties, including high strength, high specific surface area, high polymerization degree, high crystallinity, and low thermal expansion coefficient. Compared to other nanomaterials, nanocellulose has advantages such as good biocompatibility, biodegradability, and renewability. More importantly, the surface of nanocellulose contains abundant hydroxyl groups, making it highly reactive. By introducing functional groups, the functionality of nanocellulose materials can be tailored, making them more adaptable to various application fields with different needs. Nanocellulose can be widely used in applications such as paper and paperboard reinforcement, barrier properties, anti-oil, waterproof materials, resin composites, lightweight high-strength automotive parts, fully degradable high-strength materials, filtration and separation materials, green degradable barrier packaging materials, and personal care materials. It is expected to elevate the traditional papermaking industry to a fully biodegradable functional materials industry.

The close integration of other organic polymers, inorganic materials, and metals with papermaking has made significant contributions to technological advancements in China's specialty paper sector, with huge development potential. For example, by adding specialty chemicals with hydrophobic, oleophobic, hydrophilic, or oleophilic properties, and using physical–chemical methods, chemical methods, surface coating, and nanoparticle deposition techniques, paper or fibers can be modified to create hydrophobic/oleophobic/hydrophilic/oleophilic paper-based functional materials. Through flame-retardant treatment, such as adding flame retardants during the papermaking process, the resulting flame-retardant paper-based materials exhibit noncombustibility and high-temperature resistance [8]. These materials are widely used in fields such as construction, packaging, interior decoration, automotive filtration, and cable insulation. They can also be used for the long-term safe preservation of books and important documents (e.g., archives and certificates). Thermal-insulation paper-based functional materials, primarily made from fibers, are characterized by thermal insulation performance, lightweight, loose, porous structure, and low thermal conductivity. Other typical paper-based functional materials, such as paper-based analytical detection chips, paper-based ceramic carriers, paper-based flexible electronic circuits, paper-based light-emitting and heating materials, and paper-based friction materials, also have vast application prospects across multiple sectors of the national economy.

1.4 Economic and Social Value of Paper-based Functional Materials

Paper-based functional materials represent an advanced new field of materials, expanding the application scope of functional materials while playing an increasingly important role in the technological advancement of papermaking and related industries. These materials have immense economic and social values, with far-reaching significance, particularly in the areas of specialty paper and papermaking chemicals.

On the one hand, papermaking chemicals can impart various special superior properties to paper, significantly improving paper quality, enhancing paper machine efficiency, promoting the development of new and more valuable paper varieties, and effectively conserving valuable papermaking resources (including wood, water, energy, etc.). They also substantially reduce environmental pollution in the papermaking process and have strong relevance in promoting the development of downstream industries such as news, publishing, printing, packaging, electronics, automotive, food, and tobacco. With the development of China's national economy, the papermaking industry is progressing toward larger and faster paper machines, neutral papermaking processes, and closed-loop whitewater recycling. However, it faces significant bottlenecks in resources, environment, and technology. Papermaking chemicals are key materials for overcoming these bottlenecks, playing an increasingly crucial role in improving quality, transforming, and upgrading the papermaking industry, saving energy, reducing emissions, conserving resources, and developing and applying high-grade and specialty papers. For example, a national science and technology project undertaken by Xianping Yao's research team developed a series of new products using starch as the main raw material and multistage modification technology, which have been widely adopted in China's papermaking industry. These products have been proven to increase paper strength by 15% to 25%, save 10% to 35% of pulp, reduce energy consumption by 5% to 15%, improve the retention rate of fine fibers and fillers by 10% to 40%, and decrease white water concentration by 20% to 30%.

On the other hand, specialty paper differs from traditional paper and paperboard in structure and performance, offering flexible design options for properties such as optical, electrical, magnetic, thermal, and acoustic characteristics. It is widely used in various fields that are closely related to people's lives, such as cultural industries, manufacturing, agriculture, and healthcare, and has important applications in national defense, aerospace, and rail transportation. Specialty paper is one of the country's important strategic materials, and its importance and value are self-evident. In terms of economic value, specialty paper is energy efficient, environmentally friendly, high profit, and rapidly developing. Although the demand for single products may not be large, the variety of products and their high economic value make them very significant. For instance, key technologies for enhancing paper-based functional materials using specialty fibers have led to the development of high-performance products such as nanoquantitative filter paper, nanoautomotive filters, flame-retardant steel paper, and waterproof insulating steel paper. In just a few years, production and sales have reached 35,000 tons, generating an output value of 700 million RMB and tax revenue of 70 million RMB.

Nanocellulose, prepared primarily from plant fibers used in papermaking, is a cutting-edge biomass material that not only exhibits the properties of nanomaterials but also has features such as lightweight, high strength, biodegradability, good biocompatibility, high thermal stability, and strong hydrophilicity. Its density is only one-fifth that of stainless steel, while its strength is more than five times greater, making it regarded as a new material for the "post-carbon fiber era." Nanocellulose has the potential to replace metals and plastics and is a global research hotspot. According to the USDA Forest Service, the global market potential is estimated at 35 million tons. It can be widely used in ordinary paper, specialty paper, fully degradable biomass materials, and health-related fields, and its lightweight and high-strength properties make it suitable for aerospace, rail transportation, and other major engineering sectors. The current international market price of nanocellulose is over 1 million RMB per ton, demonstrating its immense economic and social values. Countries around the world are paying close attention to its development, and its progress is rapid. However, the industry is still in the early stages of commercialization, facing issues such as immature technology, small scale, and high prices. China has listed nanocellulose as a key research project in the 13th Five-Year Plan and, in collaboration with the South China University of Technology, Hangzhou Mechanical and Electrical Design Institute, and other institutions, has established the country's first pilot-scale demonstration line for the green preparation of nanocellulose at the Hangzhou Chemical Research Institute. A series of high-value application studies have been conducted, and the commercialization and high-value applications of these results are expected, with promising prospects ahead.

1.5 Development Trends of Paper-based Functional Materials

The future development trends in the field of paper-based functional materials will primarily focus on two aspects: new processing technologies and new products. A brief introduction to the development of these two aspects is provided below.

1.5.1 Development of New Technologies

1) Foam forming technology
   Foam forming is a papermaking technology developed in the 1970s that allows the production of nonwoven materials with good uniformity, high bulk density, and high porosity on paper machines. Compared to current slow-flow web formation or wet-laid nonwoven fabric production platforms, the successful application of foam forming technology in commercial production enables paper machines to produce nonwoven substitutes at a lower cost. After a series of modern technological improvements, the new foam-forming system has been applied to the commercial production of specialty paper.
2) Stretchable paper technology
   Stretchable paper technology, developed by Finland's VTT company, primarily relies on mechanical processing to achieve 20% and 16% elongation in the longitudinal and lateral directions, respectively. Stretchable paper is a novel concept designed to replace plastic with natural paper, making it more sustainable, especially for single-use packaging. This paper can be supplied in rolls and processed on traditional lines used for polymers, including printing, coating, and other processes. Potential application areas include pallet packaging, medical packaging, paper cups and other liquid containers, and furniture decoration.

1.5.2 Development of New Products

1) Carbon-fiber-reinforced thermoplastic resin molding (paper-based thermoplastic resin composite material)

Developed by Japan's Awa Paper, this technology disperses carbon fibers (with a length of 3–6 mm) and thermoplastic resin fibers (such as polypropylene (PP) and polyethylene (PE)) in water, which are then processed into continuous fiber-reinforced thermoplastic composite (CFRTP) products through papermaking, lamination, and hot pressing. The strength of CFRTP is approximately five times that of general resins like PP and PE. Particularly, CFRTP made with polyamide fibers can achieve tensile strength equivalent to aluminum alloy "A5052" at the same volume ratio, but with only one-third to one-half of its weight.

2) Energy-storing paper
Developed by Swedish researchers, this paper, made of nanocellulose and conductive polymers, can store up to 1F of electrical energy in a sheet with a diameter of 15 cm and a thickness of just a few millimeters, similar to current supercapacitors on the market. This material can be charged and discharged hundreds of times, with each charging session taking just a few seconds. It can be used as a backup energy source during peak electricity demand and as a supercapacitor at electric vehicle charging stations.
3) New degradable paper mulch
Developed by Walki, a Finnish company, this new type of paper mulch uses kraft paper as the base material, coated with a degradable coating. The product provides excellent coverage, reduces weeds, and does not affect crop yields.
4) Porous titanium electrodes for water electrolysis
German scientists are attempting to use papermaking methods to create porous titanium electrodes for water electrolysis. The process involves adding a large amount of titanium powder during papermaking, resulting in paper with more than 75% titanium powder content. This paper is then sintered at high temperatures to remove organic components and sinter the metal powder into a porous electrode material. The thickness and pore structure can be adjusted by modifying the papermaking process. This electrode material provides an optimal, low-cost gas/liquid diffusion layer for polymer electrolyte membrane (PEM) electrolysis cells, ensuring uniform water distribution and smooth gas diffusion.
5) Polystyrene nanocomposite coatings to improve antibacterial properties of packaging paper
Polystyrene, titanium dioxide nanoparticles, and silver nanoparticles are used to prepare polystyrene nanocomposite coatings. When applied to packaging paper made from straw pulp (at 5% or 10% concentration), the coated paper exhibits improved breathability, tensile strength, water absorption, and barrier properties. The coating also shows strong antibacterial effects against *Pseudomonas aeruginosa*, *Staphylococcus aureus*, Candida, and other bacteria.
6) Sharklet-microstructure-coated release paper with breakthrough technology
A North American company has developed cast-coated release paper with a Sharklet microstructure. This innovation prevents bacterial growth on the paper surface without the use of toxic additives or chemicals.

7) Replacing single-use plastics with paper-based materials
   Notably, the pollution caused by single-use plastics has become a serious threat to human health, and global efforts to combat plastic waste have gained widespread support. The papermaking industry, using plant fibers as the primary raw material, has abundant resources and the advantage of full biodegradability. With the successful development of high-strength, high-barrier nanocellulose, the use of paper to replace plastics, particularly single-use plastic products that take centuries to degrade, has seen significant breakthroughs. Once the technology matures, it is expected to drive a resurgence in the traditional papermaking industry.

In conclusion, the future development of paper-based functional materials offers vast potential. It is essential for material scientists and papermaking experts to work together to develop more advanced, high-performance, and widely applicable products to meet the growing market demand.

References

1 Chen, G. (2018). Research and application of specialty papers and paper-based functional materials. *China Pulp & Paper* 39 (23): 56–61.

2 Lin, H. (2017). Current status and development trends of paper-based functional materials. *Heilongjiang Pulp & Paper* 3: 29–31.

3 Yao, X., Yao, Z., Jiang, J. et al. (2018). Development and application of paper-based functional materials. In: *Proceedings of the 12th Academic Conference on Chemical Engineering, Metallurgy, and Materials Engineering of the Chinese Academy of Engineering*, 589–595. Beijing: Chemical Industry Press.

4 Yao, X. (2016). Thoughts and suggestions on the development of the paper chemicals industry in China during the 13th five-year plan. *China Pulp & Paper* 37 (13): 22–27.

5 Alén, R., Zhang, S.F., Zhang, B. et al. (2016). *Papermaking Chemistry*, Series on Papermaking and its Equipment Technology, vol. 8. Beijing: China Light Industry Press.

6 Yao, X. and Zheng, L.P. (2014). Research and application technology of green papermaking chemicals. In: *Proceedings of the 10th Academic Conference on Chemical Engineering, Metallurgy, and Materials Engineering of the Chinese Academy of Engineering*, 117–121. Beijing: Chemical Industry Press.

7 Wu, A.B. (2019). Development and application of paper-based functional materials. *China Pulp & Paper* 40 (13): 181–184.

8 Yao, X. (2019). *Biomass-Based Functional Materials—New Developments in the Research and Application of Starch Derivatives and Nanocellulose*. Shanghai: National Symposium on Papermaking Chemical Technology and Biomass Functional Materials.

2

High-performance Starch Paper-based Functional Materials

2.1 Overview

2.1.1 Starch and Starch-based Functional Products

Starch is widely distributed in nature and is a common component of higher plants. It constitutes a major portion of green plant fruits, seeds, tubers, and rhizomes. It is a product of the carbon dioxide and water photosynthesis process, serving as an inexhaustible and renewable natural resource. With the progress and development of technology, people have gradually realized the seriousness of issues such as energy and the environment. Therefore, the in-depth development and application of renewable resources have increasingly garnered attention. Natural starch is insoluble in cold water, and starch paste is prone to aging. Its poor film-forming and stability characteristics, as well as lack of functional properties, have limited its applications. Based on the structure and physicochemical properties of starch, people have developed modification techniques for starch. These techniques involve applying physical, chemical, or biological methods to modify native starch, imparting it with special functions and purposes. This process is known as starch modification, and the resulting products are referred to as modified starch or starch derivatives. Modified starches endowed with functional characteristics are termed starch-based functional products.

Tate & Lyle Group, based in the United Kingdom, is a major producer of starch, starch sugars, organic acids, and other carbohydrates. With over 30 production facilities worldwide, as of March 2019, Tate & Lyle achieved a total revenue of £2.9 billion. The National Starch and Chemical Company (NSCC) in the United States has established 125 customer service centers and production facilities in 36 countries around the world, with a global annual sales revenue exceeding US$3.3 billion. AVEBE, a company based in the Netherlands, is the world's largest potato starch group. It operates nine factories globally producing potato and cassava starch, located in countries such as the Netherlands, Germany, Thailand, and China. The annual starch production capacity is close to a million tons. These large international companies possess advanced research institutions, excellent research and production equipment, as well as a large number of professional technical talents. They continuously develop new products, processes, technologies, and applications.

Paper-Based Functional Materials: Preparation and Applications, First Edition.
Xianping Yao, Genrong Chen, Zhen Yao, and Liping Zheng.

From the cultivation of new starch plant varieties to starch processing, and further to the development and application of starch derivatives, a complete industrial system has been established. In recent years, these large starch and starch derivative groups have also established subsidiary branches in Thailand to develop cassava starch derivatives in the country. Currently, the global papermaking industry produces approximately 6 million tons of starch derivatives annually. The European papermaking industry utilizes around 1.6 million tons of starch derivatives per year, with a paper and paperboard consumption of 21 kg per ton. The papermaking industry in Japan consumes over 500,000 tons of starch derivatives annually.

After over 30 years of development, China's independent innovation capability has continuously strengthened. Currently, there are numerous research institutions engaged in the field of starch derivatives, such as the Hangzhou Research Institute of Chemical Industry, Guangxi University, Jiangnan University, Tianjin University, Northwestern Polytechnic University, and Dalian University of Technology. Especially noteworthy is the research and development team led by Yao Xianping at the Hangzhou Research Institute of Chemical Industry. Since 1985, the team has been engaged in the research and development of starch derivatives. It has also established an industrial base for the commercialization of starch-based functional products—Hangzhou Papermate Science & Technology Co., Ltd. (named after "Friend of Papermaking"). The team has successively established several commercialization bases for its achievements in Zhejiang, Shandong, and the northeastern region. All of the team's research achievements have been fully industrialized, and the annual production scale has reached 400,000 tons. Guangxi State Farms Mingyang Biochemical Co., Ltd. is the largest domestic producer of cassava starch derivatives. The company has established a provincial-level technology center and conducted in-depth research on cassava starch derivatives.

Changchun Dacheng Industrial Group Company LTD., as a pioneer in the industrialization of corn resources in China, is among the first batch of nationally recognized key leading agricultural industrialization enterprises. It has established itself as a prominent player in the corn starch, starch sugar, amino acid, chemical alcohol, bio-feed fertilizer, and machinery manufacturing sectors. The group has formed a comprehensive industrial cluster comprising six major corn deep-processing product categories. It has built five industrialized production bases across China in cities such as Changchun, Dehui, Jinzhou, Fuzhou, and Shanghai.

According to data compiled by the China Starch Industry Association in 2019, the total production of various types of starch in China in 2020 reached 33.8897 million tons, showing a growth of 5.46% compared to the year 2019. Among them, the production of corn starch was 32.3258 million tons, showing a year-on-year increase of 4.36%. The corn starch production of Shandong, Jilin, and Hebei provinces, the top three provinces in China, accounted for 48.30%, 6.30%, and 13.25% of the total national corn starch production, respectively. The production of value-added starch products reached 18.7416 million tons, showing a year-on-year increase of 7.88%. Among them, the production of functional starch was 1.7511 million tons, accounting for 7.33% of the total value-added starch product production, with a year-on-year increase of 9.08%. The annual consumption of functional starch in

China's papermaking industry is approximately 1 million tons. In 2020, the top three companies in terms of functional starch production capacity in China were Hangzhou Papermate Science & Technology Co., Ltd., Shandong Xilai Starch Co., Ltd., and Zhucheng Xingmao Corn Developing Co., Ltd.

2.1.2 Types and Functions of Starch-based Functional Products for Papermaking

Starch-based functional products for the papermaking industry mainly include anionic starch (oxidized starch, phosphate ester starch, carboxymethyl starch), cationic starch, amphoteric and multi-modified starch, cross-linked starch, hydroxyalkyl starch, composite modified starch, and grafted starch. Most modified starches exhibit excellent characteristics such as low gelatinization temperature, high paste transparency, high viscosity, good stability, minimal gel formation, strong film-forming ability, and resistance to acid and alkali. They can be tailored to various functional roles based on the specific requirements of different paper types.

Due to the fact that both starch and cellulose have glucose units in their molecular structures, they exhibit good compatibility and strong binding affinity. Functional starches used in papermaking are typically added in quantities of around 1%, yet they have a significant impact on the quality of paper products and the operation of paper machines. They are often referred to as the "umami of papermaking" by the paper industry. Functional starches are primarily used in the papermaking industry as wet-end additives, interlayer strengtheners, surface-sizing agents, and coating binders. They offer benefits such as enhancing papers' physical strength, aiding retention and filtration, and improving sizing effectiveness and surface properties. As a result, they contribute to improving paper quality and grade, conserving high-quality fibers and other raw materials, and achieving energy-saving and emission-reduction goals (Figure 2.1).

2.1.3 Application of Starch-based Functional Products in Papermaking

Starch-based functional products can be utilized throughout the entire papermaking process (Figure 2.2), including wet-end addition, interlayer spraying, surface sizing, and coating adhesion. On average, their usage constitutes around 2% of the paper and paperboard production, and for certain paper types, this can even reach 5–6%. They are the most extensively used, widely applied, and highly functional chemical agents in the papermaking industry. As papermaking additives, starch-based functional products have been consistently included for multiple years in the "Guidelines for National Priority Development in High-tech Industrialization" by the National Development and Reform Commission, Ministry of Science and Technology, Ministry of Industry and Information Technology, Ministry of Commerce, and the China National Intellectual Property Administration. They fall under the category of new materials technology.

The modified starches used for wet-end addition, as shown in Figure 2.2, mainly include cationic starch, anionic phosphate ester starch, amphoteric and

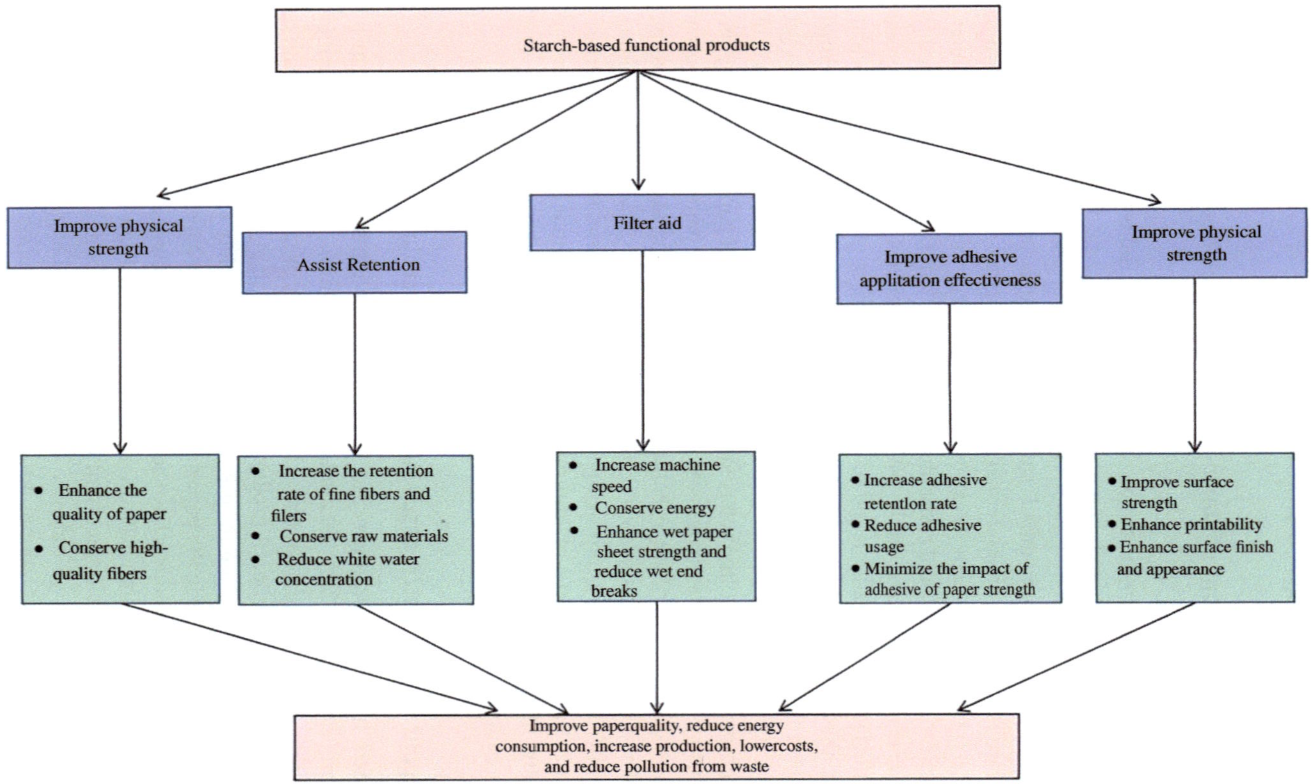

Figure 2.1 Functional roles of starch-based functional products in papermaking.

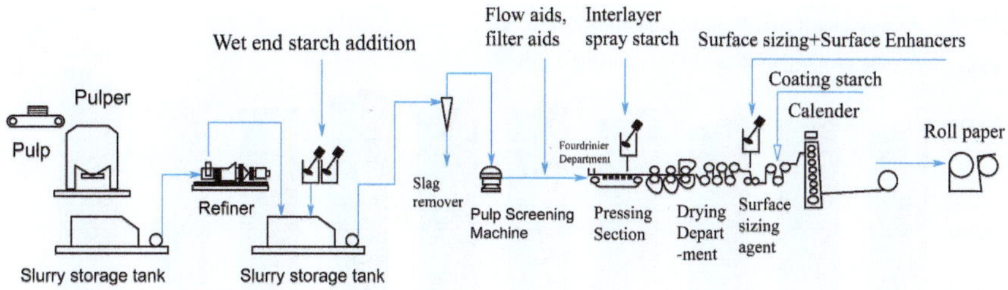

Figure 2.2 Applications of various starch-based functional products in the papermaking process.

multi-modified starch, and graft copolymer starch. These are primarily used to enhance the paper's physical strength, increase the retention of fine fibers and fillers, improve water filtration performance, and enhance sizing effects. Starches used for interlayer spraying are primarily anionic starches. They are required to have strong binding properties and a low gelatinization temperature. They are mainly used to enhance the interlayer bond strength of paperboard, while also contributing to improved stiffness, ring-crush strength, and surface strength. Starches used for surface sizing are primarily oxidized starch and enzyme-converted starch. They are used to enhance the surface strength and printability of paper, while also improving attributes such as tear resistance, foldability, tensile strength, smoothness, resistance to delamination, and ring-crush strength, among other physical strength indicators of paper. In recent years, with the advancement of digital technology, specialty surface-treated digital printing paper (inkjet paper) has seen rapid development. New high-retention surface-sizing starches are gradually becoming standard products widely adopted for digital printing paper and other premium office papers due to their specific surface properties. In Europe, currently, 75% of modified starch falls under the category of high-retention modified starch, mainly used for surface sizing and internal addition in papermaking. Common modified starches such as oxidized starch and enzyme-converted starch are rapidly decreasing in use as surface-sizing agents. Starches used for coating are primarily oxidized starch, phosphate ester starch, and hydroxyalkyl starch. They are required to have strong binding properties and low viscosity. They are typically combined with inorganic pigments like kaolin clay and are used to optimize the rheology and binding strength of coatings. Their main purpose is to improve the coatings' rheological properties, provide binding strength, and enhance the stiffness, ink absorption, and printing suitability of the finished paper.

2.1.4 The Significance of Starch-based Functional Products in Maintaining Forest Resources and Energy Conservation and Emissions Reduction

In 2019, China's paper production reached 107.65 million tons, maintaining its position as the global leader for several consecutive years (Figure 2.3). However, the industry faces serious challenges such as scarcity of wood resources, low

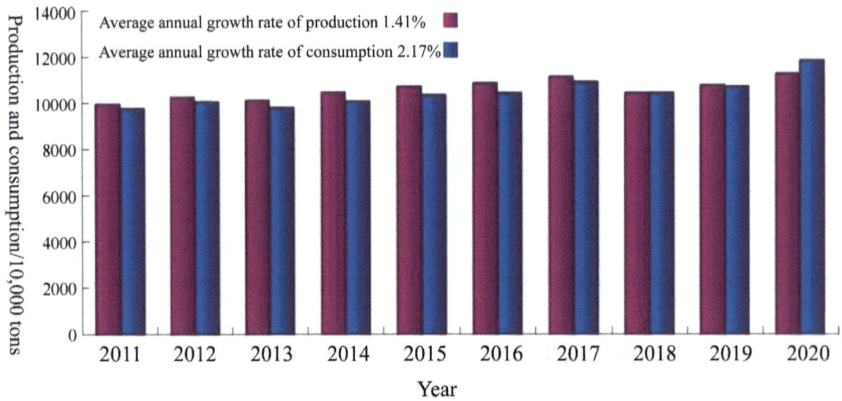

Figure 2.3 Paper and paperboard production and consumption in 2011–2020.

paper quality, and significant wastewater pollution. Research has shown that starch-based functional products can become key functional materials for breaking the bottlenecks in the papermaking industry mentioned above.

2.1.4.1 Alternative Wood-Fiber Raw Materials

China is a country with a severe shortage of forest resources, ranking 119th globally in terms of per capita forest area. Currently, the exploitable wood resources in China are far from sufficient to meet the demands of the papermaking industry. Most of the bleached wood pulp is reliant on imports. In 2017, China imported 54.29 million tons of pulp and waste paper, resulting in an expenditure of US$26.1 billion. Since the establishment of the People's Republic of China, non-wood-fiber pulping and papermaking have played a significant role in the development of the country's paper industry. China is the world's largest producer of non-wood-fiber paper and possesses the most extensive production experience in this field.

China has abundant non-wood-fiber resources, with bamboo, straw (rice straw), reed, and bagasse being the primary non-wood-fiber raw materials used for papermaking. More than a thousand years ago, China used materials such as straw, wheat straw, and bamboo as raw materials for handmade paper. The main challenges during that time included the short length and low strength of non-wood fibers, as well as the presence of various chemical impurities. Utilizing abundant non-wood resources and employing various papermaking chemicals to produce high-quality paper, thereby reducing deforestation and preserving the ecological environment, is a task of significant societal importance.

To address the aforementioned bottlenecks and constraints, the research and development team led by Yao Xianping made products (with starch as the main ingredient) that can increase the strength of straw pulp by 15–25% and reduce the wood pulp ratio by 10–30% (saving 4 to 5 m^3 of wood per ton of wood pulp). This has contributed significantly to conserving national wood resources.

The utilization of waste-paper resources is another significant feature of China's paper industry development. Currently, the proportion of recycled fiber pulp in

China has exceeded 60%. Recycling 1 ton of waste paper can produce 0.8 tons of high-quality paper; utilizing 1 ton of waste-paper pulp can save 3–5 m^3 of wood, as well as a significant amount of energy, water resources, and chemical raw materials. This practice also helps reduce air and water pollution. Many developed countries have enacted legislation to support the recycling and reuse of waste paper, and they have established legal requirements for the minimum content of secondary fibers in certain paper grades. In recent years, the global recycling rate of waste paper and its proportion in papermaking raw materials have increased rapidly, with China's growth rate being particularly notable.

The recycling of waste paper relies on specialized functional chemicals. Waste paper is subject to bacterial erosion, ultraviolet light exposure, and oxidation from air before and after recycling, which results in a reduction in fiber strength. Additionally, waste paper contains a high content of various chemical impurities, leading to potential interference. Ordinary chemicals do not provide significant enhancement effects in such cases. Therefore, the papermaking industry has an especially urgent demand for specialized enhancers for recycled paper. Calculated at 10 kg per ton of paper, the annual demand could exceed 500,000 tons, showcasing tremendous market potential and promising prospects for industrialization. The project "Interference-resistant Enhancers for Recycled Paper," undertaken by the Hangzhou Research Institute of Chemical Industry and Hangzhou Papermate Science & Technology Co., Ltd., primarily utilizes starch as the main ingredient. Using molecular design techniques, anions and cations are introduced into starch molecules, followed by interference-resistant enhancement reactions. This process effectively prevents interference from various impurities in waste-paper pulp and significantly enhances the affinity between recycled fibers, resulting in a notable strengthening effect on recycled paper.

2.1.4.2 Promote Energy Conservation and Emissions Reduction

The papermaking industry is a significant pillar industry in China, but it is also characterized by high energy and resource consumption, significant water usage, and severe pollution. Energy conservation and emissions reduction have been a hot topic and a focal point of research for the papermaking industry worldwide as well as within the environmental field for many years. The discharge of chemical oxygen demand (COD), a major pollutant in papermaking wastewater, accounts for 40% of the total industrial wastewater emissions in the country. This not only poses challenges to the sustainable development of the papermaking industry but also causes significant damage to the ecological environment. It has a serious impact on human health and is not conducive to the construction of a resource-efficient, environmentally friendly, and harmonious society. The categorization of wastewater in the papermaking industry mainly includes four major types: cooking pulp wastewater (black liquor), pulping wash water (washing wastewater), bleaching wastewater, and papermaking wastewater. Among these, the environmental pollution caused by cooking black liquor is the most severe, accounting for 90% of the overall pollution in the papermaking industry.

The research team led by Yao Xianping from the National Engineering Research Center for Paper Chemicals at the Hangzhou Research Institute of Chemical Industry has played a significant role in overcoming the bottlenecks and constraints faced by China's papermaking industry. Their team has not only successfully developed a series of new energy-saving and emission-reducing functional products, such as "new high-retention starch surface sizing agent," "high cationic network cross-linking retention and drainage agent," and "neutral sizing specialty product," but also achieved significant breakthroughs in application technology and equipment research. While significantly improving paper quality, water and energy consumption have notably decreased. These innovations have been widely adopted in China's papermaking industry.

2.2 Preparation Methods of Starch-based Functional Products

Currently, there are hundreds of starch deep-processing enterprises in China, with most of them having an annual production scale of several thousand to tens of thousands of tons. Small-scale, outdated facilities, high energy consumption, and prominent environmental issues in these enterprises have attracted close attention from the State Environmental Protection Administration and relevant departments.

Traditional starch processing methods include wet process, dry process, and semi-dry process. Among them, intermittent wet process technology is predominant, accounting for about 80% of the total processing. Modern wet and dry processing methods have seen significant improvements in equipment and automatic control technologies. However, equipment costs are relatively high. For example, the former Yushu Starch Factory in Jilin, China (acquired by Hangzhou Papermate Science & Technology Co., Ltd. in 2005, now known as Jilin Zhongyou Science & Technology Co., Ltd.), introduced a wet-process functional starch production line with an annual capacity of 5,000 tons from Germany's Stadler-Kolb company, and an enzyme-modified starch production line with an annual capacity of 10,000 tons from Denmark's Niro company, with an equipment investment of over CN¥100 million. The National Key Technological Renovation Project of Jiangxi Hongxing Starch Factory introduced a vacuum drying functional starch production line with an annual capacity of 5,000 tons from Switzerland's Bertrams company, with an equipment investment of over CN¥50 million. Domestic equipment prices are relatively lower, but there is still a certain gap in performance indicators compared to developed countries.

2.2.1 Wet-processing Technology

Wet-processing technology is a process in which starch is mixed with water or other liquid mediums to form a starch slurry, and then undergoes modification reactions with chemical reagents under specific conditions to produce functional starch. During this process, starch particles remain in a non-gelatinized state. If the

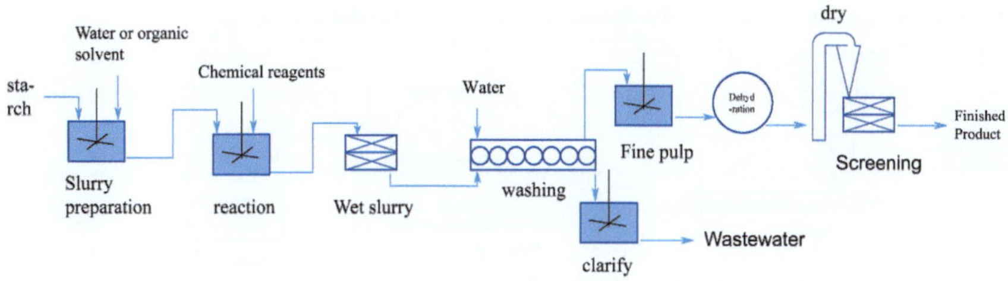

Figure 2.4 Flowchart of the wet-process preparation for functional starch.

dispersing medium used is not water but rather organic solvents or solvent-water mixtures, the wet process is referred to as the solvent method. The organic solvents used in the solvent method are expensive, flammable, and explosive, and their recovery is difficult. Therefore, this method is only used for producing high-substitution and high-value products. In the wet process, after the reaction is terminated, the process involves procedures such as filtration, separation, and drying, which are mostly carried out in a batch manner. During the reaction process, it is usually necessary to add salt-based anti-gelling agents, and the reaction temperature is controlled below the gelatinization temperature to prevent starch gelatinization. Therefore, the reaction time is relatively long, typically ranging from 4 to 20 hours. The advantages of the wet process include uniform and gentle reactions, low requirements for process equipment, and suitability for manufacturing various starch derivatives. However, its disadvantages include large amounts of wastewater (up to 3–6 tons per ton of product), high energy consumption (drying required during post-treatment), low yield, and high production costs. The wet-process preparation flowchart for functional starch is shown in Figure 2.4.

2.2.2 Dry Process

Dry process is a modification process in which starch undergoes modification reactions while in a "dry powder" state, hence the term "dry process." The dry process involves higher reaction temperatures (130–180°C) and requires a well-designed heating system for the solid-phase reactor. The reaction time is shorter compared to the wet process, usually around 1–4 hours. After the reaction is complete, the product is rapidly cooled. At this point, the material typically has a lower moisture content, ranging from 1% to 3%, and needs to be adjusted for humidity. Because the moisture content in the reaction system is very low, one of the major challenges in dry-process production is achieving a uniform mixture of starch and chemical reagents. In industrial settings, aside from using dedicated mixing equipment, a two-step process is employed: mixing in a wet state and reacting in a dry state. This approach ensures the successful production of functional starch. The variety of functional starches produced using dry-process technology is relatively limited. Among them, the most widely produced and commonly used types are pregelatinized starch, hydroxypropyl starch, and other acid-degraded functional starches. Due to the high production yield, wide application range, and lack of

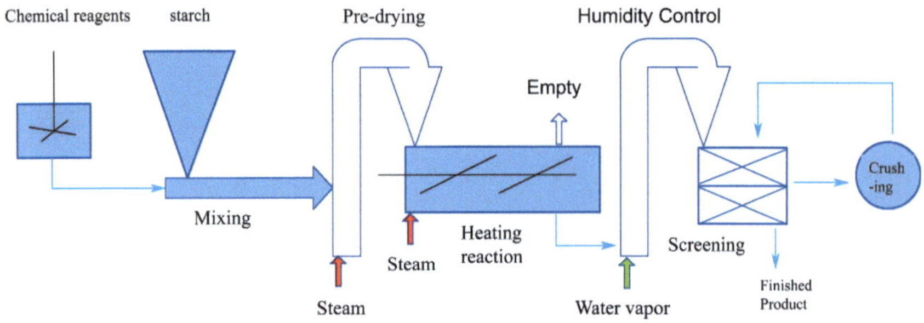

Figure 2.5 Process flow of dry method preparation of functional starch.

wastewater pollution of products like hydroxypropyl starch, pregelatinized starch, and acid-degraded starch produced through dry-process technology, the dry-process method is considered a promising approach. Its drawbacks include high equipment requirements, high energy consumption, poor product stability, and challenges in controlling dust. The process flow of dry method preparation of functional starch is shown in Figure 2.5.

2.2.3 Semi-dry Process

The semi-dry process technology is situated between the wet process and the dry process. The main distinction from the dry process technology is that the reaction temperature is lower and the moisture content is relatively higher. The production of functional starch using the semi-dry process involves short reaction times, high reaction efficiency, and generates no wastewater or exhaust gases. Taking the production of cationic starch as an example, in the wet process, the reaction time is 7–24 hours, and the typical reaction efficiency ranges from 60% to 80%. In contrast, with the semi-dry process, the reaction time is 2–5 hours, achieving a high reaction efficiency generally exceeding 80%. Moreover, there's no need to add inhibitory salts, which reduces production costs. Consequently, the semi-dry process technology is increasingly being adopted.

In the process of producing functional starch using either the dry process or the semi-dry process technology, it is essential to rigorously control the production process, reaction time, and reaction temperature based on the selection of raw materials. Additionally, it is necessary to carry out procedures such as system dewatering, temperature stability control, and cooling treatment of the output products. Furthermore, post-processing should be conducted based on the application requirements of the products, such as incorporating suitable amounts of compounded enhancing materials.

2.2.4 Continuous Flow and Pipeline-based Green Manufacturing New Process

Both domestic and international starch-based functional products commonly utilize traditional wet and dry processing methods. In the wet process, the materials react in a liquid phase, and the products need to undergo procedures such as washing, filtration, separation, and drying. The advantages include even and mild reactions, while the disadvantages involve a large amount of wastewater (3–6 m3 per ton of product), high energy consumption (0.5–0.8 tons of standard coal per ton of product), long processing time (3–10 hours), and low reaction efficiency (70–75%). The dry process involves mixing the modifying agent with starch slurry and subjecting it to a high-temperature reaction. The advantages include high efficiency (75–80%), short processing time (2–5 hours), and no wastewater generation. The drawbacks include high equipment requirements, low mixing efficiency (around 80%), high energy consumption (reaction temperature of 100–150°C), poor product stability, and challenges in dust control. Both wet and dry traditional processes have prominent issues with high energy consumption and pollution.

Many relevant enterprises in our country are facing issues of small scale and outdated technological equipment, which urgently need to be addressed. In response to the aforementioned issues and the pressing demand for high-performance starch-based functional products in industries such as papermaking in China, the team led by Yao Xianping studied the molecular characteristics of starch-based functional materials, the reaction mechanisms, and the common issues in traditional processes, and creatively developed a new process for the continuous flow and pipeline-based green manufacturing of functional starch (Figure 2.6). Additionally, the team established the world's first demonstration production line for this new process with an annual capacity of tens of thousands of tons [1].

The new process has achieved breakthroughs in the following key technologies:

1) Overcoming the rapid, continuous, and uniform mixing technology of a large amount of starch powder and trace reaction reagents

   In the wet process, the aqueous solution accounts for over 60%, making it easy to achieve uniform mixing of reagents and starch. In the dry process, starch and reagents need to be pre-mixed. When trace reagents are introduced into a large amount of powder, the reagent droplets are encapsulated by the powder particles, making it difficult to achieve uniform dispersion. Typically, the mixing efficiency is only around 80%, which can result in waste and pollution. The powder-liquid continuous high-speed mixing device, known as the High-Speed Flying Knife Mixer, independently designed and developed by the research team led by Yao Xianping, has achieved rapid, uniform, continuous, and efficient mixing. The mixing efficiency of the powder-liquid mixture can exceed 98%. Through combined optimization, it can meet the preparation technical requirements of complex multi-component process formulations and a variety of product series. Its performance surpasses that of traditional methods, serving as a core technology to ensure uniform and stable product quality. This technology is challenging, demanding high technical precision.

2) The invention of continuous flow and pipeline-based technology for starch powder

   The key to this technology is to enable the powdered reaction materials to flow and react within the pipeline, similar to a liquid. Typically, starch-based functional products require the addition of one or more liquid reagents during the

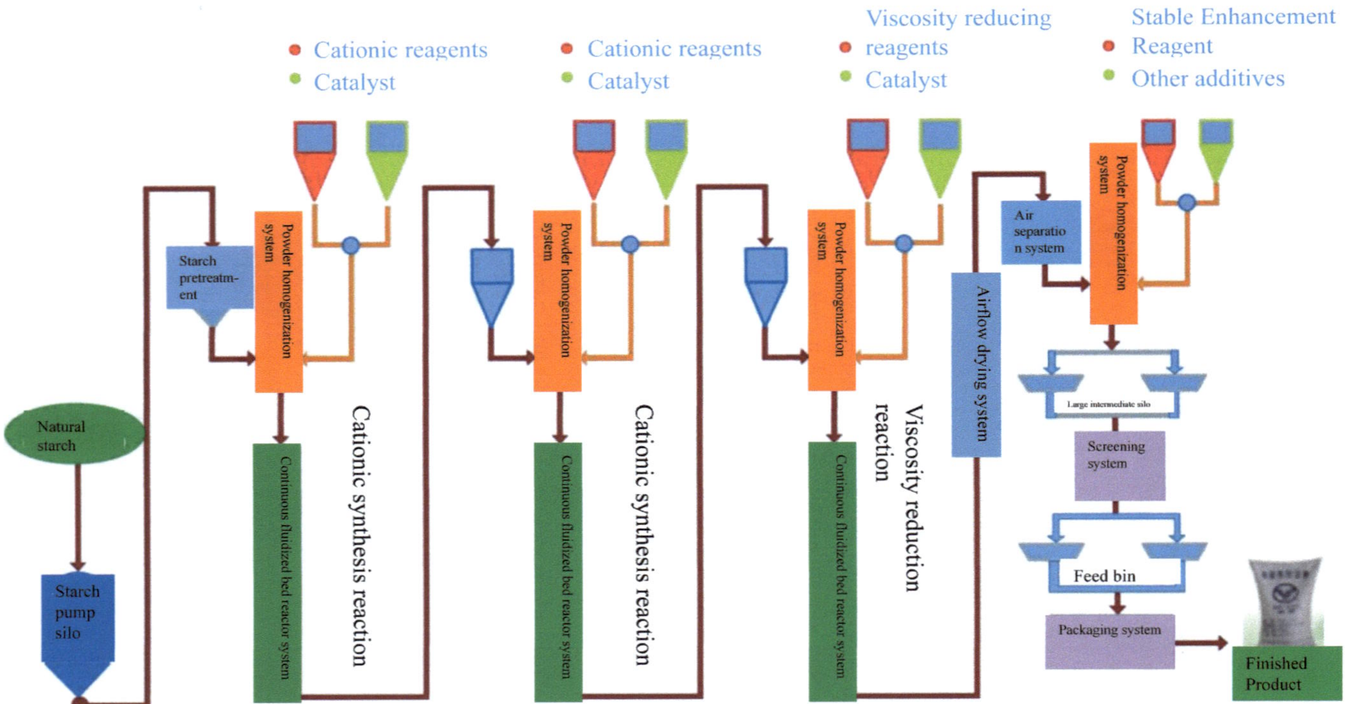

Figure 2.6 Technological roadmap of the continuous flow and pipeline-based green manufacturing new process for functional starch.

reaction process. However, when the moisture content exceeds 14%, they are prone to clumping and cannot be made to flow. Especially for complex products involving multi-component reactions, the moisture content of the mixture can sometimes exceed 20%, rendering the materials within the pipeline incapable of achieving flow. The team led by Yao Xianping developed a specialized flow aid that not only significantly enhances the flowability of wet starch but also improves the effectiveness of product applications. Even with a high water content of up to 25%, the wet powder material can still flow and react rapidly within the pipeline, effectively addressing the challenges of wet powder flow and pipeline-based reaction technology.

3) The invention of starch powder solid-phase rapid catalytic reaction technology

A highly efficient composite powder catalyst was developed that can rapidly form reactive centers, specifically designed for powder reactions. This breakthrough overcame the challenges of achieving fast and efficient reactions under the conditions of powdered flow and pipeline-based processing. Compared to traditional catalytic techniques, this technology not only meets the requirements for powder flow but also possesses high catalytic activity and efficient rapid characteristics. Taking cationization of starch as an example, the wet process requires 45–55°C for 8–10 hours, the dry process requires over 100°C for 3–4 hours, whereas the new process only requires room temperature for 0.6 hours.

Building upon the aforementioned key technologies, the team led by Yao Xianping achieved an industrial production capacity of 400,000 tons per year through a series of engineering technological innovations, including product formulation design, powder conveying, solid-liquid mixing, efficient catalysis, and continuous flow and pipeline-based powder processing. This is currently the largest new process production line in Asia. The product has been used by domestic and international papermaking companies, receiving positive feedback, and demonstrating significant economic and social benefits.

Compared to similar technologies both domestically and internationally (Table 2.1), this new process offers unparalleled advantages over traditional methods. These include minimal investment (10 times less investment for a production scale of 10,000 tons), low energy consumption (most reactions occur at room temperature, with reaction times about 1/5th of similar technologies, and energy consumption per unit output is only 1/10th of peers), high efficiency (reaction efficiency increased by over 15%), environmentally friendly practices (no wastewater, exhaust gas, or waste residue in the production process), and stable product quality. It is particularly well-suited for modern large-scale high-speed paper machines and continuous automation applications. More than 50 high-performance products have been developed using the new process, including 10 products that have achieved national-level recognition. The new products have been applied in the world's most advanced paper machines and premium paper grades through global bidding processes. They have demonstrated significant competitiveness by enhancing paper strength (15–25%), saving wood pulp fibers (10–35%), reducing papermaking white-water concentration (20–30%), lowering energy consumption (5–15%), and promoting paper recycling. Up to now, a cumulative total of 1.1 million tons of

Table 2.1 Comparison of technological parameters between the new process and foreign companies' "Wet Process" and "Dry Process" technologies.

	Process		
Project	**Wet Process Technology of a German Company**	**Dry Process Technology of a Swiss Company**	**New Process**
Investment (CN¥100 Million/10,000 Tons)	1	0.5	0.059
Reaction Time (Hours)	3–10	2–5	0.6
Reaction Temperature (°C)	45–55	>100	Room Temperature
Water Consumption (Tonnes Per Ton of Product)	3–10	Small Amount	Small Amount
Steam Consumption (GJ Per Ton of Product)	0.7	1	0.25
Electricity Consumption (kW·h Per Ton of Product)	120	300	30
Wastewater (Tonnes Per Ton of Product)	3–6	None	None
Reaction Efficiency (%)	70–75	75–80	85–90
Production Cost	High	Medium	Low
Reaction Uniformity	Good	80%	98%
Product Stability	Quite good	Poor	Good

these related new products have been promoted, used in hundreds of millions of tons of paper and paperboard production. This has resulted in significant savings of wood resources, water resources, and energy consumption for the paper industry, along with reduced wastewater discharge. The economic and social benefits are remarkable.

2.3 High-performance Starch-based Functional Products for Papermaking

2.3.1 Special Modified Starch for Non-wood Fiber

China's forest resources are severely inadequate, with a per capita forest stock volume of only 12.6% of the global average. It is a major non-wood-fiber papermaking country. Until the late 1990s, the proportion of non-wood-fiber usage in the paper industry was still around 80%, and bleached wood pulp was heavily reliant on imports. In contrast, the global paper industry's proportion of non-wood-fiber usage is only 7%, with wood fiber accounting for 93%. Utilizing non-wood fiber for papermaking holds significant importance in conserving China's forest resources.

Table 2.2 Technical achievements of related projects.

Project Name	Appraisal Evaluation	Invention Patent Number	National Key New Product Approval Number
Specialized Enhancer for Non-wood-fiber	Internationally Leading	CN99113918.6	20023300J011
Specialized Wet-end Chemicals for Newsprint	Internationally Leading	CN99113920.8	2002ED700072
Retention and Drainage Aid for Straw Pulp Papermaking	Internationally Leading	CN99113919.4	2002ED700019

Due to the short length, low strength, poor drainage characteristics, and the presence of numerous impurities, non-wood fibers not only severely affect the paper machine's forming process and the quality of the finished paper but also lead to significant pollution from waste products. Moreover, due to the inherent characteristics of non-wood fibers and the interference caused by the large amount of impurities, the use of conventional papermaking chemicals yields limited results. The papermaking industry is in urgent need of specialized chemicals for non-wood-fiber papermaking.

Against this backdrop, Yao Xianping's research team undertook a project that included three subjects: "Specialized Enhancers for Non-Wood Fiber," "Retention and Drainage Aids for Straw Pulp Papermaking," and "Specialized Wet-End Chemicals for Newsprint." The achieved technological level is at or above the international leading or advanced level. The related products are detailed in Table 2.2.

Due to the presence of various impurities in non-wood fibers, which can significantly interfere with the effects of general chemicals, the chemicals used in wood pulp papermaking are not suitable for application in non-wood-fiber papermaking. At that time, the development of enhancement, retention, and drainage functional products for papermaking was mainly focused on wood pulp papermaking. Research on chemicals for non-wood-fiber papermaking and their application performance is still lacking both domestically and internationally. The project achieved multiple technological innovations.

2.3.1.1 Product Structural Innovation—Pioneering the Molecular Structure Modification of Natural Polymer (Starch) with Multifunctional Groups

The project invented a multi-element molecular structure design, overcoming the technical challenges of mutual interference between different ion-substituted molecules. It successfully achieved the incorporation of both cationic and anionic substituents, as well as other functional substituents, within the same starch molecule chain. By utilizing the mutual coordination between anionic and cationic groups and the synergistic effect of functional groups, the product becomes suitable for high impurity content and low-quality fiber non-wood fibers and waste-paper fibers [2]. This addresses the challenge of interference caused by numerous impurities in non-wood and mechanical pulp during the application process. The technical achievements have not been reported domestically or internationally. Moreover, different functional substituents and molecular ratios can synthesize numerous excellent new products with various functionalities. Apart from applications in the papermaking industry, these innovations can also be extended to industries such as food, oilfields, cosmetics, pharmaceuticals, and more, offering a wide range of potential applications. This technological breakthrough has opened up a new development space in the field of starch derivatives, carrying profound significance for the future.

2.3.1.2 Technological Process Innovation—Inventing the Catalytic Continuous Powder Pipeline Reaction Clean Production Process

Through innovative synthesis processes, the technical challenge of mutual interference between substituent groups with different charged ions during the reaction process has been resolved. Moreover, through innovative production processes and equipment, the key technologies of catalytic reactions and continuous powder pipeline reactions have been successfully addressed. The new process has developed a continuous powder pipeline reaction clean production process along with the primary reaction equipment. Additionally, it has achieved automated process control, which is a challenging and technically complex endeavor [3]. No similar reports have been found domestically or internationally.

2.3.1.3 Application Technology Innovation—Pioneering Application Technology for Non-Wood-Fiber Pulp Papermaking

For the first time internationally, enhancement, retention, and drainage technologies were applied to non-wood-fiber papermaking. Additionally, wet-end chemicals were first used in newsprint production, leading to a significant improvement in the quality of newsprint. The successful development of the project has had a significant impact on the transformation and enhancement of the traditional papermaking industry through the utilization of advanced technologies.

Taking cultural paper as an example, the addition of non-wood-fiber enhancers can increase paper strength by 15–25%, and it can lead to a reduction in wood pulp usage by 10–35%. By adding grass pulp retention and drainage aids, pulp consistency can be lowered by 20–30%, resulting in a decrease in energy consumption by 5–15%. Moreover, the retention of fine fibers and fillers can increase by 10–40%, and white-water consistency can be reduced by 20–30%.

Taking the example of wet-end chemicals for newsprint, the resulting products can elevate the quality of recycled newsprint to the level of imported high-speed color printing newsprint, while significantly reducing production costs. Successive products have been adopted by numerous domestic recycled newsprint mills. Utilizing 10,000 tons of this product can produce 1 million tons of recycled newsprint, which is equivalent to saving 4–5 million cubic meters of wood and 1.2 million tons of standard coal, conserving 600 million kilowatt-hours of electricity and preserving 600 million tons of water.

By utilizing specialized chemicals for non-wood fiber, the quality of non-wood-fiber paper is enhanced, leading to a reduction in the consumption of wood fiber and the preservation of forest resources. This, in turn, contributes to the protection of the ecological environment and aligns with the principles of sustainable development and the scientific outlook on development advocated by the nation.

2.3.2 High-retention Surface-sizing Starch

Surface sizing refers to the process and technique of applying adhesive material onto paper and paperboard, with the aim of maintaining the dimensional stability of the paper, reducing its water absorption and ink absorption, enhancing its printing adaptability and smoothness, and improving the quality of both paper and paperboard. Modified starch accounts for over 60% of the total amount of papermaking chemicals, with surface-sizing starch being the most extensively used. Due to the weak bonding between regular oxidized starch and fibers in surface sizing, it is difficult to meet the quality requirements of high-grade paper. Moreover, during paper damage or recycling, more than 60% of the starch may enter the white-water system, leading to wastewater pollution and affecting the wet-end potential and paper machine operation in the papermaking process. Simultaneously, with the development of digital technology and inkjet printing, new requirements for paper surface properties have emerged. Therefore, there is an urgent need to develop high-retention surface-sizing starch.

Currently, only a few foreign companies possess high-retention surface-sizing starch, while in China, the majority still use regular oxidized starch. From the perspective of production process technology and equipment, domestic production lines are mostly intermittent wet or dry processes, with small scales and outdated facilities, making it difficult to produce high-quality and stable products. In terms of technical and economic indicators, the retention rate of products using regular oxidized starch in practical applications is much lower than that of high-retention surface-sizing starch, leading to significant environmental concerns.

Yao Xianping's research team, through technical innovations in product, process, equipment, and application, has developed a high-tech product with independent intellectual property rights—high-retention surface-sizing starch [4]. This innovation not only provides essential functional materials and application technology for the production of high-quality paper and paperboard in the paper industry, offering an effective solution to papermaking wastewater pollution, but also breaks the foreign technological monopoly. It addresses technical challenges in product molecular structure design, environmentally friendly preparation, and the development of application technology and equipment. Through integrated innovations, it elevates the technological level of papermaking chemicals in the field of fine chemical engineering in China, placing it among the world's leading advancements.

2.3.2.1 The Main Differences Between High-Retention Surface-Sizing Starch and Regular Oxidized Starch

The main difference between high-retention surface-sizing starch and regular oxidized starch lies in their affinity for the paper surface in terms of chemical

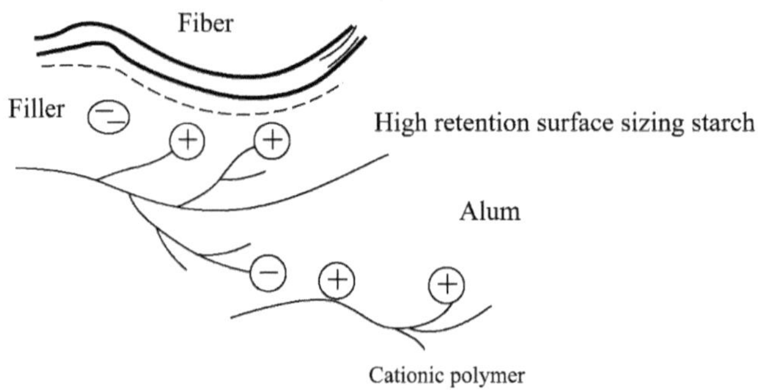

Figure 2.7 Illustration of the working principle of high-retention surface-sizing starch.

technology principles and performance. Considering that papermaking itself is a complex chemical system, different types of pulp, variations in production processes, and diverse chemical additives all form distinct papermaking systems. These systems contain various chemical substances such as fibers, fillers, and wet-end chemicals, all with different ions. Therefore, the Yao Xianping team chose to use multifunctional substituents with high affinity, employing a polymeric molecular structure. This involved attaching cationic substituents and anionic substituents within starch molecules. This strategy enhances the affinity of starch for negatively charged paper fibers and fillers and positively charged chemicals. Through processes such as viscosity reduction and stabilization, the product becomes suitable for surface sizing on different paper machines and paper types. This enhances the bonding strength with fibers, achieving the goal of high retention (Figure 2.7) [5].

Comparisons were made between the high-retention surface-sizing starch product and commonly used domestic oxidized starch in terms of surface-sizing effectiveness and surface strength (Table 2.3 and Figure 2.8). It was found that the actual usage of the high-retention surface-sizing starch is 75% less than oxidized starch, yet the surface strength is doubled, showcasing its exceptional performance. Under equivalent sizing conditions, the surface strength of paper treated with this product consistently surpasses that of paper treated with oxidized starch. Particularly noteworthy is the

Table 2.3 A comparison of surface-sizing effects between oxidized starch and high-retention surface-sizing starch.

Varieties of Starch	Application Rate of Sizing (kg/t)	Fuzzing Speed (cm/s)		Breaking Length (m)
		纵	横	
Blank	-	34	51	3321
Oxidized Starch	40	116	73	3460
High-retention Product	30	212	181	3490
Comparison of Results	Reduced Usage 25%	Enhancement 82.8%	Enhancement 148%	Enhancement 0.9%

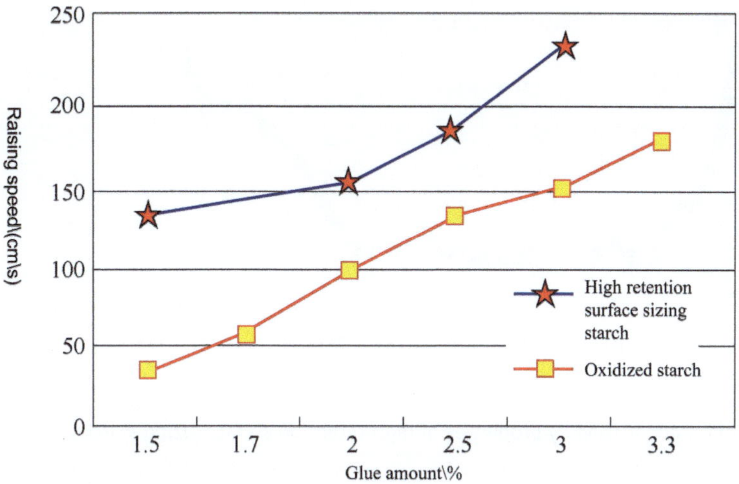

Figure 2.8 Influence of starch-sizing amount on the surface-sizing effect.

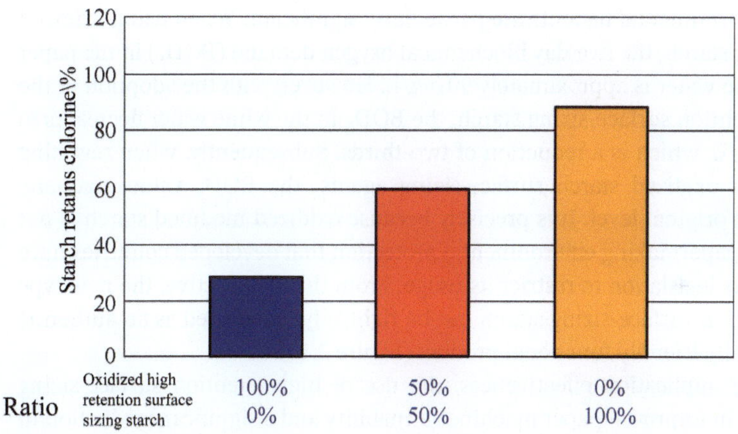

Figure 2.9 Retention rates of different varieties and ratios of surface-sizing starches during recycled paper repulping.

fact that this product, while achieving the same application effects, incurs lower additive costs compared to oxidized starch, highlighting its significant superiority.

Furthermore, in the recycling of damaged paper, the retention rate of high-retention surface-sizing starch is significantly higher than that of oxidized starch (Figure 2.9). Due to the presence of reactive groups in the product's molecules that exhibit excellent affinity for fibers, the measurement using iodine colorimetry spectrophotometry reveals that almost 100% of this product remains attached to fibers during the recycling of damaged paper. In contrast, the retention rate of regular oxidized starch surface-sizing agents is less than 40%. This means that at least 60% of the starch will enter the papermaking white-water system, leading to an increase in biochemical oxygen demand (BOD) and chemical oxygen

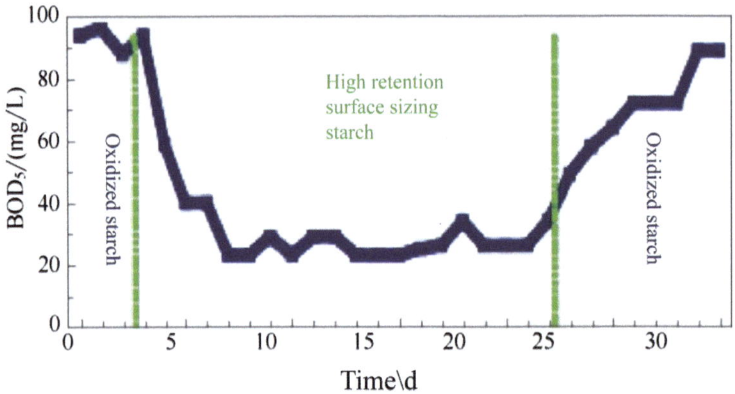

Figure 2.10 Environment-friendly properties of high-retention surface-sizing starch.

demand (COD) values. This fluctuation in Zeta potential subsequently affects the operational performance of the paper machine [6].

Due to the inherently high retention rate of the high-retention surface-sizing starch, its environmental benefits are particularly significant. When using oxidized surface-sizing starch, the five-day biochemical oxygen demand (BOD_5) in the paper machine white water is approximately 90 mg/L. However, with the adoption of the new high-retention surface-sizing starch, the BOD_5 in the white water decreases to around 30mg/L, which is a reduction of two-thirds. Subsequently, when reverting to the use of oxidized starch surface-sizing agents, the BOD_5 value gradually returns to the original level. It is precisely because oxidized modified starch is not conducive to papermaking environmental protection that developed countries have begun to enact legislation to restrict its usage. From this perspective, the new type of high-retention surface-sizing starch can be rightfully recognized as an authentic environmentally friendly functional product (Figure 2.10).

In terms of application effectiveness, the use of high-retention surface-sizing starch results in improved paper machine runnability and a significant reduction in head breaks (Table 2.4). Head breaks are the most troublesome issues in paper mills, leading to reduced production output and increased labor intensity for workers. Minimizing head breaks is one of the crucial objectives pursued by paper mills. Under the conditions of a sizing concentration of 6–8%, oxidized starch experiences 20 head breaks per day, whereas this product hardly experiences any head breaks.

Table 2.4 A comparison of daily head breaks in the paper machine under the same sizing concentration conditions.

Project	Oxidized Starch	High-retention Surface-sizing Starch
Sizing Concentration/%	6–8%	6–8%
Daily Head Breaks/Occurrences	20	0

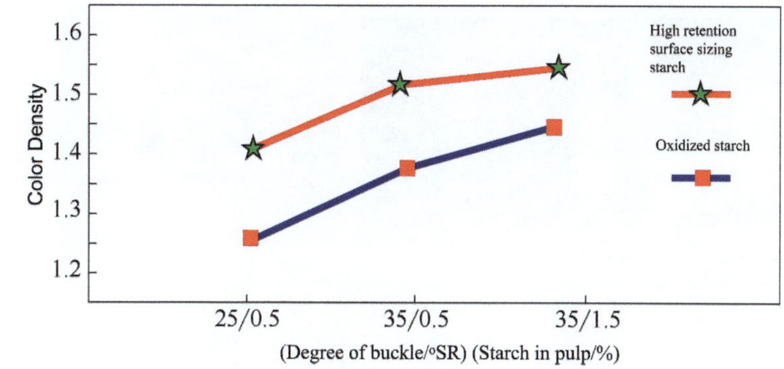

Figure 2.11 Improvement of printing effects by high-retention surface-sizing starch.

Under equivalent conditions, compared to oxidized starch, the high-retention surface-sizing starch product enhances the vibrancy of printed colors and provides more distinct outlines. The printing results are significantly better than those of paper treated with regular products (Figure 2.11).

2.3.2.2 Application of High-Retention Surface-Sizing Starch

The application of high-retention surface-sizing starch products has been adopted by dozens of paper mills in provinces such as Zhejiang, Jiangsu, Sichuan, Henan, and Guangxi. These applications span across various types of paper machines, ranging from low-speed to high-speed machines, and utilize both conventional sizing equipment and advanced film-sizing systems. This approach has resulted in numerous successful experiences. Moreover, the product has passed global bidding processes and has been introduced to renowned papermaking enterprises worldwide (Figure 2.12). As evidenced by user experiences, this product demonstrates a significantly greater enhancement in surface strength compared to regular starch. While maintaining the existing strength levels, it is possible to reduce starch sizing amounts by at least 15%. Simultaneously, this allows for increased filler usage and decreased wood-fiber consumption, leading to a noteworthy reduction in papermaking costs. In the subsequent sections of the paper machine, it is evident that steam consumption can be significantly reduced, leading to energy savings. Additionally, it can notably enhance the printability of the paper. Furthermore, it can improve paper machine operational performance and significantly reduce the occurrence of head breaks. In repulping, the high-retention surface-sizing starch retains over 90% on the fibers and serves as an aid for retention and filtration, enabling a reduction in retention-agent usage by 20–30% and a 10% reduction in sizing-agent usage. This significantly improves the white-water circulation system, reduces papermaking water consumption, and effectively decreases the pollution load in the papermaking white water (with a reduction of over 50% in BOD). It truly lives up to its name as a resource-saving and innovative environmental product, aligning with the requirements of the national circular economy and resource-efficient industries.

Figure 2.12 Application of high-retention surface-sizing starch on modern large-scale paper machines.

2.3.3 Interference-resistant Specialized Enhancer for Recycled Paper

Due to the increasing scarcity of virgin papermaking fiber resources worldwide and the growing emphasis on environmental awareness, recycled paper has become a significant source of papermaking fiber materials. Recycling waste paper can save a substantial amount of wood resources, energy, water, and chemical raw materials. In recent years, the global recycling rate of waste paper and its proportion in papermaking raw materials have rapidly increased, with China's growth rate being particularly notable. The application of recycled fibers and the manufacturing of recycled paper have gained increasing attention in China [7]. The consumption of waste paper in China has grown rapidly, reaching 63.02 million tons in 2017, accounting for 63% of the total national pulp consumption and ranking first globally.

As previously mentioned, the manufacturing of recycled paper relies on specialized chemicals and holds immense market potential, offering extensive prospects for industrialization [8]. In response to China's high proportion of waste-paper pulp in the papermaking industry, the research and development team led by Yao Xianping has invented an interference-resistant specialized enhancer for recycled paper. This enhancer effectively combats various impurities in waste-paper pulp while significantly enhancing paper strength [9]. This product primarily utilizes starch as its main ingredient. Through molecular design methods, it is produced using a multi-functional modification technique and interference-resistant enhancement composite reaction. The product is manufactured using a proprietary developed green preparation process with a continuous flow fluidized powder pipeline technology. The overall technology has reached an advanced level, with key aspects of the technology being at the forefront internationally.

Table 2.5 Changes in paper strength after adding interference-resistant specialized enhancer for recycled paper.

Indicator name	Blank	With enhancer	Growth rate/%
Breaking Length/m	3506	4519	28.89
Tear Index/($Pa{\cdot}m^2/g$)	2.52	3.42	35.71
Folding Endurance/Cycles	32	58	81.25

Application has demonstrated that compared to commercially available regular enhancers, the interference-resistant specialized enhancer for recycled paper shows significant enhancement effects (Table 2.5).

Compared to existing regular enhancers, this interference-resistant specialized enhancer for recycled paper not only enhances paper strength but also effectively resists the interference of impurities and ions. This characteristic allows it to be directly applied in waste-paper papermaking, unlike conventional enhancers that require the use of impurity-controlling agents to counteract interference. Therefore, this product simplifies the paper production process, enhances production efficiency, reduces production costs, and simultaneously improves paper properties, thereby enhancing product quality as well (Table 2.6).

2.3.4 Environmentally Friendly Paper Coating Adhesive

Paper surface chemicals are a crucial component in improving the quality of raw paper. Paper surface chemicals account for over 70% of papermaking chemicals, with adhesives being the largest category within coating chemicals. Apart from pigments, adhesives hold the greatest proportion and play a crucial role in terms of performance.

In traditional coating adhesives, emulsion-based chemicals exhibit excellent film-forming properties, ensuring a coating surface with good ink compatibility and sufficient bonding strength. They are the most widely used type of coating adhesives. However, their non-biodegradability and inherent chemical toxicity, coupled with the environmental pollution caused by their manufacturing processes, constrain their further development. Therefore, there is an urgent need to develop an environmentally friendly coating adhesive to fully or partially replace traditional emulsion-based products.

Modified starch-based coating adhesives are a type of adhesive chemical that achieves the optimal balance between bonding strength and coating cost. They are also the most suitable chemicals, apart from emulsions, for metered pressurized application in coating. In light of this, countries around the world are focusing their attention on using natural starch-based materials to synthesize adhesive products that outperform conventional coating starches, are more cost-effective than synthetic emulsions, and are better suited to the unique ionic and rheological characteristics of coatings. In Hu Zhiqing's experiment of replacing a portion of

Table 2.6 A comparison between the patent product of interference-resistant specialized enhancer for recycled paper and similar products.

Compare items	Colloidal substance control agent	Dry strength agent for papermaking	Amphoteric starch dry strength agent	This patented product
Ability to Resist Interference from Impurities	Carries a significant amount of positive charges, which effectively neutralize the negative charges in impurities	In dry strength agents, positive charges are easily neutralized by impurities, leading to the loss of functionality.	Positive charges are easily neutralized by impurities, resulting in a loss of functionality.	The interference-resistant agents within the product can adsorb impurities while also enhancing its effect.
Enhancement Effect	Lacks a noticeable enhancement effect	When used effectively, it demonstrates a notable enhancement effect.	When used effectively, it demonstrates a significant enhancement effect.	The product demonstrates a noticeable enhancement effect.
Retention Aid Effect	Cannot improve the retention aid effect	It might lead to deterioration in the retention aid effect.	It might lead to a decrease in the retention aid effect.	It significantly improves initial retention.
Cost	Higher	Higher	Lower	Low
Technological Advancement	Formulated product, simple process, lack of specificity for the product	Prepared through organic monomer polymerization, the product lacks specificity.	The product has weak interference resistance, and the preparation process is conventional.	Designed specifically for recycled paper regeneration, the product exhibits good compatibility and interference resistance, along with an advanced preparation process.

emulsion with starch, phosphate ester starch was used to substitute 25% of the emulsion. The experimental results revealed a slight increase in paper surface strength compared to using emulsion, indicating the feasibility of the starch substitution approach [10]. Wang Jiafu et al. prepared ultra-fine grinding starch particles (UGSP) using a mechanical grinding method and studied the effects of partially replacing styrene-butadiene emulsion with UGSP on coating performance and coated-paper properties. The results showed that when 2% UGSP was used to replace 2% styrene-butadiene emulsion, the coating exhibited excellent water retention properties and favorable rheological behavior. This ultra-fine grinding starch can partially replace synthetic emulsion in coatings, effectively saving costs, reducing carbon emissions, and the application of the coating has no significant impact on the properties of the finished paper [11]. Wang Xuqing et al. utilized coating starch as an auxiliary adhesive in composite adhesives and found that it significantly enhances the stiffness and opacity of paperboard, improves the printability of the finished paper, reduces costs, and optimizes the rheological behavior and water retention of the coating. The usage of this starch as an auxiliary adhesive constitutes only 10–30% of the total adhesive amount [12].

Currently, regular modified starch available on the market cannot meet the development requirements of high-speed coating. In order to meet the development demands of modern high-concentration and high-speed paper coating, the market has introduced composite-modified, diverse-modified, or enhanced-functional compound-modified starches. In 2009, the American company Ecosynthetix introduced a patented product designed to replace petroleum-based synthetic latex adhesives widely used in the coating of paper and paperboard. The product involves mixing starch raw materials with plasticizers and additives such as TiO2 in a Brabender high-shear mixer, and under conditions of high shear forces and temperatures exceeding 100°C, melting and blending them into a thermoplastic polymer-additive composite. Subsequently, crosslinking agents like ethylene glycol are introduced into the composite, and bio-latex nanoparticles are prepared using a specific extruder.

The research and development team led by Yao Xianping utilized starch as the main raw material and employed techniques such as phosphate esterification for viscosity reduction, etherification, and cationic complex modification to innovate and develop an environmentally friendly coating adhesives. This adhesive possesses multiple ionic groups and characteristics of balanced charge, resulting in high bonding strength, low migratory properties, the formation of flexible and resilient films, as well as excellent dispersibility and compatibility with pigments [13]. The active functional groups present can react with ethylene glycol, formaldehyde, melamine-formaldehyde (MF) resin, and urea-formaldehyde (UF) resin to achieve excellent water resistance. It can produce high-solids-content coatings with excellent flowability, stability, and water retention. It can be used for blade coating, resulting in coated paper with good gloss, smoothness, pull strength, and ink absorption, while also improving paper stiffness. It has low cost, is virtually non-toxic, and can significantly reduce the COD and BOD values of papermaking wastewater after pulping. It possesses environmentally friendly properties such as biodegradability and recyclability [14]. This high-performance, cost-effective modified starch can replace or partially substitute synthetic latex-based coating adhesives.

The environmentally friendly coating adhesive mentioned above is synthesized through a multifaceted modification process, including viscosity reduction through phosphate esterification and etherification with chloroacetic acid, followed by cationic modification. This process imparts the adhesive with various ionic groups such as phosphate ester, carboxyl, and cationic groups, endowing the product with distinct ionic and adhesive properties. No waste pollution is generated during the product synthesis process, making it an environmentally friendly product. The relevant process has been searched for in the technology search center of the Zhejiang Provincial Academy of Science and Technology Information, and it has not been mentioned in other literature sources both domestically and

Table 2.7 Comparison of common starch viscosity reduction techniques.

Viscosity reduction methods	Advantages	Disadvantages
Enzymatic Transformation	On-site Modification, Low Usage Costs, Suitable for Large-scale Industrial Applications	Unstable Viscosity, Not Suitable for Coating Adhesives
Oxidation	Wet Reaction, Mild Process Conditions, Stable Quality	High Pollution, Low Yield, Unstable Oxidation Degree, Lack of Incorporation of Special Functional Groups in Molecular Structure, Poor Heat Resistance, Prone to Yellowing
Hydrochloric Acid Degradation	Wet Reaction, Mild Process Conditions	High Pollution, Unstable Viscosity, Susceptible to Aging
Gelatinization	Low Cost, Environmentally Friendly Process	Poor Water Retention in Coatings, Low Strength
Phosphate Esterification	Moderate Cost, Environmentally Friendly Process, Stable Viscosity, Strong Bonding Strength, Good Compatibility	

internationally. The following is a brief overview of the determination of the relevant process route.

2.3.4.1 Viscosity Reduction Process

Common starch viscosity reduction techniques are shown in Table 2.7. After undergoing any one or a combination of the chemical or biological modifications listed in the table, starch can achieve the required viscosity range. However, phosphate esterification not only reduces viscosity but also decreases the gelatinization tendency of starch, thereby enhancing the stability of paste and even coating viscosity.

The laboratory prepared modified starches using various viscosity reduction techniques. The viscosity of these modified starches was determined at different temperatures using a Brookfield viscometer (Figure 2.13). The results show that the viscosity stability of the starch modified through phosphate esterification has been significantly improved. Particularly, the low-temperature flowability of the paste has been enhanced. This indicates that when this modified coating starch is applied in coatings, the viscosity of the coatings will be more stable, making it a relatively ideal coating adhesive.

2.3.4.2 Improve the Viscosity Stability of Starch, Especially Under High-Concentration and Low-Temperature Conditions

Esterification or etherification modification of starch is the primary approach to further enhance the viscosity stability of starch. After improving the viscosity stability of starch, the average molecular weight of starch can be maintained at a higher level.

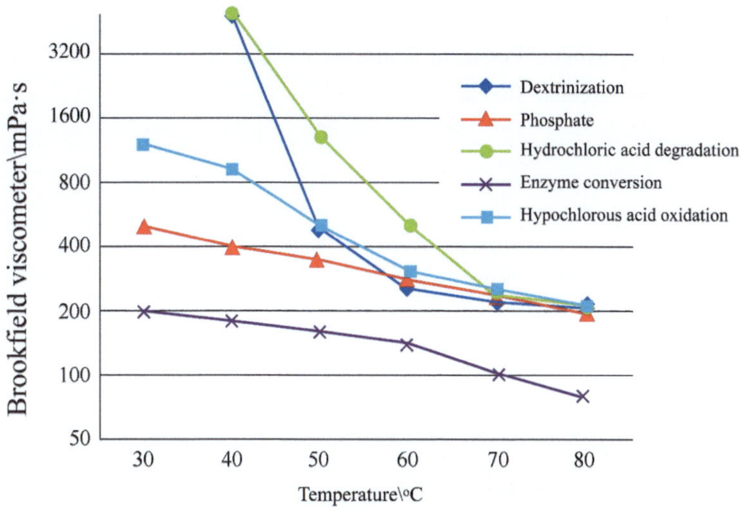

Figure 2.13 Variation in the Brookfield viscosity of coating starch with different viscosity reduction techniques.

As a result, the water retention performance and bonding strength of the adhesive will also increase accordingly. Coating glossiness, smoothness, and other indicators will also improve accordingly.

Acetic anhydride and vinyl acetate are two commonly used and highly effective starch esterification reagents. However, both of these reagents are highly volatile and irritating chemicals. They can only be used in wet starch processing and require a closed environment, demanding high environmental conditions for their use. Chloroacetic acid, as an anionic etherification agent, introduces carboxyl groups into starch through its etherification process. This results in starch paste that is more stable and forms a more resilient film. The experiment compared the enhancing effects of acetate-esterified starch and carboxymethyl starch on the paper surface. It was found that, at different starch dosages, the starch treated with chloroacetic acid resulted in higher surface strength of the paper compared to the acetate esterified starch (Figure 2.14).

2.3.4.3 Enhancing the Modification of Starch Through Cationization

While conventional coating starch applied in coating systems can provide sufficient bonding strength, it may not meet the desired requirements for water retention and ink adhesion. Therefore, it is necessary to further enhance the modification to bestow the starch with excellent water retention properties on top of its existing capabilities. The relevant process involves introducing cationic groups as specific enhancing functional groups to improve the water retention and ink adhesion capabilities of the product. From Figure 2.15, it can be observed that the water absorption of water-absorbent paper treated with enhancing starch is lower. As a result, more moisture is retained in the coating, and with an increase in starch dosage, the amount of moisture retained in the coating also increases. From Figure 2.16,

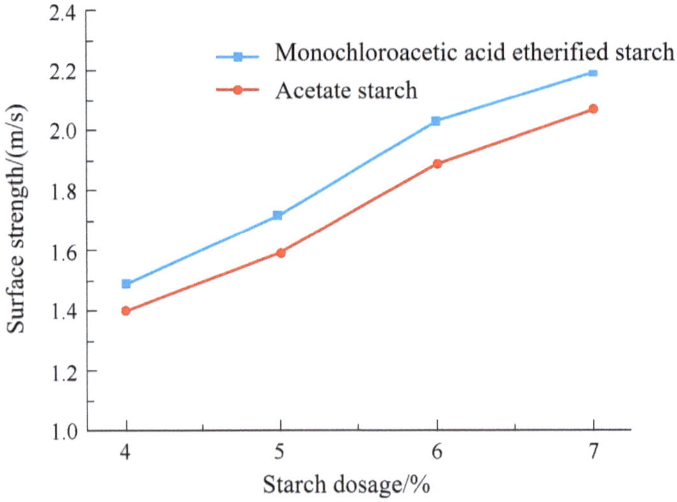

Figure 2.14 Changes in paper-surface strength at different starch dosages.

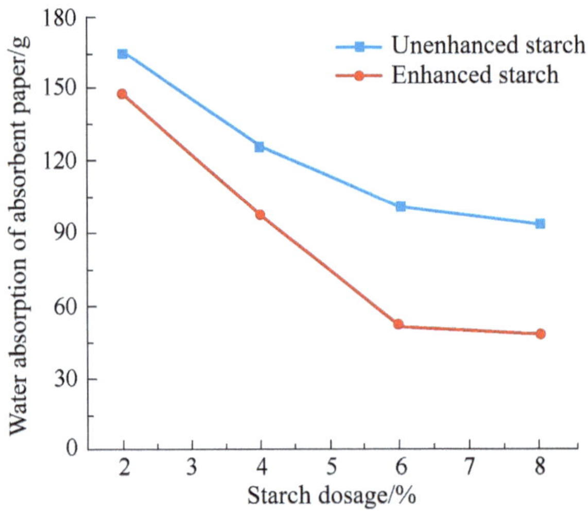

Figure 2.15 Changes in water retention of the coating at different starch dosages.

it can be observed that for different dosages, the K&N values of paper treated with enhancing starch are consistently higher than those of paper treated with non-enhancing starch. This indicates that paper treated with enhancing starch exhibits better ink absorption properties.

Currently, the product has been tested in multiple paper mills, and application performance testing as well as large-scale paper machine trials have yielded positive results. It has demonstrated effective potential to replace certain latex adhesive products.

For example, in a Zhejiang-based company's coating formulation, the environmentally friendly coating adhesive CPT was used to replace 50% of the latex adhesive. The new coating was applied on a production line for producing 250,000 tons of

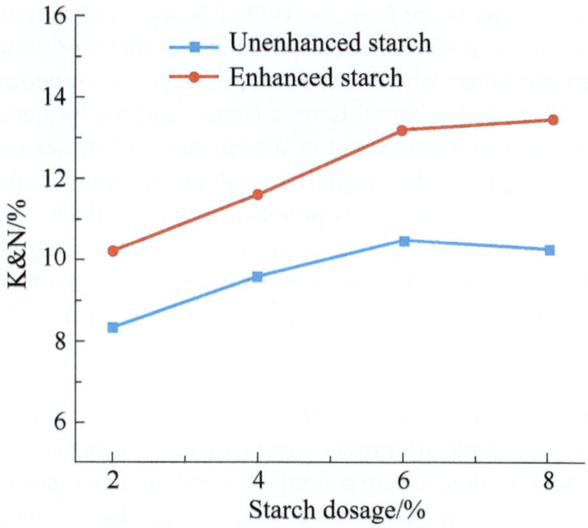

Figure 2.16 Changes in paper K&N value at different starch dosages.

Table 2.8 Comparison of effects before and after the application of environmentally friendly coating adhesive CPT in a certain company.

Sample name	Surface strength/(m/s)	Smoothness/s	Whiteness/%
Company's Latex	0.63	1150	88.8
Environment-friendly Coating Adhesive	0.87	1300	90.0

coated white paper annually, with a machine speed of 800 m/min. The performance of the produced paper with different coating applications is outlined in Table 2.8.

The data indicates that compared to latex-based products, the environmentally friendly coating adhesive shows significant improvements in various aspects of paper performance. The product can effectively reduce costs and is more environmentally friendly.

2.3.5 Specialized Functional Starch for Heat Transfer Printing Paper

Digital transfer printing technologyoriginated in the United States and has also seen rapid development in countries like South Korea and Italy. In recent years, due to the continuous growth of the domestic consumer market, China has become the world's largest consumer and processing hub for digital transfer printing products. Many international businesses have gradually entered the Chinese market. Therefore, thermal transfer printing paper, which is one of the essential consumables for this technology application, is highly favored by businesses. Currently, China imports approximately 10,000 tons of thermal sublimation transfer printing paper annually from Europe, North America, and other regions. Due to price considerations, the main imports are from the United States, Europe, and South Korea. In terms of quality, imported thermal sublimation transfer printing paper is only recognized by consumers when it comes to products from American companies like AW and ATT, as well as South Korea's Hansol and the Netherlands' Jetcol [15]. With the continuous development of digital thermal transfer printing technology and the growing demand for digital thermal transfer products, thermal sublimation transfer printing paper has a very promising market outlook.

Transfer printing refers to the process of transferring dyes from transfer paper onto fabric in the printing process. Transfer printing is a process based on the sublimation properties of certain disperse dyes. Disperse dyes that sublime at temperatures ranging from 150°C to 230°C are selected. These dyes are mixed with a paste to create a "color ink." According to various design requirements, this "color ink" is printed onto transfer paper (a specially designed paper known as transfer paper). Subsequently, the transfer paper with the printed pattern is closely placed in contact with the fabric. Under controlled conditions of temperature, pressure, and time, the dye diffuses into the interior of the fabric, achieving the purpose of coloring [16].

Currently, domestic manufacturers of this type of paper keep their production techniques, materials, and processes confidential. Especially in the preparation and usage of coatings for heat-transfer-printing coated paper, each papermaking company has its unique confidential formulas and process workflows.

In recent years, the research and development team led by Yao Xianping has continuously explored and developed a specialized modified starch [17] specifically designed for coatings used in heat transfer printing paper. The production process flow is shown in Figure 2.17. This product primarily utilizes starch as its main raw material. It undergoes a wet modification process involving crosslinking to achieve special modifications of the starch. Compounds with binary or multiple functional groups are introduced, and they react with the hydroxyl groups of starch to form a multidimensional network structure [18], enhancing molecular interactions and bond strength. Through extensive processing, when the degree of crosslinking reaches 1/100 AGU (Anhydroglucose Unit), functional products that are heat-resistant, non-expanding, non-gelatinizing, and devoid of viscosity can be obtained [19]. During the reaction, as the starch crosslinking degree increases, the product's temperature resistance, acid and alkali resistance, and stability against

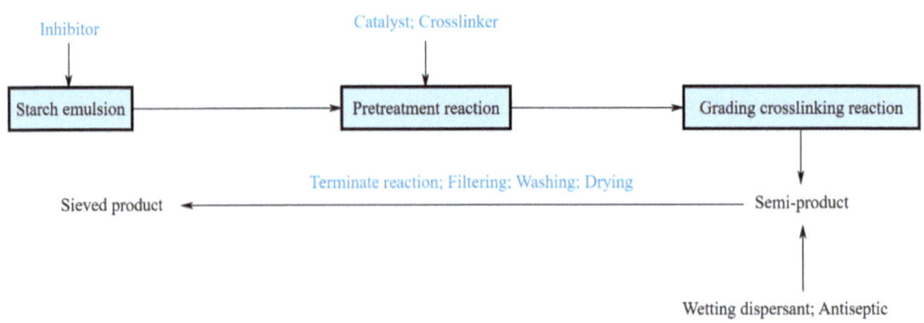

Figure 2.17 Process flowchart for the production of specialized functional starch for heat transfer printing paper.

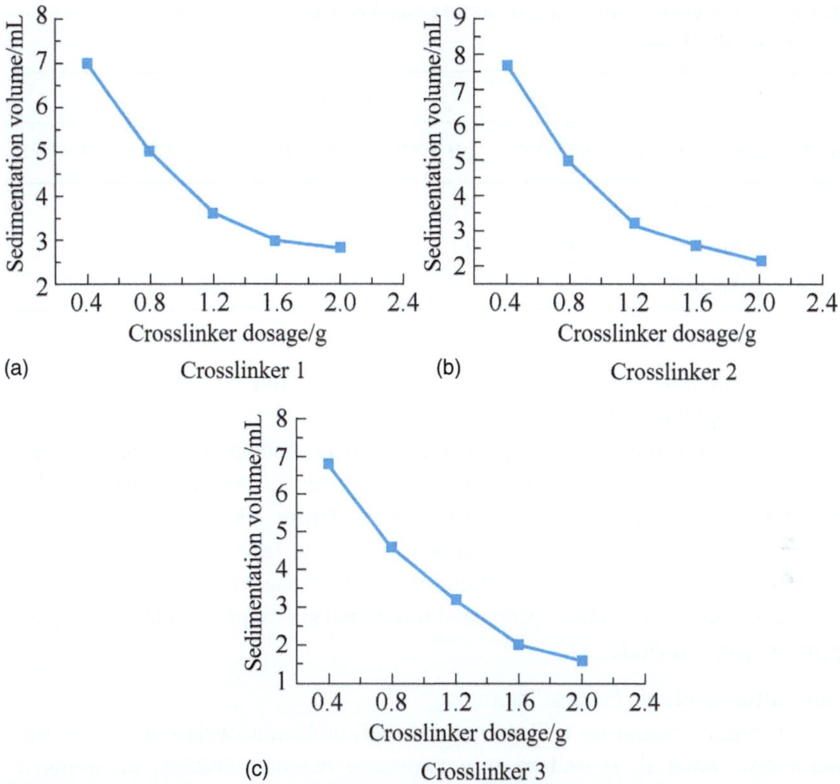

Figure 2.18 Sedimentation volume of starch at different crosslinking agent dosages.

high shear and enzyme actions, all increase. This also maintains the integrity of the expanded particles [20]. Additionally, the product maintains good dispersibility and free-flowing characteristics, making it suitable for various applications in the production process of heat transfer printing paper coatings.

The project conducted research on the crosslinking effects and product stability of starch produced with different crosslinking agents and different dosages. Using the sedimentation volume of the obtained starch samples after the reaction as the judging criterion, it can be observed from Figure 2.18 that under the same dosage, crosslinking agent 3 has the highest degree of crosslinking, while crosslinking agent 1 has the poorest crosslinking effect. The starch with the deepest crosslinking degree prepared using the above three crosslinking agents was selected for stability evaluation. The starch was gelatinized under different pH conditions, and the obtained starch viscosity data are shown in Table 2.9 (viscosity data measured under conditions of \NDJ-1, 1#, 60 r/min, 6%, and 50°C).

From Table 2.9, it can be observed that starch samples modified using crosslinking agent 1 and crosslinking agent 2 exhibited varying degrees of viscosity changes under different acidic and alkaline conditions. Only the starch samples modified with crosslinking agent 3 maintained an extremely stable state under various

Table 2.9 Starch gelatinization of products modified with different crosslinking agents under different pH conditions.

Type of crosslinking agent	Viscosity/MPa·s				
	pH = 1.0	pH = 2.0	pH = 7.0	pH = 11.0	pH = 13.5
Crosslinking Agent 1	5.6	27.2	55.0	42.3	9.6
Crosslinking Agent 2	6.8	12.7	23.0	15.8	11.7
Crosslinking Agent 3	3.4	3.4	3.5	3.4	3.4

conditions, with almost no gelatinization occurring. This indicates that crosslinking agent 3 has a high level of crosslinking efficiency.

Furthermore, the project investigated the effects of reaction pH, reaction time, reaction temperature, and crosslinking agent dosage on crosslinking efficiency, starch thermal stability, and acid-alkali resistance. Ultimately, the optimal conditions were determined as follows: reaction pH of 13.0, reaction time of 5 hours, reaction temperature of 50°C, and crosslinking agent dosage of 2g.

The main innovations of this specialized functional starch product for heat transfer printing paper include:

1) Innovative Synthetic Process Technology

   The product primarily utilizes natural high-molecular-weight starch as the main raw material. It undergoes a two-stage reaction process, including a shallow-level and deep-level reaction, under high pH conditions (pH > 13). This process imparts the product with properties such as high temperature resistance and resistance to strong acids and bases. In the later stages, high-molecular-weight wetting and dispersing agents, as well as sterilization and preservatives, are introduced to further improve the starch's dispersibility, free-flowing characteristics, and overall performance.

2) Innovative Crosslinking Formulation Design

   By optimizing the formulation design and selecting special crosslinking agents, an efficient, cost-effective, and stable crosslinking agent is developed. The crosslinking agent compound with binary or multiple functional groups reacts with the hydroxyl groups of starch, forming diether or diester bonds. The reaction with starch results in a high-strength, stable multi-dimensional network structure. The starch product obtained in this way exhibits high stability under conditions of high heat (above 130°C) and strong acids or bases (pH values below 1.0 or above 14.0). It maintains the basic stability of physical and chemical indicators and performance, effectively addressing technical challenges such as slow drying speed of heat transfer paper printing inks, poor heat transfer printing effects, low color yield, poor ink utilization efficiency, and difficulty in accurately controlling drying time.

Currently, this specialized functional starch product for heat transfer printing paper is being widely promoted and applied in specialty paper manufacturing

Table 2.10 A comparison of the application effects of related patented products with similar products available domestically and abroad.

Sample of paper	Ink drying time (Minutes)	Ink transfer rate (%)	Heat transfer printing effect (Using black Ink density as an example)
Patented Product	<1	>85	1.43
Domestically Available Products	2~3	70~75	1.28
Imported Products	<1	>85	1.35

enterprises in provinces such as Zhejiang, Jiangsu, and Guangdong. Its performance fully meets the requirements of similar foreign products and customer demands. In particular, the product has outstanding performance in improving ink transfer rates, drying speed, and paper smoothness, significantly improving the situation where domestic manufacturing enterprises previously relied on imported foreign products.

The data in Table 2.10 indicates that the heat transfer printing effect using functional starch for heat transfer printing paper is significantly better than domestic products and is comparable to imported products, making it a viable substitute for imports.

2.3.6 High-substitution High-molecular-weight Cationic Starch

Cationic starch refers to a category of starch where natural starch molecules are etherified using cationic agents to impart them with cationic properties. Common cationic groups include tertiary amine groups and quaternary ammonium groups. Its cationic properties make it highly attractive for various applications such as papermaking, textiles, adhesives, and more. Since the first patent for cationic starch was introduced in the 1960s, the number and variety of cationic starches have continued to grow. Cationic starch differs from natural starch mainly in two characteristics: (1) significantly lower gelatinization temperature, and (2) a shift in Zeta potential to a positive value.

In the papermaking industry, cationic starch is used not only for paper strength but also exhibits excellent retention and drainage effects. It is also significant for reducing white-water concentration and facilitating closed-loop water systems in paper mills. Common commercial and large-scale cationic starch products include tertiary amine alkyl starch ethers and quaternary ammonium alkyl starch ethers. Among them, quaternary ammonium alkyl starch ethers have gained recognition and attention from paper manufacturers due to their stronger cationic properties and broader pH adaptability, leading to rapid development. Cationic starch typically used as a papermaking strength agent has a degree of substitution ranging from 0.015 to 0.030. Its primary purpose is to enhance the comprehensive physical strength of paper, including tensile strength, burst resistance, folding endurance, and ring-crush strength. It also allows for increased filler usage while reducing the consumption of high-quality fiber materials [21].

High-substitution, high-molecular-weight cationic starch is a new type of wet-end papermaking chemical that has been introduced in recent years. Its cationic degree is nearly double that of regular papermaking cationic starch, and its viscosity is more than twice as high as that of ordinary cationic starch. These characteristics make it particularly suitable for high-filled paper and paper with a high proportion of recycled fibers, significantly improving the first pass retention and increasing the coating pick strength.

Most of the existing production methods for high-substituted cationic starch and high-molecular-weight starch are wet processes. After the reaction is terminated, they require procedures such as washing, filtering, and drying. These processes are intermittent, resulting in low reaction efficiency, long reaction times, high energy consumption, large wastewater volumes, and significant pollution. The team led by Yao Xianping has developed a new type of high-substituted, high-molecular-weight cationic starch product, which is produced through a continuous, green manufacturing process using a fluidized bed reactor (Figure 2.19). The production process is free from wastewater pollution, has low energy consumption, short reaction times, high efficiency, and continuous product output, which significantly reduces production costs. The product has a high cationic degree and some cross-linking. The starch paste cooked using continuous starch cooking equipment has high viscosity, which, when applied in high-speed paper machines, can significantly improve the retention of fine fibers and fillers, increase ash content and surface strength, save energy, reduce consumption, and mitigate pollution from the paper mill's waste streams.

The research team prepared various high-molecular-weight cationic starches with different levels of cationic substitution. They studied the impact of products with different substitution levels on the performance of recycled paper. By comparing

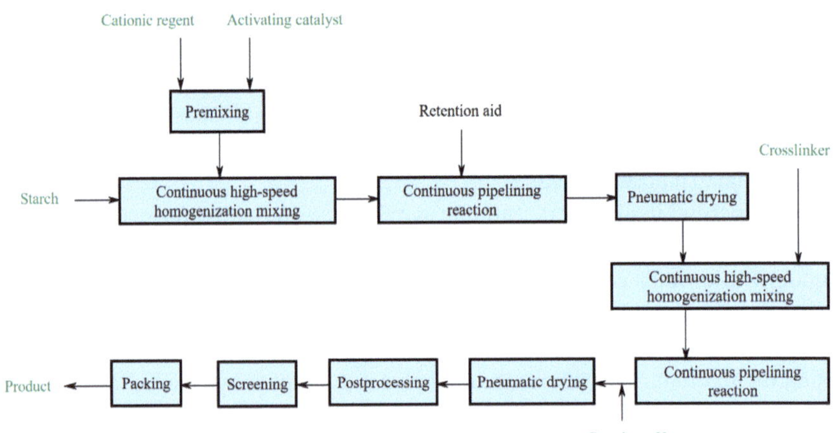

Figure 2.19 Process flow for the production of high-substitution, high-molecular-weight cationic starch.

the application effects of products with different substitution levels at the same starch dosage, they determined that the optimal substitution level is approximately 0.045.

In addition, the research team investigated the impact of different degrees of cross-linking on paper formation and paper properties. Under the optimal cationic substitution degree, various degrees of cross-linking were achieved by adding different amounts of cross-linking agents to prepare high-substitution, high-molecular-weight starch products. The application effects of starch products with different degrees of cross-linking were compared, including properties like filtration rate, first-pass retention, ash content, tensile strength, and surface strength. The best cross-linking-agent dosage was determined to be approximately 0.1%.

The main innovations of this high-substitution, high-molecular-weight cationic starch product include:

1) Innovation in the preparation technology pathway. This project innovatively employs a two-step dry process to produce high-substitution, high-molecular-weight cationic starch: first, an intermediate product with a cationic substitution degree of 0.04 to 0.06 is produced, and then a high-molecular-weight cationic starch is obtained through a cross-linking reaction.
2) Adoption of a new starch powder continuous fluidized pipeline green manufacturing process. This process allows for the rapid, continuous, and uniform mixing of a large amount of starch powder with small amounts of reaction reagents. It addresses the technical challenges of continuous-flow reactions with wet powder materials within pipelines. Additionally, most products can react rapidly at room temperature, resulting in significant energy savings and reduced consumption.

The application results of the developed high-substitution, high-molecular-weight cationic starch product in Paper Mill No. 3 are shown in Table 2.11. The operating conditions were as follows: After continuous cooking, 50% of the high-substitution, high-molecular-weight cationic starch was added to the filler (calcium carbonate), and the remaining 50% was added to the waste-paper pulp. The total addition rate was 5 kg/ton of paper.

Table 2.11 Application effects of the high-substitution, high-molecular-weight cationic starch product in paper mill No. 3.

Testing parameters	Blank	Addition	Before and After Comparison
Burst Length (m)	3180	3747	+17.83%
First Pass Retention (%)	74.7	87.6	+17.27%
Ash Content (%)	10.4	13.5	+29.80%
Printing Surface Strength (m/s)	1.6	2.1	+31.25%
Smoothness (s)	18	30	+66.67%

Based on the application data, the addition of high-substitution, high-molecular-weight cationic starch significantly improves paper quality, greatly increases ash content, reduces production costs, and also improves the retention system to some extent, leading to a further increase in the first pass retention rate.

There are many functional starch-based products for papermaking, and the ones listed above are just some of the important ones with significant quantities and wide applications. Due to space limitations, we won't go into detail on each of them individually.

2.4 Application Technologies of Starch-based Functional Products in Papermaking

The application technology of starch-based functional materials is a high-tech practical technology that spans the chemical materials and papermaking industries. The application effects and benefits vary greatly depending on different paper machines, paper types, fiber ratios, chemical combinations, application methods, and even water quality. For a new starch-based functional material, customers often lack initial application knowledge, and the developers also have limited understanding of the product's performance. Therefore, there is inevitably an exploratory and refining process. While the application process may not be very complex, it involves cross-disciplinary aspects and typically goes through three stages: simple application, coordinated application, and comprehensive application. The so-called simple application technology addresses questions such as where to add the material in the entire papermaking process, how much to add, and what effects to achieve. Coordinated application technology is developed on the basis of simple application, emphasizing the synergistic effects between different materials added in the papermaking process. It aims to achieve cost reduction and efficiency improvement through shared techniques by ensuring that different additives complement each other in the papermaking process. Whether it's a simple application or a coordinated application, the effectiveness of the application is closely related to the production process conditions and management level of the factory itself. To achieve better application results, it is necessary to comprehensively optimize various factors within the factory, maximize the overall efficiency of functional materials, reduce production costs, improve product quality, and enhance economic and social benefits. This is the essence of integrated application technology. Coordinated application technology is developed based on the foundation of simple application technology and serves as an important intermediate step toward achieving integrated application technology. Paper mills can achieve the best results in the use of starch-based functional materials only when they have a correct understanding of the integrated application technology of these materials.

Each product has its specific application range. Some are suitable for wet-end addition, some are suitable for inter-ply spraying, some are suitable for surface sizing, and some are suitable for coating and bonding. Even products intended for wet-end addition can have multiple functional effects such as retention, drainage aid, and strengthening. Even within the category of strengthening agents, there can be various grades and types tailored to different paper types and machine conditions. With numerous manufacturers and a wide range of product names for starch-based functional materials, how should paper mills choose the product that suits their specific needs? Firstly, it is important to thoroughly understand the functional characteristics of the product. Identify the product's category and its ionic characteristics. You should also learn about its main physicochemical indicators and understand its application in similar types of paper. Additionally, you should gather information about the technical management level of the manufacturer and the stability of their product quality. Afterward, you can organize application trials. Considering these factors, paper mills should pay attention to the following points before conducting trials:

1) Work closely with the manufacturing unit, carefully and meticulously develop a well-planned application strategy, and make thorough preparations.
2) Organize a dedicated trial team and ensure clear responsibility assignments. Assign specific individuals to oversee various aspects of the trial, from on-site supervision to paper sample analysis and paper machine processes.
3) After the trial is completed, conduct a thorough economic and technical evaluation, and establish standard operating procedures for regular application.
4) After normal application, it is important to promptly identify any issues and continuously work on improvement and enhancement.

2.4.1 Wet-end Application Technology

The wet-end of the papermaking process is characterized by a complex chemical environment, including multiple components such as fibers, fillers, adhesives, as well as resins present in the fibers, residual chemicals from the pulping stage, various ions, microorganisms, and more. The Zeta potential of the wet-end system can indirectly reflect the electrical potential balance of these components, providing a reference point for controlling paper quality and maintaining the paper machine system. Extensive experiments have shown that by controlling the Zeta potential of the wet-end, it is possible to optimize the paper machine's wet-end forming system. As wet-end additives, starch-based functional materials primarily include cationic starch, anionic phosphate ester starch, amphoteric and multifunctional modified starch, and graft copolymer starch, among others.

Due to the negatively charged surface of fibers and commonly used fillers like precipitated calcium carbonate, talc, kaolin, etc., also having a negative charge, it's not difficult to understand that cationic starch is the most commonly added functional starch in the wet end. Cationic starch, due to its positive charge, can undergo electrostatic adsorption with fibers, thereby enhancing the retention of fine fibers and fillers. Additionally, it can improve wire section filtration and reduce energy consumption.

When anionic starch is added to the wet end, it can lead to charge repulsion, which is not conducive to starch adhesion. Therefore, it is necessary to add cationic bridging agents such as alum to improve adhesion.

Amphoteric and multi-functional modified starches contain anionic, cationic, or non-ionic functional groups. They can adhere to the fiber surface, adsorb fillers, and eliminate the interference of impurities in the system. They have superior reinforcing and retention effects. Their inherent potential is neutral, so they do not affect the Zeta potential of the wet-end system, making them highly adaptable.

In papermaking, single chemicals often fail to achieve good results. Fillers have difficulty adhering to fibers without retention aids, and rosin size cannot be fixed on fibers to achieve sizing effects without alum, which are just the simplest examples. Modern papermaking technology demands higher quality and functional characteristics for paper, and there are more and more papermaking chemicals with various functional properties. The coordination between different chemicals often results in a multiplying or additive effect. Using a single chemical is difficult to achieve at the same additive cost. Paper mills can achieve the best application results at the lowest additive cost only by fully utilizing the synergistic effects between chemicals. This is also the fundamental reason why we need to pay full attention to the technology of synergistic application of chemicals.

No matter what kind of paper is being produced, paper mills have their own corresponding manufacturing processes, such as pulp ratio, pulp consistency, adhesive dosage, paper machine speed, steam pressure, basis weight, moisture content, and so on. The process parameters should all be controlled within a certain range. However, the application of functional starch leads to corresponding changes in the above system. Therefore, to achieve the best application results, in addition to adopting coordinated application technology, it is also necessary to make corresponding adjustments to papermaking process parameters. For example:

1) The beating degree of the pulp will significantly decrease after using retention and drainage aids. A decrease in the beating degree implies an improvement in drainage performance, leading to phenomena such as forward movement of the waterline on the paper machine, reduced vacuum pressure at the couch roll, and decreased paper moisture content. For example, when using high-substitution cationic starch in a high-furnish pulp, the decrease in beating degree can reach around 10°SR. In this case, measures such as increasing machine speed (which can also increase paper production) or reducing the steam pressure in the dryer section (to decrease energy consumption) should be taken.
2) After using a strengthening agent, strength indicators such as tensile strength, tear resistance, and fold endurance significantly improve, while the tear index may sometimes decrease. This is because the tensile strength, tear resistance, and folding endurance of paper primarily depend on the bonding strength of fibers (generally increasing with higher pulp beating degree), followed by the average length of the fibers. When pulping reaches a certain degree, and the average fiber length significantly decreases, a turning point of decline will occur. The addition of functional starch increases the bonding strength between fibers, thereby significantly improving physical indicators such as tensile strength, tear resistance, and folding endurance. However, tear resistance is different; the primary factors affecting tear resistance are average fiber length, followed by fiber

bonding strength and fiber strength. In the early stages of pulping, as the beating degree increases, fiber bonding strength plays a dominant role, leading to a significant decrease in tear resistance. Moreover, tear resistance is more likely to reach the turning point of decline compared to tensile strength, tear resistance, and folding endurance. The application of functional starch improves the retention of fine fibers and fillers, leading to a decrease in average fiber length. Therefore, the decline in tear resistance is a natural outcome. Ensuring tear resistance as an effective measure, while significantly improving other strength indicators, is achieved by reducing pulping time. Shortening the pulping time not only enhances tear resistance and saves electricity but also reduces the content of fine fibers.

3) If the paper strength decreases or does not significantly improve after using retention and drainage aids, you can consider the following methods:
   1) Move the location of adding starch functional materials toward the pulping direction while simultaneously reducing the addition concentration (<1%) to ensure even contact between the additive and fibers, preventing localized excessive flocculation.
   2) While ensuring an increase in ash content, reduce the filler dosage appropriately.
   3) Choose additives that have both strengthening and retention effects simultaneously.
4) If you find that paper brightness decreases after using starch functional materials, the main solutions are as follows:
   1) Place the whitening agent and functional starch as far apart as possible during addition to reduce the chances of interaction. It is recommended to add the whitening agent during pulping.
   2) You can use composite fillers to improve brightness.
   3) You can add a whitening agent during surface sizing to avoid mutual interference between additives.
5) After a significant increase in paper strength, you can reduce the cost of papermaking by increasing the proportion of grass fibers or secondary fibers and decreasing the use of wood pulp.

2.4.2 Surface-sizing Technology

The amount of surface-applied functional products accounts for approximately 60–70% of the total functional products used in papermaking. The main types include functional starch, carboxymethyl cellulose, polyvinyl alcohol, alginates, paraffin, stearic acid/chromium chloride complexes, chromium fluorides, alkyl ketene dimer (AKD), silicone resin, acrylic-maleic anhydride copolymer, and styrene-maleic anhydride copolymer, as well as pigments, latex, and other coating chemicals. In a strict sense, surface application technology is a form of advanced paper processing technology. Paper that has undergone surface treatment is far superior in terms of appearance, quality, printability, and other aspects compared to untreated paper. This is especially crucial for certain specialty papers, such as banknote paper, security paper, drawing paper, high-quality writing paper, and offset printing paper, all of which have high-quality requirements. Surface sizing is an extremely important step in the production process. It not only enhances the surface strength of the paper and imparts excellent physical properties such as improved tear resistance, folding endurance, tensile strength, ring-crush strength, and so on, but it also allows for the selection of specific sizing agents to give the paper certain characteristics as per requirements, such as water resistance, oil resistance, acid resistance, and alkali resistance.

With the development of modern technology, there has been a significant increase in the demand for specialty papers. Surface application technology has increasingly demonstrated its importance in the field of papermaking chemicals and applications. Surface sizing, in essence, is the process of applying sizing agents to the surface of paper to bond fibers with the paper and create a nearly continuous thin film on the paper's surface. Surface sizing is typically achieved through a process called size press application, which allows almost all of the sizing solution to be retained on the paper's surface. This overcomes the drawback of additives being lost within the pulp and increases the efficiency of sizing solution utilization. It also contributes to the closed-loop circulation of white water in the paper machine [22]. This characteristic makes the cost of surface sizing much lower than internal sizing (when using the same sizing agents).

2.4.2.1 The Selection of Surface-Sizing Agents

The functional starches suitable for surface sizing primarily include: (i) heat or thermally converted starch, (ii) acid-modified starch, (iii) oxidized starch, (iv) enzyme-converted starch, (v) acetylated starch, (vi) hydroxyalkyl (propyl or ethyl) starch, (vii) anionic dual-modified starch, (viii) cationic starch, and (ix) octenyl succinic anhydride modified starch.

Due to the large quantities of starch used in surface sizing, most paper types produced in China employ functional starches that are convenient to convert and cost-effective. Products such as the ones mentioned in (i) to (iv) fall into this category. In particular, oxidized starch remains the mainstream product in China.

Thanks to the invention and widespread adoption of continuous-cooking equipment for starch, modern large and medium-sized paper mills have increasingly integrated the gelatinization and addition processes of starch into the design of their equipment and paper machines. This has made the application of starch simpler, more practical, and amenable to automatic control.

In recent years, with the development of digital technology, specialty digital printing papers (inkjet printing papers) with unique surface properties have seen rapid growth. New types of high-retention surface-sizing starch are gradually becoming essential products for digital printing papers and other high-quality office papers.

2.4.2.2 The Selection of Surface-Sizing Equipment

There are primarily three types of surface-sizing equipment: size press application, gate roll sizing, and film sizing (in Figure 2.20). Gate roll sizing and film sizing have emerged as new developments in recent years.

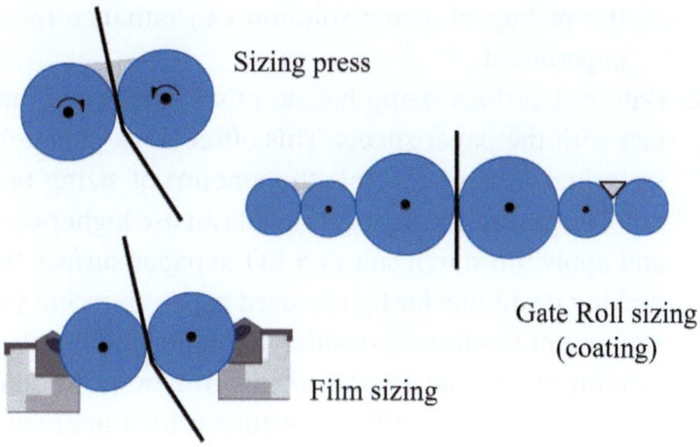

Figure 2.20 Various types of surface-sizing equipment.

1) Size press application

Size press application is a relatively traditional sizing equipment and comes in various forms such as vertical, inclined, or horizontal. Its operating principle involves the paper web passing through a sizing solution bath before entering the nip between the press rolls. This allows the sizing agent to be applied to the paper's surface. Subsequently, the paper web passes through the press rolls, which press the sizing agent into the paper while removing excess sizing material from the paper surface. Size press application equipment is often chosen for common cultural papers and medium to low-grade paper machines.

The main factors affecting size press application include:

1) The sizing degree of the base paper

Paper with a low sizing degree absorbs a large amount of sizing solution. Paper that has not been sized at all absorbs 100% more sizing solution than sized paper. However, once a certain level of sizing degree is reached, the impact on sizing solution absorption becomes less significant. In the case of neutral paper, it's different. Due to the lag effect of sizing degree in neutral paper, at a certain machine speed, when the basis weight of the paper increases, it often absorbs more sizing solution.

2) Machine speed of the paper machine

Between a machine speed of 100 m/min and approximately 300 m/min, the pickup of sizing solution significantly increases. However, after the machine speed reaches 650 m/min, there is no significant change in the pickup of sizing solution.

3) Basis weight of the paper

For various basis weights of paper sized with rosin, between a machine speed of 200 and 360 m/min, the basis weight and machine speed have almost no impact on the pickup of starch.

4) Other properties of the paper

The smoother the paper surface, the lower the pickup of sizing solution, while higher paper porosity results in a higher pickup of sizing solution. When the viscosity is less than 20 mPa·s, increasing the starch solid content can increase the pickup of sizing solution. For coated paperboard, increasing the pickup of sizing solution can enhance the ring-crush strength of the paperboard.

2) Gate roll surface sizing has an offset stock pool that does not come into contact with the paper sheets. This offset stock pool delivers sizing material to the metering nip and controls the amount of sizing material entering the second nip. The gate roll sizing equipment can use higher-concentration starch solutions and apply the starch solution to the paper surface through a "coating" process, making it suitable for lightly sized paper. In recent years, gate roll surface-sizing equipment has found considerable application in the light coating of newsprint, yielding significant improvements in newsprint quality. However, it has drawbacks such as high initial investment, high maintenance costs, and long downtime during roll replacement.
3) Film sizing utilizes a metering bar, which allows for precise control of the surface-sizing amount. By adjusting the pressure on the metering bar, the sizing quantity can be easily modified. This feature enables control over two independent variables: pickup amount and solid content. Another characteristic of film sizing is its similarity to doctor blade coating, which allows for increased solid content and reduced sizing solution volume. This means that a certain level of surface strength can be achieved with less starch, leading to improved operational efficiency.

2.4.2.3 Shared Chemical Technology in Surface Sizing

In typical surface sizing, functional starch is usually the only chemical used. However, to meet the performance requirements of different types of paper, shared chemical technologies are increasingly being used in surface sizing, such as

1) To enhance the paper's water resistance, starch can be used in combination with chemicals that improve water resistance such as UF resin, AKD, alkyl ketene dimer (ASA), or dispersing rosin size. This approach yields significant results and is often employed for highly water-resistant papers.
2) To improve the paper's resistance to oil, starch can be combined with fluorinated oil-resistant resins, polyvinyl alcohol (PVA), and other chemicals. Oil-resistant paper is used more frequently in food packaging.
3) To enhance heat resistance, starch can be used in conjunction with heat-resistant chemicals like styrene-maleic anhydride copolymers.
4) To improve surface strength, starch can be combined with high-molecular-weight polymers like acrylic-maleic anhydride copolymers. This is quite common in cultural papers.
5) Other aspects such as improving brightness, opacity, and smoothness, can involve adding ingredients like whitening agents and fillers to the starch adhesive.

The development of shared surface-sizing technology is advancing rapidly, and colored inkjet chemicals are an example of a combination of multiple chemicals used together. By employing film sizing, the goal of producing colored inkjet printing paper can be achieved.

Surface sizing is a complex papermaking technology with a rich content. Here, we can only focus on the main aspects for introduction.

2.4.3 Interlayer Application Technology

Interlayer strengthening chemicals mainly include interlayer spray starch. Their development began in the early 1970s. However, their growth rate has been rapid.

Spray starch was initially developed to enhance the interlayer bond strength of paperboard. In recent years, its development has been particularly rapid, and its application has expanded beyond paperboard, moving to thicker paper grades. Its application purpose has also expanded beyond just improving interlayer bond strength to include enhancing stiffness, ring-crush strength, and surface strength widely. The application results demonstrate that the extent of improvement in paper strength significantly depends on the method of starch addition. Using spray starch and its application technology has proven to be the most effective method for enhancing the stiffness, ring-crush strength, and interlayer bond strength of paperboard and thick paper.

Compared to wet-end addition technology, while starch spray may not have the same ability to aid in the retention and filtration of fine fibers and fillers, its strengthening effect is no less than that of internal addition. This is especially true for pulp with a higher content of ions, such as groundwood pulp, recycled paper pulp, and straw pulp. The strengthening effect of starch spray is significantly better than wet-end addition in such cases, and spray starch has a higher retention rate than starch added internally. Compared to surface sizing, although it may not match surface sizing in improving paper and paperboard surface properties, it excels in enhancing the internal bond strength of paper, and it does not require a drying section after the sizing press, making it particularly suitable for paper machines without a sizing press. Certainly, by combining wet-end addition, starch spray, and surface sizing, and leveraging their respective strengths, more ideal results can be achieved. This is a commonly used technique in countries with developed papermaking.

2.4.3.1 Application Mechanism and Performance Requirements

The application mechanism of spray starch is as follows: When paper and paperboard are formed in the wet end of the paper machine, starch particles are dispersed in water to form a suspended slurry, which is then sprayed onto thick paper or paperboard through a spray system. Starch particles are retained through mechanisms such as adsorption by pulp fibers, obstruction by the fiber network, intermolecular attraction and binding, and encapsulation by the fibers that make up the paper sheet. Subsequently, the retained starch receives heat in the drying section, causing it to gelatinize and enhance the interlayer bond strength of the paperboard through cross-linking and hydrogen bonding, among other mechanisms [23].

Spray starch and wet-end starch have many differences in performance and application methods, as outlined in Table 2.12.

Table 2.12 Differences between wet-end starch and spray starch.

Project	Wet-end starch	Spray starch
Applicability Pre-addition Treatment Addition Location Dosage Addition Method Primary Application Effects Starch Retention Rate	Generally suitable for most paper grades	Only suitable for thick paper and board paper
Applicability Pre-addition Treatment Addition Location Dosage Addition Method Primary Application Effects Starch Retention Rate	Starch solution needs to be pre-gelatinized	No need for pre-gelatinization
Applicability Pre-addition Treatment Addition Location Dosage Addition Method Primary Application Effects Starch retention rate	Added in the pulp before the headbox (i.e., before the wire section)	Applied in the wire section or between layers
Applicability Pre-addition Treatment Addition Location Dosage Addition Method Primary Application Effects Starch Retention Rate	Typically added at a consistency of 0.5–2.0% dry pulp weight	Typically controlled at around 1g/m2, but the dosage should be doubled when improving stiffness and ring-crush strength
Applicability Pre-addition Treatment Addition Location Dosage Addition Method Primary Application Effects Starch Retention Rate	Added either by metering or in a one-time addition	Applied by metered spraying
Applicability Pre-addition Treatment Addition Location Dosage Addition Method Primary Application Effects Starch Retention Rate	Enhances paper tensile strength, tear resistance, and surface strength	Enhances stiffness, ring-crush strength, and interlayer bonding strength
Applicability Pre-addition Treatment Addition Location Dosage Addition Method Primary Application Effects Starch Retention Rate	Increases retention of fine fibers and fillers	Almost 100% retention rate

Due to the special application method and mechanism of action of spray starch, spray starch must possess the following conditions:

1) It must have good adhesive properties with fibers.
2) It should have a relatively high initial retention rate, especially when used in paper recycling; it should not have a negative impact on strengthening, retention, and papermaking.
3) It should have a lower gelatinization temperature, allowing it to gelatinize promptly and effectively as paper passes through the drying cylinder.
4) It should have low viscosity and high adhesive strength.
5) It should have a small particle size, capable of producing fine spray through the spraying system, distributing evenly on the paper, and not clogging the nozzles.

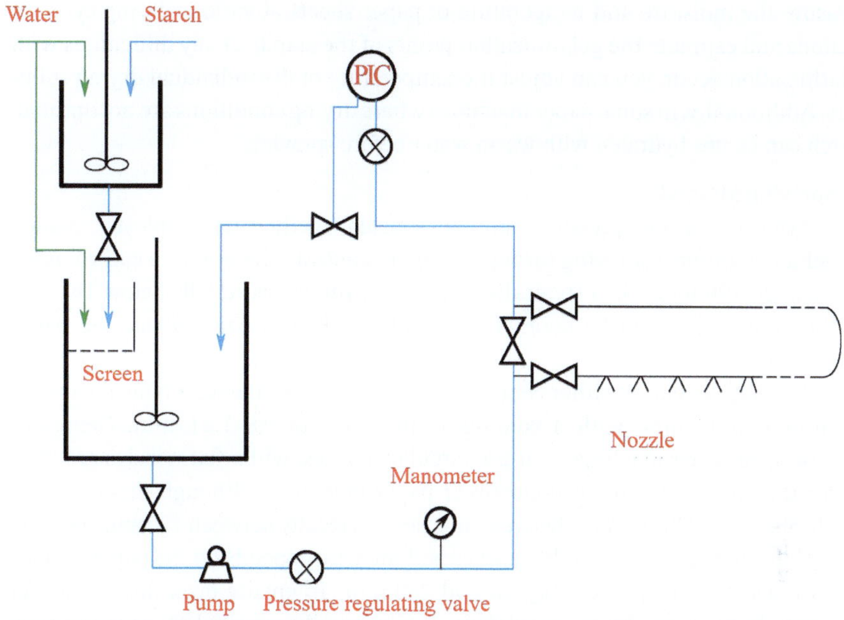

Figure 2.21 Spray starch application process flowchart.

2.4.3.2 Spray Technology

The application process of spray starch varies depending on the type of paperboard and the conditions of the paper machine. The basic process flow is as shown in Figure 2.21.

Spray starch is first prepared by mixing it with cold water to form a certain concentration suspension under agitation. After the suspension is filtered through a 120-mesh screen, it is pumped through a high-pressure metering pump (to generate a certain pressure) and sprayed onto the paperboard using a spraying device. The concentration of the starch solution, the spray flow rate, and the paper machine speed determine the amount of spray starch added. Different quality requirements and paper machine conditions allow for the selection of different addition amounts. Factors such as different spray points, spray pressure, spray state, and spray distance all affect application effectiveness, and these are the key technical aspects of the process. Through diligent efforts, China has gained a lot of successful application experience in various types of paper (such as whiteboard paper, kraftliner, white top linerboard, corrugated paper), different paper machines (long wire, long and round wire, multi-ply wire), and various quality requirements (interlayer bonding strength, surface strength, stiffness, etc.).

It is worth noting that if you can see water droplets, stripes appear on the paper, and there are significant fluctuations in paper moisture and temperature in the drying section compared to before use, these issues may be caused by poor atomization. The quality of atomization directly affects the effectiveness of use, so controlling atomization is crucial. Gelatinization is also crucial. Before use, it's important to measure the moisture and temperature of paper sheets at various drying cylinder sections and estimate the gelatinization points of the starch. If any difficulties with gelatinization occur, you can adjust the temperature of the individual drying cylinders. Additionally, in some paper machines where drying conditions are not optimal, starch can be pre-hydrated with warm water before spraying.

1) Spraying Method

   Typically, a low-pressure, non-air-assisted method is employed, which achieves uniform spraying through pressure control. The pressure control is set between 196 to 392 kPa (typically 294 kPa). If the pressure falls below 196 kPa, the atomization will be poor. If it exceeds 392 kPa, it can damage the paper formation.

   The nozzles can be either circular or flat. The nozzle diameter typically ranges from 1 to 1.5 mm, with a volumetric flow rate of 0.8–1.2 L/min. Generally, long-wire paper machines can use circular nozzles, while flat nozzles are used for interlayer spraying in Fourdrinier paper machines, although this is not an absolute rule. The spacing between nozzles is typically between 120 and 160 mm, and the distance between the nozzles and the wire depends on the paper surface coverage by spraying, usually around 250 mm. To ensure even distribution of spray droplets on the paper surface, it is generally required to have multiple cross-overlapping (as shown in Figure 2.22). To minimize the loss of atomized starch and ensure uniform spraying, you can adjust the angles of the nozzles and spray pipes as needed.

2) Spray Location

   The spray points for starch vary depending on the conditions of the paper machine and the intended application. For long-wire paper machines, round-wire paper machines, as well as long and round-wire combined paper machines, typical addition locations are shown in Figure 2.23. For long-wire paper machines, because the substance being sprayed is essentially an uncooked starch slurry, the retention mechanism involves the filtration of starch particles. Therefore, the spray location is crucial: If the nozzle is positioned close to the

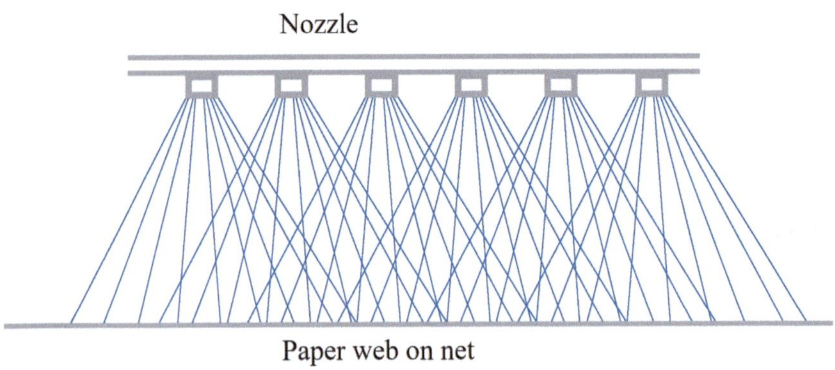

Figure 2.22 Starch fog liquid coverage diagram.

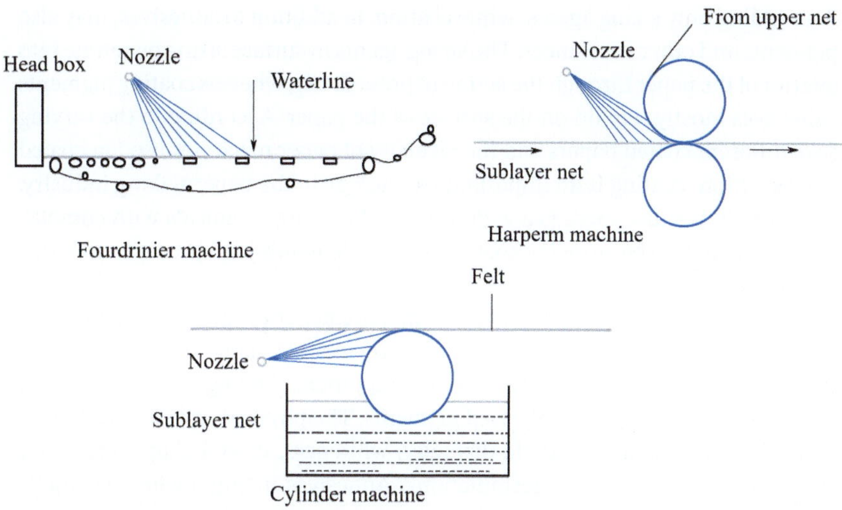

Figure 2.23 A schematic diagram illustrating the application of sprayed starch on different paper machines.

wire, the majority of the starch will be distributed on the paper's surface, which is highly effective in improving surface performance. If the nozzle is moved in the direction of the headbox, the starch will be evenly distributed in the Z-direction of the paper sheet, which is highly effective in improving paper-fiber bonding, increasing tear strength, and enhancing tensile strength, among other physical properties. But it should not move too far away from the waterline or exceed the waterline, otherwise it will affect the actual application effectiveness. Therefore, relevant experiments should be conducted to determine the optimal position when using it. For long-elliptical-wire paper machines or circular-wire paper machines, starch is typically sprayed onto the composite point between the two layers of paper.

3) Spray Volume

The amount of starch sprayed is determined based on actual requirements. When the paper machine conditions remain constant, it is typically adjusted through the starch slurry concentration, which is generally maintained between 1 and 10%. When used as an interlayer enhancer, the quantity can be lower (slurry concentration approximately 2%). If it is used to improve internal bond strength or surface strength, a slightly higher quantity is appropriate. If stiffness needs to be increased, the slurry concentration should be higher.

2.4.4 Coating Application Technology

Coating and surface sizing both involve applying chemicals to the surface of paper, but there is a significant difference between the two: surface sizing is typically done in the press section, while coating is done in the dry section. Surface sizing typically involves adding only sizing agents, while coating, in addition to adhesives, may also use pigments and other substances. The sizing agents in surface sizing penetrate into the interior of the paper through the action of press sizing, whereas coating pigments and adhesives mostly remain on the surface of the paper. According to the varying needs of different coated papers and the diversity of paper machines used in coated paper production, coating is an important technology in the papermaking industry. This is not only because there is a wide variety of coating chemicals with complex formulations, but also because the coating process demands high precision in terms of technology.

Coating chemicals mainly fall into three major categories: pigments, adhesives, and other functional chemicals. Pigments mainly include kaolin, China clay, satin white, calcium carbonate, titanium dioxide, and others. Among these, kaolin and China clay are the most commonly used pigments. The physical and chemical properties of pigments, such as particle diameter, whiteness, crystal shape, etc., have a significant impact on coating formulations. Adhesives mainly include synthetic latex (carboxylated styrene-butadiene latex, styrene-butadiene latex, polyvinyl alcohol, etc.), modified starch, casein, soy protein, and others. Functionality starch, commonly used for surface sizing, can generally also be used for coating, but with lower viscosity and better stability requirements. Due to high costs and environmental issues such as strong odor and poor biodegradability associated with synthetic latex, in recent years, Hangzhou Zhiyou Technology Co., Ltd. has successfully developed a fully biodegradable and high-performance bio-latex. This bio-latex can replace two to three parts of synthetic latex in formulations, significantly reducing costs and environmental pollution. It has been successfully applied in multiple large paper mills. The other functional additives mainly include coating dispersants, waterproofing agents, lubricants, preservatives, defoaming agents, etc., and are not further elaborated upon due to space limitations.

The core technology of coating lies in the formulation of coatings, and different types of paper and different companies all have their own formulation research. In coating formulations, commonly used types of functional starch include thermally modified starch, oxidized starch, enzyme-modified starch, hydroxyethyl starch, esterified starch, and so on. In coatings, functional starch typically acts as an adhesive, exhibiting excellent bonding properties, moisture retention, and a wide viscosity range, as well as characteristics of safety, environmental friendliness, and biodegradability.

Fuyuqing et al. used surface coating with cornstarch adhesive to enhance the strength of corrugated base paper. By increasing the coating amount, they were able to gradually improve the corrugated base paper's ring-crush index and tensile strength but slightly decrease its foldability. Its improvement in the ring-crush index is significant. Through surface coating with starch adhesive, it is possible to achieve a ring-crush index for low basis weight base paper comparable to that of high basis weight base paper, providing a new pathway for the papermaking industry to conserve fiber raw materials and reduce costs [24].

2.4.5 Relevant Green Technology Applications and Equipment

2.4.5.1 High-Temperature Jet Cooking Technology and Equipment for Starch Continuous Cooking

The high-temperature jet cooking technology for starch is a new technique that involves the instantaneous high-speed mixing of starch slurry and steam through a special device, causing rapid gelatinization of starch under high pressure and high temperature. The special device used for this purpose is referred to as a high-temperature jet cooking starch continuous cooker. The world's first high-temperature jet cooking starch continuous cooking device for commercial applications was developed in the 1950s. Compared to conventional batch tank cooking technology, high-temperature jet cooking starch continuous cooking technology has distinct advantages. It serves as the foundation for a range of starch application technologies that require temperatures exceeding 100°C, including high-temperature starch hydrolysis, continuous enzymatic starch conversion, high molecular weight starch, high amylose starch, cross-linked starch, and polymerized starch, among others. In other words, without this technology, these starches would be either unusable or unable to realize their full potential. As a significant application equipment for starch derivatives, the high-temperature jet cooking starch continuous cooker is a standard component of large, modern paper machines. Its importance in papermaking can be easily understood.

As is widely known, the application of starch derivatives involves processes such as heating for gelatinization, insulation, and dilution. Currently, the majority of papermaking companies in China still primarily employ traditional batch tank cooking methods. However, batch tank starch cooking has many drawbacks, including low production capacity, high energy consumption, high labor intensity, difficulty in controlling cooking conditions, starch agglomeration, viscosity fluctuations, low starch gelatinization efficiency, and compromised paste quality and application performance. Particularly noteworthy is that traditional batch tank cooking systems have long cooking times (typically requiring around 1.5–2 hours per batch of starch paste), low daily processing capacity (mainly in the range of 1–5 tons per day), and are unable to achieve mechanized operation.

With rapid improvement in the technological equipment level of paper mills, both the individual machine output and operational efficiency of paper mills have significantly increased. Previously, there were no specialized manufacturers in China producing continuous starch cooking equipment, and papermaking companies had to spend millions of yuan to import jet-cooking starch continuous cooking equipment. In recent years, Hangzhou PaperMate Technology Co., Ltd. has successfully developed various models of high-temperature jet-cooking starch continuous cooking equipment. Its continuous cooking process is depicted in Figure 2.24, and it has been applied by dozens of paper mills. The design capacity can be tailored to specific requirements, with cooking capacities ranging from 0.15 to 80 tons per day, as shown in Figure 2.25.

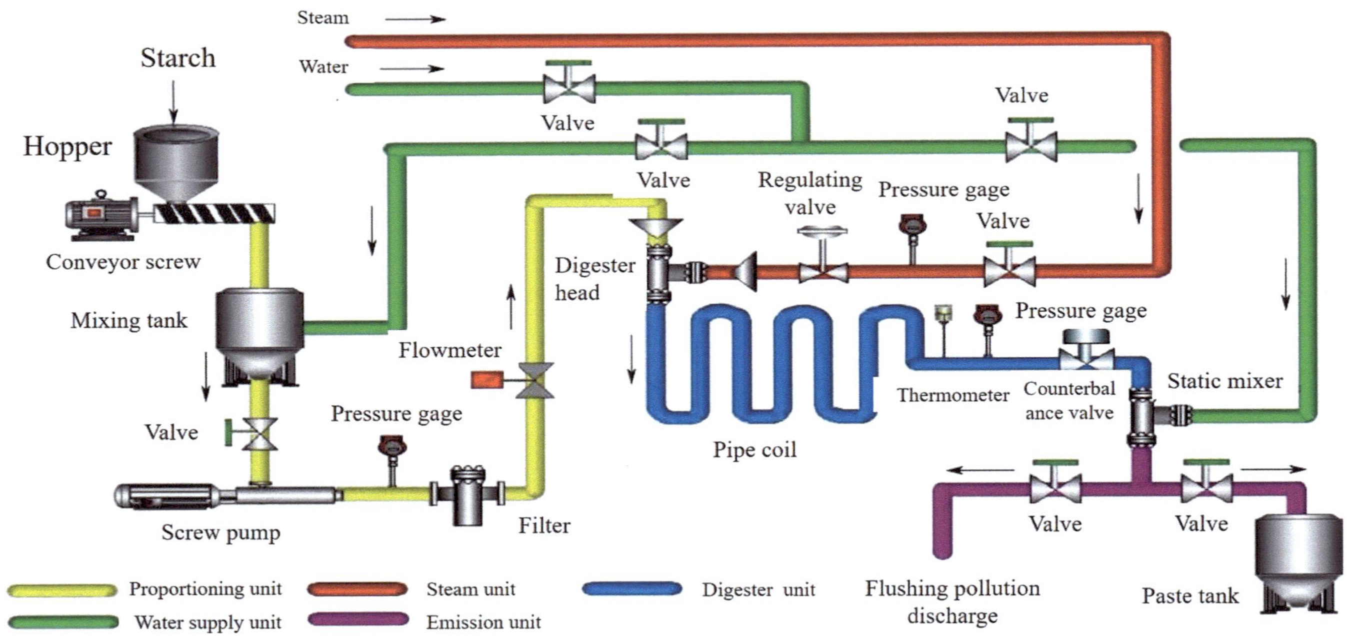

Figure 2.24 Process flowchart of high-temperature jet cooking starch continuous cooking.

Steaming and boiling starch (0.1~0.2t/d)
(a)

Steaming and boiling starch (20~30t/d)
(b)

Figure 2.25 Different types of high-temperature jet cooking starch continuous cookers.

The main advantages of high-temperature jet cooking starch continuous cookers are as follows:

1) Due to the instantaneous high-speed mixing of starch slurry and steam, it enables efficient, energy-saving continuous production and significantly increases production capacity. Through calculations, it has been determined that high-temperature jet cooking starch continuous cooking can save approximately 50% of electricity and about 10% of steam compared to batch tank cooking. This is mainly due to the substantial increase in starch cooking concentration and significant reduction in cooking time.
2) Due to its full automation and modularity, it can be operated with high mechanization, resulting in extremely low labor costs. When integrated with a distributed control system (DCS), there is almost no need to hire additional personnel, leading to a significant reduction in labor expenses.
3) High gelatinization concentration and short cooking time allow for on-demand production, reducing preparation time, and promoting environmental protection and energy efficiency.
4) Due to complete and uniform starch gelatinization, it significantly enhances the uniformity of starch paste, improves application effectiveness, and maximizes the functionality of modified starch. Especially for cationic starch used in the wet end, it can enhance the efficiency of the papermaking wet-end circulation system, reduce the usage of cationic starch, lower the load on white-water consistency, and consequently reduce production costs.
5) Starch can undergo enzymatic conversion, allowing for the substitution of relatively inexpensive starch products for commercial starch.
6) Thanks to fully automated and intelligent equipment, it can prevent errors caused by human factors.

7) High-temperature jet cooking starch continuous cooking serves as the foundational application for a range of high-end starch derivatives, including high-temperature starch hydrolysis, enzymatic starch conversion, high-molecular-weight starch, high-amylose starch, cross-linked starch, and polymerized starch.

The high-temperature jet cooking starch continuous cooking technology for papermaking can significantly enhance the modernization level of chemical application equipment in paper mills, improve the application effectiveness of wet-end, surface, and coating starch, promote energy conservation and environmental protection, reduce overall operational costs, and enhance paper machine operational efficiency, among other benefits. Due to the invention and widespread adoption of high-temperature jet cooking starch continuous cookers, modern large and medium-sized paper mills have incorporated this equipment as standard for their large paper machines. Configuration for this equipment is now considered during paper machine design. The high-temperature jet cooking starch continuous cooking technology simplifies and makes starch applications more practical, allowing for automated control. It is a green technology worth vigorous development and widespread promotion.

2.4.5.2 Papermaking Fillers Online Surface Modification and Jet Mixing Technology and Equipment

In the production process of filled paper sheets, fillers are typically added to the paper pulp system before the headbox. Single or dual-component retention aids are often added before and after the filler addition point to improve the retention of fillers on the forming wire. Adding fillers to the paper pulp not only helps improve physical properties such as opacity, brightness, texture, and printability of the paper but also accelerates the drying rate of wet paper sheets, saving on steam consumption during drying. Furthermore, fillers are typically much cheaper than papermaking fibers, so increasing the use of fillers helps reduce paper production costs and lessen the ecological environmental impact.

However, at present, the addition of fillers to paper can only be maintained at relatively low levels, with one significant limiting factor being that increasing the amount of fillers in paper can lead to a decrease in paper strength. For example, this decrease in paper strength can manifest as reduced tensile strength, fold resistance, tear resistance, and compression resistance. It can also lead to increased linting and dusting during cutting and printing processes, among other issues. One crucial prerequisite for increasing the ash content in paper is to enhance the initial retention rate of fillers. This necessitates that fillers exhibit high retention and filtration performance under the shearing, turbulent, and vacuum conditions during the papermaking process. Retention aids typically play the role of retaining fillers and are generally added at single or multiple points both before and after the filler addition point. One drawback of this system is that retention aids are prone to being captured by long fibers, small fibers, anionic impurities, or other chemicals before coming into contact with the fillers. This reduces the chances of retention aids interacting with the fillers. One undesirable solution is to increase the dosage of retention aids, but this can lead to the formation of larger pulp flocs with a detrimental effect on paper uniformity, which is undesirable in the papermaking process.

An increase in the ash content of paper typically leads to a decrease in paper strength. On one hand, this is because a reduction in the proportion of fibers results in a decreased number of hydrogen bonding sites between fibers. On the other hand, fillers themselves do not possess hydrogen bonds, and they cannot form hydrogen bonds between fillers or between fillers and fibers. Furthermore, fillers placed between fibers also hinder the formation of hydrogen bonds between fibers due to spatial constraints.

Is it possible to enhance the adhesion between fillers and fibers by modifying the surface of the fillers, thereby increasing filler retention while minimizing strength loss? Yao Xianping's team has proposed a new method for continuous on-site surface modification of fillers in papermaking. The team uses a specialized jet mixing device to apply modified starch and retention aids for the surface modification of papermaking fillers [25]. The results have demonstrated that this method, while maintaining paper strength, allows for an increase in the usage of papermaking fillers, consequently reducing the consumption of fibers in the papermaking process. It represents a more favorable green technology.

The aforementioned specialized jet mixing device is the key to achieving online surface modification of fillers. The online jet mixing device is a coaxial distributed Venturi jet mixing device (see Figure 2.26). The three coaxial injectors, A, B, and C, divide the space into four chambers, labeled as a, b, c, and d. In chamber a, the fluid line velocity of the Venturi injector ranges from 10 to 25 m/s, with the preferred range being 16–20 m/s. A higher injector flow rate results in a greater axial distribution of the mixed fluid. The chemical flow rate in chamber b of the Venturi injector depends on the dosage and concentration of the chemical used. In chamber c, the flow rate of the turbulent mixing flow ranges from 0.5 to 5 m^3/h, with a larger turbulent flow rate leading to a wider distribution of the mixed fluid in the cross-section. Chamber d serves as the turbulent mixing zone, where the jet flow, chemical flow, turbulent flow, and pulp are uniformly mixed within the area.

The preferred choice for jet flow and turbulent flow is using papermaking white water or paper machine white water. This helps conserve freshwater resources and reduces the energy required for heating clean water. To achieve the desired jet pressure and flow rate, centrifugal pumps or screw pumps can be used to pressurize the jet flow and turbulent flow. To accurately meter the chemical flow, metering pumps or screw pumps can be used to measure and add chemicals.

When the specialized jet mixing device is installed at two different positions, one before and one after the paper machine wet-end pressure screen, the jet mixing device before the pressure screen meters the addition of papermaking fillers into the jet flow of the paper machine white water or dilute white water, forming the primary jetting medium together. Modified starch, as the main medium in the chemical flow, mixes with the fillers in the area where the jet flow and chemical flow intersect. After further mixing through the pressure screen, the jet mixing device located after the pressure screen encapsulates retention aids, fiber pulp, papermaking fillers, and modified starch within a relatively stable micro-flocculated particle.

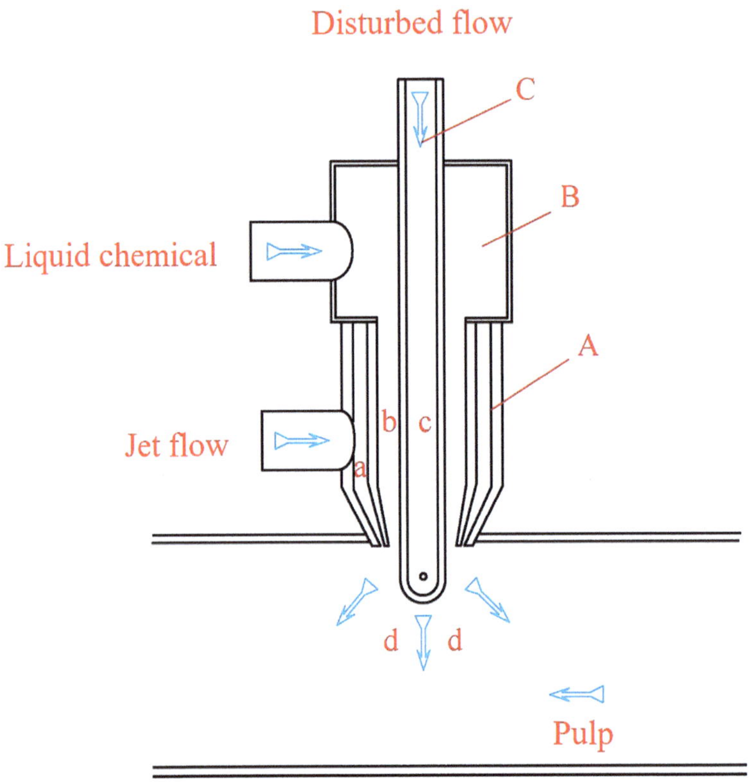

Figure 2.26 Online papermaking chemical injection mixing device.

The turbulent flow is a fluid ejected from holes evenly distributed on the bottom side of the turbulent flow injector. It forms a “water curtain” that disrupts the jet flow, chemical flow, and pulp flow, creating turbulent fluid, and evenly distributes the surface-modified filler micro-flocs within the pulp before being sent to the headbox for papermaking. These micro-flocs are no longer subjected to strong shear forces from pulp pumps, pressure screens, and the like before entering the headbox, which results in a higher initial retention rate. The primary advantages of this application technology and equipment are the improvement of paper surface and internal bonding strength, reduction in fiber consumption, increased ash content in paper, cost savings in paper production, and the simplicity and stability of practical operation. Compared to traditional chemical injection devices, the key feature of this equipment is a significant enhancement in the instantaneous and uniform mixing and modification effect between chemicals and papermaking fillers.

2.4.6 Comprehensive Economic and Technical Evaluation

Whether it is a simple application, a collaborative application, or a comprehensive application, economic and technical evaluation is required. In economic and technical evaluations, many paper mills often only conduct a direct benefit analysis. In reality, this is not sufficient. We advocate conducting a comprehensive economic and technical evaluation, which generally requires analysis and evaluation in the following aspects.

2.4.6.1 Retention Rate Analysis

1) Retention of starch itself—the starch content in white water below the wire can be measured using the I2-KI reagent, which can then be used to calculate the retention rate in paper and paperboard.
2) Increased retention of fine fibers and fillers—these benefits can be estimated by measuring the decrease in white-water concentration.

2.4.6.2 Quality and Benefits Analysis

1) Economic benefits resulting from improved product qualification rates, enhanced printing performance, and so on.
2) Benefits arising from upgrading product grades and obtaining higher prices for higher-quality products.

2.4.6.3 Analysis of Raw Material Cost Reduction Benefits

1) Savings in pulp, especially high-grade wood pulp.
2) Savings in other chemicals, such as reduced usage of additives like rosin, alum, and AKD fillers.

2.4.6.4 Analysis of Other Benefits

1) Benefits in terms of power and energy, such as reduced pulping time and decreased steam consumption.
2) Benefits in terms of increased labor productivity, such as reduced downtime, improved papermaking efficiency, increased machine speed, and higher production yield.
3) Other societal benefits, such as reducing the need for high-grade paper imports and mitigating pollution from the papermaking industry, are also noteworthy.

The comprehensive application of functional starch is a low-cost, high-benefit technology that involves various categories and disciplines of knowledge. It has a strong scientific and practical nature, and truly mastering it is quite challenging. It is worth our continuous exploration and discovery.

2.5 Outlook and Recommendations

China's paper industry has entered a period of rapid development, and the research and development of starch-based functional products should closely follow the development of the paper industry.

First, China's starch-based functional products should accelerate serialization and scale development. Currently, China leads the world in paper and paperboard production, but per capita paper consumption is much lower than the global average. Therefore, China's paper industry is expected to continue its rapid growth. At the same time, China's starch-based functional products are still in the early stages of development, with low production, unstable quality, limited variety, and poor specificity being common issues. In order to meet the demands of the rapidly growing paper industry, it is essential to transition from the current small-scale and limited variety pattern toward large-scale, specialized, and diversified production.

Second, it is essential to emphasize research and development in application technology and specialized equipment for starch-based functional products. In a certain sense, the application technology of starch-based functional products is more important than synthetic technology. China's papermaking raw materials and paper machine processes differ significantly from developed countries. For example, the presence of various chemicals in straw pulp can severely disrupt the normal application of chemicals, and without corresponding application technology solutions, it will be difficult to achieve the desired results. We not only need to conduct in-depth research on the pulp's potential, chemical composition, and compatibility of chemicals but also select some typical paper machines, paper types, and companies for promotion. Additionally, we should enhance technical consultation and service efforts to ensure that papermaking enterprises can truly achieve benefits. On the other hand, we also need to strengthen research on specialized equipment for starch-based functional products that align with our national conditions. Developed countries have highly advanced papermaking processes and equipment that are increasingly focused on high automation, large-scale production, and high-speed operations. From pulp preparation and chemical addition to surface sizing and coating, all processes are carried out on a single production line. The paper width can exceed 10 meters, and the machine speed can reach over 1000 meters per minute, or even around 2000 meters per minute. The technological equipment of large papermaking enterprises in China has already achieved a basic alignment with international advanced levels. However, numerous small and medium-sized papermaking companies in China still lag in terms of technology and equipment. They often have low production capacity, slower processing speeds, and lower levels of automation. Currently, the equipment and processes for developing starch-based functional products in China are quite outdated, with intermittent methods being the main approach, heavily reliant on manual operations. This is not well-suited for meeting the current quality requirements for paper production. Therefore, there is a need to accelerate the development, application, and promotion of starch-based functional product application technologies and equipment.

Furthermore, there should be strengthened domestic and international technical cooperation and exchanges. As early as the 1960s, developed countries conducted extensive research on starch-based functional products. Functional starch, neutral sizing agents, polyacrylamide, and so on have been successively applied in the papermaking industry and have gradually developed into a series of products. However, in China, research efforts are scattered, and large-scale research and development companies in this field have not yet been established. The overall level of research and development of functional starch products in China is still relatively low. When compared to large and advanced paper machines from abroad, there are still certain gaps in terms of production capacity, quality, service, and technology. Therefore, we should place great importance on communication and collaboration with counterparts from abroad, enhancing the depth and breadth of cooperation, and expediting the development of China's functional starch materials industry to the greatest extent possible. On the other hand, the development of functional starch products should also prioritize collaboration with domestic chemical and papermaking companies.

Many renowned experts in the domestic chemical and papermaking fields have made significant contributions to the development of papermaking functional products in China. It can be said that it is through the collective efforts of these leaders and experts that China's paper-based starch functional materials have achieved rapid growth today.

China is a major player in the papermaking industry, and the times call for the rapid and scientific development of starch functional materials to make papermaking more resource-efficient, improve paper quality, diversify paper types, and make the papermaking process more energy and water-efficient. This can transform papermaking into an environmentally friendly industry and position the invention of paper as a leading force in the world. As scientists and researchers in the field of starch-based functional materials for paper, we have a duty that cannot be shirked.

References

1 Yao Xianping, Zheng Liping, Gong Guanshan, et al. A continuous fluidized bed cationic starch efficient and clean manufacturing method: CN, 200810305318.5. 2010-09-15.

2 Liping, Z. and Xianping, Y. (1994). Application of grass and wood pulp reinforcing agent HC-3 in industrial packaging paper. *Zhejiang Pulp & Paper* 4: 35–40.

3 Xianping, Y. and Zheng, L. (2012). Clean preparation technology of starch derivative powder in continuous fluidized pipeline. In: *Proceedings of the 147th China Engineering Science and Technology Forum - Forum on the Development of Light Industry Technology*. China Association of Chemical Industry: Beijing.

4 Yao Xianping, Zheng Liping, Tian Qingquan. Production process for high retention surface sizing starch: CN, 200410016436.6. 2006-03-08.

5 Liping, Z. and Xianping, Y. (2003). Development of a new high-retention surface sizing starch. *China Pulp & Paper* 22 (2): 24–26.

6 Borg, Y.X. and Liping, Z. (2001). Application technology of new high-retention surface sizing starch. *Paper Chemicals* 13 (2): 2–6.

7 Xianping, Y. and Liping, Z. (2002). Development and application of specialty chemicals for recycled paper. *Paper Chemicals* 14 (1): 3–8.

8 Xianping, Y. (2017). Green Development and Application of Multifunctional Modified Starch Papermaking Products. In: *Proceedings of the 2017 (30th) National Seminar on Paper Chemicals Development and Application of New*

Papermaking Technologies, 1–4. Hangzhou: China Paper Chemical Industry Association.

9 Yao Xianping, Zheng Liping, Gong Guanshan, et al. A Method for Manufacturing an Anti-Interference Recycled Paper Strength Enhancer: CN, 201110457570.X. July 9, 2014.

10 Zhiqing, H. (2004). Coating experiment using starch instead of partial latex. *Paper Science and Technology* 23 (6): 82–84.

11 Yanfen, D., Jingang, L., and Jiafu, W. (2017). Study on the properties of starch-based bio-latex prepared by different methods. *China Pulp & Paper* 36 (1): 1–8.

12 Xuqing, W. and Xiaofan, Z. (2011). The influence of partial replacement of latex by starch on the properties of coated paper. *Journal of Nanjing Forestry University (Natural Sciences Edition)* 35 (1): 79–82.

13 Xianping, Y., Zhen, Y., Jianmei, J. et al. (2018). Development and application of paper-based functional materials. In: *Proceedings of the 12th Academic Conference on Green Chemical Engineering, Metallurgy, and Materials Engineering* (ed. L. Jiongtian), 589–595. Beijing: Chemical Industry Press.

14 Zhen, Y., Jianmei, J., Xiaolei, H. et al. (2018). Development and application of paper-based functional materials. *Paper Chemicals* 30 (5): 1–7.

15 Xie Qin, L. and Yaming, X.G. (2015). Development of heat transfer printing paper. *Paper and Papermaking* 34 (2): 57–59.

16 Bofeng, C. (2007). Some insights into transfer printing. *Scientific and Educational Technology Information* 20: 303.

17 Yao Xianping, Yao Zhen, Yang Yong, et al. A Method for Producing Modified Starch for Heat Transfer Printing Paper: CN, 201711254668.9. 2021-01-08.

18 Xiong, F. and Qiang, H. (2016). *Modified Starch for Food Use*. Beijing: China Light Industry Press.

19 Xianping, Y. and Liping, Z. (1999). *Starch Derivatives and Their Application Technology in Papermaking*. Beijing: China Light Industry Press.

20 Yuzhong, W., Xiuli, W., and Fei, S. (2015). *Starch-Based New Materials*. Beijing: Chemical Industry Press.

21 Xianping, Y. and Liping, Z. (2003). Principles and application technology of papermaking wet-end chemistry. *Paper and Papermaking* 22 (1): 9–12.

22 Jinding, Z. (2018). Surface sizing of paper. *Heilongjiang Papermaking* 46 (3): 31–33.

23 Renvall, S., Kim, J.D., and Tynkkynen, T. (2013). Spraying starch to improve strength of container board produced from weak recycled furnish. *Tappi Journal* 12 (5): 11–16.

24 Qiuying, F.W., Dawei, L. et al. (2019). Influence of starch surface sizing on the strength properties of corrugated base paper. *Today's Printing* 6: 66–68.

25 Xianping, Zheng Liping, Tian Qingquan. A Method for In-line Surface Modification of Papermaking Fillers: CN, 201110296959.0. 2013-06-05.

3

Functional Materials Based on Nanocellulose Paper

3.1 Overview

Natural cellulose is the most abundant biomass resource on Earth, widely present in plants, animals, and some bacteria and minerals. Cellulose is a linear natural polymer compound composed of β-D-pyranose glucose units linked by 1,4-β-glycosidic bonds. Fibrous materials prepared from cellulose as raw materials through chemical, physical, or biological treatments with one-dimensional dimensions below 100 nm in the form of rod-shaped, fibril-shaped, or filamentous cellulose are collectively referred to as nanocellulose (NC) [1].

In 1947, Nickerson and Habrle obtained cellulose nanocrystals (CNC) through sulfuric acid hydrolysis of cellulose raw materials. Later, Filson and others successfully prepared CNC by hydrolyzing secondary fibers with endoglucanase. Inorganic strong acid hydrolysis and enzymatic hydrolysis have long been considered the most common methods for preparing CNC [2].

The term "original fibrous cellulose" was first proposed in patents and papers filed by ITT Rayonier [3]. In the early 1980s, Turbak and Herrick successfully prepared cellulose nanofibrils (CNF) for the first time using a homogenizer. The diameter of NC prepared by this method is generally 20–100 nm, with a length of tens of micrometers. However, there are still some issues in the preparation of NC using this method, such as homogenizer blockages, high energy consumption (up to 30,000 kW·h/t, 10 cycles), and strong mechanical damage to the fiber crystal structure, making proper pretreatment essential [4].

With the development of pretreatment technologies such as chemical or enzymatic methods (such as carboxymethylation), the energy consumption for preparing NC has also sharply decreased (500–2000 kW · h/t). Currently, NC can be prepared through enzymatic hydrolysis, carboxymethylation, TEMPO (2,2,6,6-tetramethylpiperidine-1-oxyl)-mediated oxidation, grafting-carboxymethyl cellulose method, cationization method, acetylation method, phosphorylation method, sulfonation method, and combined with mechanical treatments (such as grinding). In recent years, the commercial production of NC has gradually evolved from small-scale laboratory settings to industrial-scale production [5].

Paper-Based Functional Materials: Preparation and Applications, First Edition.
Xianping Yao, Genrong Chen, Zhen Yao, and Liping Zheng.

In February 2011, Sweden's Innventia commissioned the world's first experimental device for producing CNF (daily output of 100 kg). Canada's CelluForce company pioneered a pilot production line for CNC based on sulfuric acid hydrolysis in 2012, with a capacity of 1 ton per day. In 2012, the US Forest Service opened the first NC production plant in Wisconsin, mainly producing CNC based on sulfuric acid hydrolysis (10 kg/d) and CNF based on mechanical grinding (1000 kg/d). In 2014, the University of Maine established a pilot production line for CNF based on mechanical disk refining, with a capacity of 1 ton per day. Pilot and demonstration production lines for NC are currently mainly concentrated in developed countries such as the United States, Canada, Japan, Sweden, and Finland.

In recent years, Japan has made rapid progress in the research and industrialization of NC, with the TEMPO catalytic oxidation method developed by Professor Akira Isogai of the University of Tokyo being a representative example. Japan's Nippon Paper has begun to establish a production line based on this method, with a designed capacity of 500 t/a of CNF. In January 2017, Oji Paper in Japan announced that they had begun producing CNF with high viscosity and high shear thinning properties based on a method combining phosphorylation pretreatment and mechanical processing, with a capacity of 40 ton per annum.

However, these scaled-up production lines still primarily rely on sulfuric acid hydrolysis, TEMPO oxidation, and mechanical methods, leading to issues such as difficult chemical recovery, high costs, large water usage, or high energy consumption. Therefore, further efforts are needed from researchers to develop green, efficient, and sustainable methods for preparing NC.

China began researching NC as early as the 1980s, but there is still a significant gap in large-scale production compared to developed countries. With the active promotion of major research institutions in China, NC has attracted attention at the national level and been taken seriously by relevant enterprises. For example:

On November 7, 2015, the Nanocellulose and Materials Professional Committee of China Papermaking Society was established at the National Center for Nanoscience and Technology.

From May 20 to 22, 2017, the China Papermaking Society and Hangzhou Research Institute of Chemical Industry successfully hosted the "First International Conference on Nanocellulose Materials" in Hangzhou, Zhejiang Province, greatly promoting exchanges and cooperation between domestic scientific research and industry and international peers.

In 2017, in the key special project of the 13th Five-Year Plan of the Ministry of Science and Technology—"Efficient Utilization Technology of Fiber Raw Materials Based on Papermaking Process," it was clearly stated that the establishment of a pilot demonstration line for the industrialization of plant micro-nanocellulose is required (with a capacity of no less than 100 kg/d). The pilot demonstration line project brings together the scientific research strengths of South China University of Technology, Hangzhou Research Institute of Chemical Industry, Hangzhou Institute of Light Industry Design and Research, and other units, with the construction task undertaken by Yao Xianping's nanocellulose research group. It passed acceptance in 2021 and the pilot demonstration line has been built and put into operation, with a capacity of 210 kg/d, featuring advanced characteristics such as green, continuous, and low cost.

In addition, some domestic technology companies are also dedicated to the research and commercial application of NC.

With the increasing investment from the Chinese government, research institutions, and the business community, and the joint promotion of production, academia, research, and application, it is believed that China will be able to achieve the large-scale preparation and commercial application of NC more quickly, narrowing the gap with European and American countries.

3.1.1 Classification of Nanocellulose

There are two ways to classify NC: one is to categorize NC into three types based on the different raw materials and functional characteristics - nanofibrillated cellulose (NFC), nanocrystal cellulose (NCC), and bacterial nanocellulose (BNC) [6]; the other is based on the characteristics of NC by the Technical Association of the Pulp and Paper Industry (TAPPI), which divides NC into nanofibers (CNC and CNF) and nanostructured materials [cellulose microcrystal (CMC) and cellulose microfibril (CMF or MFC)] [7]. The specific classification is shown in Figure 3.1.

Currently, there is no unified naming convention for the classification of NC. Chen Wenshuai and others from Northeast Forestry University in the book "Fundamentals of Mechanical Preparation and Application of Nanocellulose" unifiedly named rod-like, whisker-like, fibrous, and linear nanomaterials with diameters below 100 nm as NC [8]. This chapter will adopt this expression and uniformly refer to it as NC.

3.1.1.1 Cellulose Nanocrystals

CNC, also known as nanocrystalline cellulose, are generally obtained by strong acid hydrolysis, presenting a needle-like crystal whisker structure with a small aspect ratio. They have a diameter of 5–70 nm and a length of 100–250 nm. Due to the preservation of the amorphous region during acid hydrolysis, CNC has a high degree of crystallinity, typically ranging from 60% to 90%, and exhibits high mechanical properties, with a Young's modulus of approximately 150 GPa and tensile strength of around 10 GPa [9].

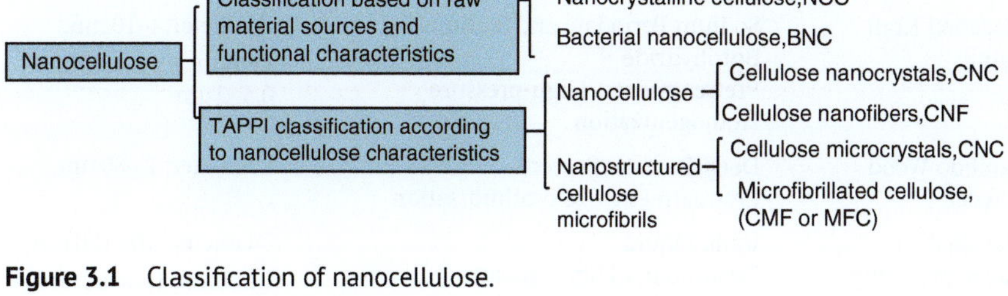

Figure 3.1 Classification of nanocellulose.

Table 3.1 Size of CNC prepared from different raw materials.

Raw material source	Length *L* (nm)	Diameter *D* (nm)
Softwood pulp	100~300	3~5
Cotton fiber	100~300	7
Bacterial cellulose	>100	5~50
Tunicate	>100	10~20
Oak shell	100~200	20
Sisal	250	4

Source: Adapted from [10].

The dimensions, morphology, and crystallinity of CNC obtained from different fiber raw materials and reaction conditions vary. Table 3.1 provides size data for CNC prepared from several different raw materials.

3.1.1.2 Cellulose Nanofibrils

Cellulose nanofibrils, also known as nanofibrillated cellulose or microfibrillated cellulose, are mainly obtained by the mechanical strong shear force treatment of cellulose raw materials, presenting a fibrous shape with a diameter of 5–60 nm and a length of 1000–10,000 nm. During the processing, the amorphous region of cellulose is typically not removed, resulting in CNF consisting of both crystalline and amorphous regions, with a high aspect ratio and good flexibility [11].

The dimensions, morphology, and crystallinity of CNF obtained from different fiber raw materials and preparation methods vary. For specific comparisons, refer to Table 3.2 [9].

Table 3.2 Characteristics of CNF obtained from different raw materials and methods.

Raw material	Method	Characteristics
Bleached Eucalyptus Pulp	Grinding	Diameter: 4~30 nm
Bleached Kraft Pulp	TEMPO Oxidation Pretreatment + Mechanical Treatment	Diameter: 3~4 nm, Several μm
Bleached Softwood Pulp	Formic Acid Pretreatment + High-pressure Homogenization	Diameter: 5–20 nm, Length: 300–1200 nm, Crystallinity 52.9%
Bleached Kraft Pulp	Sodium Periodate and Sodium Borohydride Pretreatment + High-pressure Homogenization	Diameter:4–10 nm, Length: Approximately 0.5–2 μm
Bamboo Wood Powder	Deep Eutectic Solvent Pretreatment + Microfluidization	Diameter: 2~80 nm
Bleached Eucalyptus Pulp	Ionic Liquid Treatment + High-pressure Homogenization	Diameter: 20~100 nm, Cellulose II Structure, Crystallinity: 34.43%

3.1.1.3 Bacterial Nanocellulose

Bacterial nanocellulose is typically synthesized by specific bacteria (such as Acetobacter xylinum), where glucose chains are produced inside the bacterial cell and extruded through pores in the cell membrane. Several glucose chains further combine to form microfibrils, which intertwine to create a three-dimensional nanonetwork of bacterial cellulose. Compared to plant cellulose, BNC does not contain lignin, pectin, or hemicellulose, resulting in higher purity, high crystallinity (above 95%), high polymerization degree, variable length, and a diameter of 20–100 nm. It possesses high water absorption capacity, high tensile strength, good shape retention ability, and excellent biocompatibility [12].

3.1.2 Cellulose Nanofibrils - CMF (or MFC)

Microfibrillated cellulose (MFC) is a novel cellulose product developed in the 1980s by Turbak and others at the American company ITT. It is a colloid-like cellulose product with nanoscale dimensions (typically 5–60 nm in diameter) obtained by repeatedly homogenizing a suspension of plant cellulose fibers. Although the term "micro" does not refer to micrometer-scale dimensions, it has been commonly used to describe MFC. Many studies also refer to this material as CMFs, and due to its nanoscale diameter, it is sometimes called nanofibrillated cellulose in literature. MFC typically consists of a random network of highly aspect ratio fibrous microfibrils. Compared to conventional cellulose and some cellulose derivatives, MFC has a larger surface area, exposing numerous hydroxyl groups, leading to high water retention value, good dispersibility, and stability. Additionally, the high surface area and hydroxyl content make MFC easily combinable with other materials, providing a wider range of options for the preparation of new multifunctional materials. This chapter will focus on introducing CNF and CNC materials derived from plants.

NC, as a new type of biomaterial, possesses unique nanoscale structure, mechanical properties, and optical properties, making it a forefront and hotspot in cellulose research. The main characteristics of NC are as follows:

a) Nanoscale size, high specific surface area, and three-dimensional network structure
b) Sourced from renewable plant resources, biodegradable
c) High crystallinity, high purity
d) Lightweight, high Young's modulus, high strength - only 1/5 of the weight of steel but with strength over five times that of steel, comparable to Kevlar aramid fibers
e) High hydrophilicity, ultrafine structure, and high transparency
f) Low thermal deformation, with a thermal expansion coefficient of approximately $0.1 \times 10^{-6} K^{-1}$, only 1/50 of glass, similar to quartz glass
g) Good biocompatibility and other characteristics.

The numerous excellent properties of NC make it have enormous potential applications in areas such as papermaking, construction, automotive, food, cosmetics, electronics, medicine, and more. The high-value utilization of NC can generally

Table 3.3 Applications of NC.

High-volume applications	Low-volume applications	Novel and emerging applications
Papermaking	Coating	Sensor
Degradable bioplastics	Thermal insulation materials	Water treatment
Automotive interior materials	Aerogel	Air purification
Lightweight materials for automobile bodies	Aerospace	3D printing
Building cement	—	Organic LED light-emitting diode electronic panel
Textile	—	Solar photovoltaic
Medical and health care -	—	Drug-sustained release

be roughly divided into three major parts, as shown in Table 3.3: high-volume applications, low-volume applications, and novel and emerging applications [13].

With the continuous decline in reserves of fossil resources such as petroleum and coal, the prices of petrochemical raw materials continue to rise. Coupled with the increasing global attention to environmental pollution issues and the call for green chemistry, NC, as a widely used biomaterial, holds infinite business opportunities and promising development prospects. This chapter will focus on the preparation and application of NC-based functional paper materials.

3.2 Preparation and Characterization of Nanocellulose

3.2.1 Preparation

NC is extracted and isolated from natural cellulose under the action of various chemical reagents or mechanical forces. Cellulose consists of amorphous and crystalline regions, where the loose arrangement of molecules in the amorphous region can be degraded, while the structure of the crystalline region is preserved, resulting in nanoscale cellulose with high crystallinity. Currently, the preparation of NC mainly involves mechanical methods, chemical methods, and biological methods.

3.2.1.1 Mechanical Methods

In 1983, Turbak et al. and Herrick et al. first used a high-pressure homogenizer to prepare CNF with a diameter below 100nm. Subsequently, other methods have been gradually developed, such as high-pressure homogenization, microfluidization, fine grinding, ultrasonication, cryo-crushing, PFI refining, twin-screw extrusion, emulsification, steam explosion, ball milling, and fluid impact methods [2].

Among these methods, high-pressure homogenization, microfluidization, and fine grinding are the most commonly used, with corresponding production equipment being high-pressure homogenizers, microfluidizers, and colloid mills (as shown in Figure 3.2).

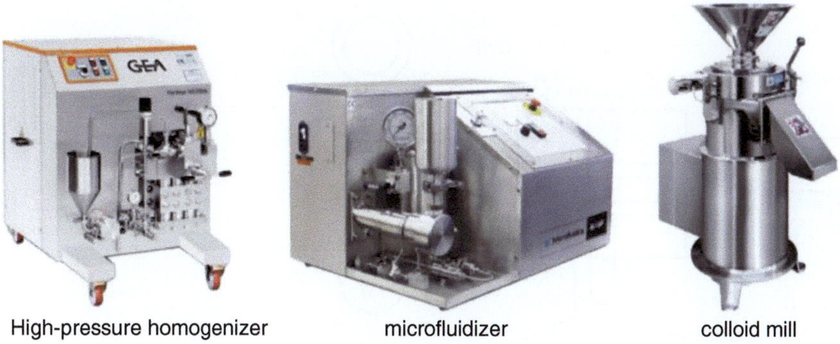

Figure 3.2 Main equipment for preparing CNF through mechanical methods *Source:* GEA Group Aktiengesellschaft.

Figure 3.3 Working principle of the high-pressure homogenizer.

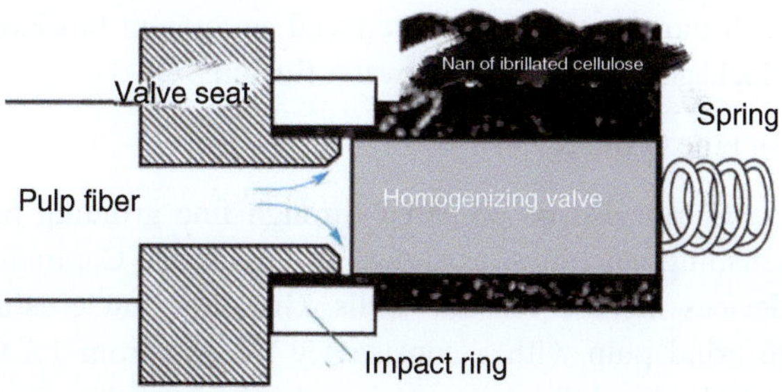

1) High-Pressure Homogenization

High-pressure homogenizers are one of the most effective production equipment for preparing nanomaterials, as shown in Figure 3.3. Powered by a high-pressure reciprocating pump, fibers are transported to the homogenization valve. As the fibers pass through the homogenization valve, a sudden pressure drop creates a cavitation effect and high-speed impact, generating strong impacts, cavitation, shearing, and turbulent vortex effects. This process leads to the ultrafineization of fibers in the suspension, resulting in the production of NC [9].

CNF prepared by homogenization method has characteristics such as large specific surface area, high aspect ratio, and wide diameter distribution. Additionally, this method is simple to operate and can easily achieve industrial-scale continuous production. However, this method also faces challenges such as high energy consumption and susceptibility to blockages during the preparation of NC, which to a certain extent hinders its large-scale production.

2) Microfluidization

Microfluidizers, commonly used in cosmetics, biotechnology, and pharmaceutical industries, are another type of equipment similar to high-pressure homogenizers. In recent years, they have been used to separate pulp fibers and prepare CNF. The working principle of this equipment is shown in Figure 3.4.

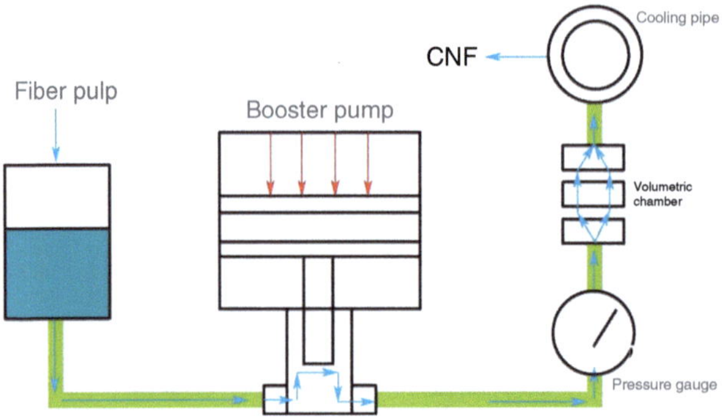

Figure 3.4 Working principle of the microfluidizer.

Compared to high-pressure homogenizers, microfluidization can produce CNF with more uniform sizes, reduced equipment blockages, and easier resolution of blockages through reverse water flushing.

3) Fine Grinding

NC can also be prepared through fine grinding methods using grinding and crushing machines, as shown in Figure 3.5. Common ultrafine particle grinding devices include Masuko mills. Grinding and crushing machines can be used to grind pulp with a consistency ranging from 1% to 5% (mass fraction) without the need for prior "micronization" treatment of fibers. Through multiple

Figure 3.5 Grinding and pulverizing machine.

cycles of grinding, nanoscale dimensions can be achieved, with no equipment blockages and convenient disassembly and cleaning. However, issues such as significant heat generation in the grinding disc, expansion of the grinding disc material leading to imprecise control of the grinding disc gap, and immature technology for continuous production of equipment pose challenges.

4) Pretreatment

The mechanical preparation of CNF from plant cellulose typically requires significant energy consumption, with energy consumption values ranging from 20,000 to 30,000 kW·h/t being common, and reports of energy consumption as high as 70,000 kW·h/t. High energy consumption has been a significant factor limiting large-scale production of CNF. To reduce the energy consumption of preparing CNF, fiber pulps are generally subjected to pretreatment, such as chemical or enzymatic pretreatment. Reports indicate that pretreatment can significantly reduce energy consumption values to below 1000 kW · h/t.

Furthermore, purely mechanical methods for fiber fibrillation have low efficiency and can cause severe damage to the cellulose structure, resulting in uneven CNF particle sizes, low crystallinity, and poor dispersion. Pretreating the fibers by loosening or breaking down the fiber cell wall structure through pretreatment methods before mechanical processing is an effective way to achieve green, efficient preparation of NC. In summary, the combination of chemical or enzymatic pretreatment with mechanical methods is expected to be the mainstream trend for CNF preparation. Currently, pretreatment methods mainly include alkali pretreatment, TEMPO oxidation pretreatment, and enzymatic pretreatment.

Alkali Pretreatment: Alkali pretreatment mainly aims to break down lignin components in fibers and help dissolve lignocellulosic bonds. Mild alkali treatment can dissolve lignin while preserving pectin and hemicellulose. Alkali extraction needs to be strictly controlled to avoid cellulose degradation and ensure that hydrolysis reactions occur only on the fiber surface to obtain complete NC.

TEMPO Oxidation Pretreatment: TEMPO-mediated oxidation is an effective method for surface modification of natural cellulose fibers, introducing carboxyl and aldehyde functional groups in natural cellulose under aqueous conditions and mild temperatures. After TEMPO-mediated oxidation, the fiber morphology is mostly maintained, with oxidation occurring only on the surface of the microfibrils, giving them a negative charge that repels nanofibers, making fibrillation easier.

The mechanical energy consumption for preparing CNF after TEMPO oxidation is approximately 570 kW·h/t, significantly lower than 70,000 kW·h/t before treatment. However, TEMPO catalysts are expensive and challenging to recycle.

Enzymatic Pretreatment: Cellulase pretreatment can improve the accessibility and reactivity of fiber cell walls, effectively increase the fibrillation degree of fibers during subsequent mechanical processing, and greatly reduce the energy consumption. By adjusting the degree of enzymatic pretreatment and the mechanical processing process, CNF with different average diameters can be prepared. Compared to TEMPO oxidation pretreatment, enzymatic-assisted mechanical processing significantly reduces energy consumption during the preparation

Figure 3.6 The 210kg/d nanofibrillated cellulose pilot demonstration production line established by Hangzhou Chemical Research Institute.

process, has milder conditions, and requires less chemical reagents, making it one of the green and efficient methods for preparing CNF. The world's first pilot plant for CNF production (Lindström's Group from Innventia) has adopted a method combining enzymatic pretreatment with mechanical processing, indicating the broad industrial prospects of this approach.

Based on extensive research, Yao Xianping's research team at the Hangzhou Research Institute of Chemical Industry has completed the construction of a 210 kg/d NC pilot production line (see Figure 3.6). This pilot production line has four main features: (i) It uses enzymatic pretreatment and a mechanical combination method, with the entire process producing no wastewater, waste residue, or chemical emissions, making it environmentally friendly; (ii) It incorporates a fully automatic computer control system to achieve continuous production; (iii) Enzymatic pretreatment significantly reduces the energy consumption of preparation, lowering production costs and providing cost advantages; and (iv) It is equipped with concentration technology equipment that can increase the product concentration from 2% to nearly 30%, facilitating further product modification and widespread application.

3.2.1.2 Chemical Methods

Acid hydrolysis is the most common method for preparing CNC, as this method effectively removes the amorphous regions of cellulose, reducing the size of the cellulose while producing high crystallinity CNC.

Common acids used in acid hydrolysis include sulfuric acid, hydrochloric acid, phosphoric acid, with sulfuric acid being the most commonly used, resulting in CNC with good size controllability and water dispersibility. During acid hydrolysis, inorganic and organic acids diffuse into the interior of the cellulose network under hydrothermal conditions, degrading the noncrystalline regions [14]. Although the experimental process of acid hydrolysis for preparing CNC is relatively advanced, it still generates a large amount of waste acid and impurities, leading to issues such as high posttreatment costs and equipment corrosion.

It is estimated that producing 1 kg of CNC requires the consumption of 9 kg of concentrated sulfuric acid (used in the acid hydrolysis process) and 13 kg of sodium sulfate (used for neutralizing the waste acid solution) [15]. Due to the presence of sulfate ions, CNC prepared by sulfuric acid hydrolysis has poor thermal stability. Therefore, for current practical production, large-scale production of CNC using high concentrations of inorganic acids is still not sustainable environmentally and economically [5]. This has forced researchers to seek environmentally friendly and mild reaction methods, such as solid acid methods, ionic liquid methods, and biological methods.

3.2.1.3 Biological Methods

Biological methods mainly involve enzymatic hydrolysis. Enzymatic hydrolysis is a primary method for preparing CNC, where cellulase enzymes catalyze the hydrolysis to remove the amorphous regions of cellulose, retaining the dense and crystalline regions with a certain aspect ratio.

Cellulase enzymes are not single enzymes but a multienzyme system, mainly consisting of endoglucanases, exoglucanases, and β-glucosidases. Endoglucanases attack the low crystallinity regions of cellulose fibers to create free chain ends; exoglucanases remove cellulose disaccharide units from the free chain ends, further degrading cellulose molecules; and β-glucosidases hydrolyze cellulose disaccharides to produce glucose.

The enzymatic hydrolysis process has relatively mild conditions and high specificity, maintains the basic chemical structure of cellulose, improves product purity, and significantly reduces energy consumption, making it a green and sustainable preparation method. However, there is ongoing research on how to effectively increase the yield of CNC and control the extent of cellulase hydrolysis.

3.2.2 Modification Techniques

NC, as a representative of cellulose-based nanomaterials, has attracted attention from scholars both domestically and internationally. It possesses unique properties that are significantly different from natural cellulose, such as high strength, large specific surface area, and high crystallinity, making it widely applicable in many fields.

Despite having many excellent properties, NC still has several shortcomings: (i) The numerous hydroxyl groups on the surface of NC prevent it from dissolving well in weakly polar solvents and polymer matrices; (ii) NC has a large specific surface area and high thermodynamic potential, making it prone to aggregation between crystals, with aggregation becoming more irreversible at higher temperatures; and (iii) NC lacks various target properties of polymeric compounds [16]. Surface modification of NC will expand its application areas, thus realizing the high-value utilization of biomass resources.

There are mainly two methods for surface modification of NC: one is small molecule chemical modification, including adsorption of surfactants or

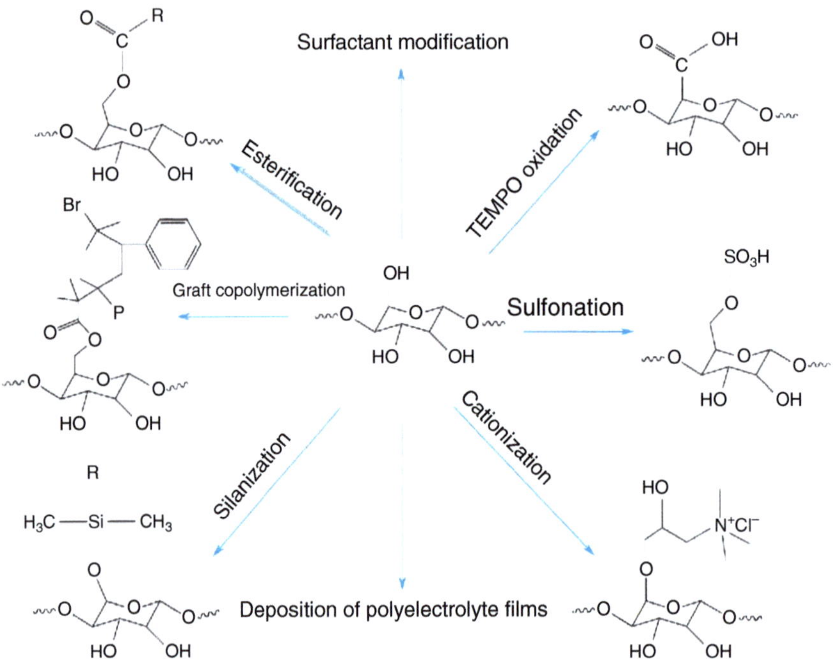

Figure 3.7 Modification methods of nanofibrillated cellulose.

polyelectrolytes, esterification, etherification, oxidation, silanization, etc., and the other is graft copolymerization. Please refer to Figure 3.7 for more details.

3.2.3 Main Characterization Methods

The differences in preparation methods of NC result in significant variations in its size, morphology, and surface properties. As a potential high-value functional material, the size and distribution of NC are closely related to product quality, directly impacting the physical properties of the prepared materials. Consistent, reliable, and accurate characterization methods are particularly urgent for understanding the performance of NC and establishing industry standards.

Characterization of NC and its derivatives mainly focuses on surface morphology and internal structure. The following key indicators are discussed in detail:

1) Microscopic Morphology Characterization

The microscopic morphology of NC can be characterized using transmission electron microscopy (TEM), scanning electron microscopy (SEM), field emission scanning electron microscopy (FE-SEM), and atomic force microscopy (AFM).

a) Scanning Electron Microscopy

Scanning electron microscopy allows for the visual observation of the appearance and aggregation state of NC samples. Software such as Image J and Nano Measurer can be used to statistically determine the width of NC. However, due to the entanglement between cellulose fibers, it is difficult to identify the two ends of a single fiber, making it challenging to determine the length of NC.

b) Transmission Electron Microscopy

Transmission electron microscopy uses an electron beam with an extremely short wavelength as the illumination source and focuses transmitted electrons through an electromagnetic lens to achieve high-resolution, high-magnification imaging. When observing the morphology of NC, cryo-electron microscopy is generally selected to reduce damage to the sample by the electron beam, minimize sample deformation, and obtain a more authentic sample morphology.

c) Atomic Force Microscopy

Atomic force microscopy studies the surface structure and properties of materials by detecting extremely weak atomic interactions between the sample surface and a micro-force-sensitive component. By scanning the sample, changes are detected by the sensor, providing information on force distribution, thus obtaining surface morphology structure information and surface roughness information with nanometer-level resolution. Surface topography can be more accurately reflected by AFM morphology images combined with phase images.

2) Structural Characterization

Characterization methods or indicators such as X-ray diffraction (XRD), degree of polymerization (DP), and thermal gravimetric analysis (TGA) can be used to characterize the internal structure of NC. When characterizing NC aerogels, specific surface area (BET) measurements are also necessary.

a) X-ray Diffraction

X-ray diffraction is commonly used to determine the crystallinity of cellulose, known for its direct and convenient characteristics.

b) Degree of Polymerization

The degree of polymerization represents the number of glucosyl units connected in a molecular chain. In the molecular formula $(C_6H_{10}O_5)_n$, n represents the degree of polymerization, which can be used to calculate the molecular weight (molecular weight = 162 × DP). DP is a physical and chemical indicator representing the degree of cellulose degradation. The determination of the degree of polymerization of NC requires the use of the national standard method, the cupriethylenediamine viscosity method. DP is closely related to the aspect ratio of NC fibers, with longer fibers having higher degrees of polymerization.

c) Thermal Gravimetric Analysis

Thermal gravimetric analysis equipment is an important material research tool that analyzes the combustion characteristics of biomass by measuring the relationship between the mass of the test sample and temperature changes under controlled temperature conditions. By analyzing the thermal gravimetric curve, information related to the composition of the sample and its possible intermediate products, thermal stability, thermal decomposition, and the products generated can be obtained, all linked to mass-related information.

d) Specific Surface Area

The specific surface area, adsorption-desorption isotherms, and pore size distribution curves of samples can be determined using an automatic specific surface area adsorption instrument. According to the definition of the International Union of Pure and Applied Chemistry (IUPAC), pores smaller than 2 nm are referred to as micropores, pores larger than 50 nm are referred to as macropores, and pores ranging from 2 to 50 nm are referred to as mesopores. By analyzing the adsorption–desorption isotherms and pore size distribution curves, the type, pore volume, and average pore size of pores in the aerogel can be determined.

3.3 Functional Application of Nanocellulose in PaperMaking

China is not only the invention of papermaking technology, but also the world's papermaking power. In 2019, China's paper production of 12,515 million tons has been ranked first in the world for many years. The main raw material of papermaking is cellulose, but unlike ordinary cellulose, NCs have many excellent properties, such as nano size effect, high strength, high hydrophilicity, high crystallinity, high barrier property, and high Young's modulus, and can have special properties such as high water resistance and high charge density after chemical modification. In particular, NC is combined with papermaking chemicals such as starch derivatives, which can play increasingly effective functional roles and can prepare new paper-based functional materials with better performance, higher added value, and improved environmental protection, such as improve paper strength, reduce beating energy consumption, and improve the retention of fine fibers and fillers, which can be used as high barrier packaging materials, oil-proof materials, high water resistance materials, and so on, continuing to expand the terminal application field. At the same time, NC can replace the existing chemical synthetic papermaking additives, making the papermaking system more clean and environmentally friendly. Therefore, NC is a promising new paper-based functional material, which will strongly promote the technological progress and industry development in the field of paper-based functional materials.

Due to the unique properties of NC, domestic and foreign scientific workers on its application in papermaking have made a lot of technical achievements; however, in addition to Canada, Japan, and some paper mills in Europe, the real investment in the use of paper mills is relatively small, and the domestic NC production application is almost blank. Therefore, for the application of NC in paper-based functional materials, people with lofty ideals at home and abroad need to continue to work hard. At present, although the preparation technology of NC has become increasingly mature, the scale is still small, the cost is high, the product is not strong, and the application technology research is obviously insufficient, which restricts the application and promotion of NC in papermaking.

The following will be combined with our research results and domestic and foreign application research status, and the main functional applications of NC are briefly introduced.

3.3.1 Paper Reinforcement

Strength is one of the most important functional indicators of paper. There are many ways to improve the strength of paper, usually by increasing the content of softwood pulp, improving the beating degree, and adding papermaking enhancer and other methods to achieve. The most common way to strengthen paper is to add chemical reinforcers. Commonly used dry strength agents can be divided into natural polymers, synthetic resins, and water-soluble cellulose derivatives into three categories. Starch is the most typical use of natural polymers. Starch is cheap and effective, but starch needs gelatinization, and it will bring environmental problems in the recycling process of waste paper. Synthetic resins include polyacrylamide, polyvinyl alcohol, phenolic resin, and other substances, among which polyacrylamide is the most widely used. It cannot only be used as an enhancer, but also has the effect of retention and filtration, surface sizing, and dispersion. But such ingredients are relatively expensive and often have high levels of toxic elements. NC, as a derivative of natural cellulose, is derived from papermaking raw materials and has natural green characteristics. At the same time, NC, as a paper reinforcing agent, has a higher strengthening effect than most commonly used reinforcing agents and has great potential in the field of papermaking.

1) Principle of action

NC is an excellent bio-based paper reinforcing material, and it is generally believed that its action principle is hydrogen bonding. CNF bridges adjacent fibers through a large number of exposed hydrogen bonds, shortening the distance between fibers to the hydrogen bond action distance and forming a tangled three-dimensional network structure between fibers, and the strengthening effect increases with the increase in binding area. Second, CNF can be embedded in the gap between the fiber networks to improve the binding force of the paper fiber.

2) Application example

Example 3.1 In 2013, FPInnovations reported a revolutionary natural fiber filament, CF (cellulose filament). This material is a filamentous fiber that is stripped directly from a single wood pulp fiber by mechanical action, and its strength is retained as much as possible. Its length-diameter ratio can reach about 1000. In the ultra-lightweight wrapping paper project led by CF inventor Makhlouf Laleg, the core technology is to compensate for the poor strength of waste pulp fibers by adding CF, while improving the stiffness and water resistance of the resulting paper.

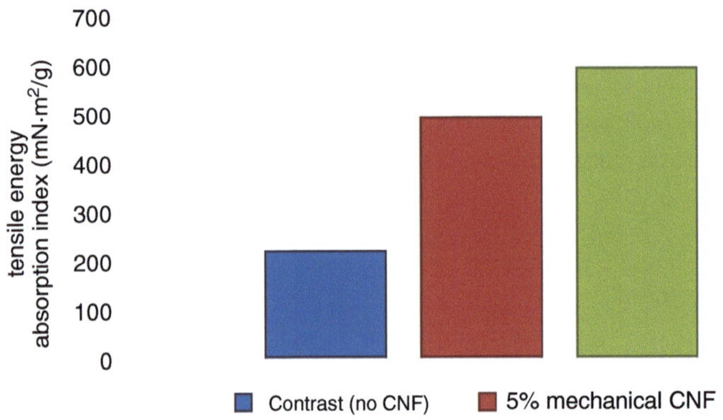

Figure 3.8 Effect of CNF on tensile energy absorption of paper.

The experimental results show that the comprehensive contribution of CF to the strength of formed paper can be converted into the reduction ratio of paper weight: adding 1%CF can reduce the weight of base paper by 7%.

Example 3.2 Hua Xujun et al., Canadian Forest Products Innovation Research Institute, disclosed a preparation method of NC with a high length-to-diameter ratio in patent CN 103502529A, and the NC prepared by this method can effectively increase the dry strength of paper. Figure 3.8 shows the effect of prepared CNF on the tensile energy absorption of the paper. When 5%CNF is added, the tensile energy absorption index of paper is increased by 150% compared with blank space.

The above patent also pointed out that the addition of CNF can improve the strength of wet paper, reduce the rate of broken paper, improve the operating performance of the paper machine, and increase the speed. For enhancing the strength of wet paper, NC has incomparable advantages over chemical additives. European researchers have also found a similar rule. In Finland, UPM, a high-speed medium paper test machine tests, found that adding 1% to 2% NC can improve the running performance of the paper machine and increase the dry and wet strength of the paper. This has also been verified in the pilot process of paper machines at the VTT Technical Research Center of Finland.

Example 3.3 China's Yao Xianping R & D team found that when CNF nanocellulose products were applied to decorative base paper, after the paper quantity was reduced from 70 to 60 g/m^2, in addition to the decrease in paper permeability, the smoothness of paper was greatly increased, the dry tensile strength and wet tensile strength of paper were not reduced, and the ash content and paper opacity were basically the same. See Figure 3.9 for details.

Example 3.4 Wang Aijiao from Yao Xianping's R&D team formed a binary strengthening system of polyamide polyamine epichlorohydrin resin (PAE) and

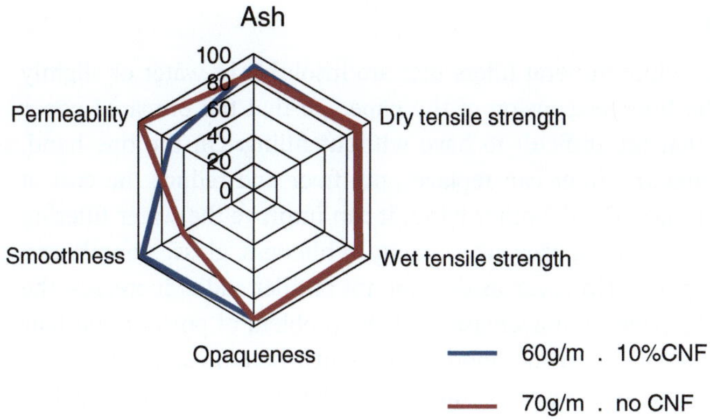

Figure 3.9 Different quantitative paper properties comparison diagram.

cationic starch (CS) with CNF, respectively, to investigate their strengthening effects on paper pages. The results showed that the PAE/CNF system had an obvious effect on paper wetting strength. When the dosage of PAE was 0.5% (the mass percentage of the absolute dry amount of additives to the absolute dry amount of pulp) and the amount of CNF was 0.3%, the wet tensile strength of paper was 6.2 times that of an untreated blank paper and 1.76 times that of a PAE-treated paper alone (PAE dosage 0.5%). The CS/CNF dual system has a more obvious effect on the drying strength of paper. When the amount of CS is 2.0% and the amount of CNF is 0.3%, the dry tensile strength of paper is 1.65 times that of an untreated blank paper and 1.26 times that of a CS alone treated paper (2.0% CS). The specific performance of the experiment is shown in Table 3.4.

The application test of NC as a paper reinforcing agent has also been carried out in paper mills in China. By adding 1~2% NC and 1% cationic starch, the tensile strength of the paper is obviously improved, and the light transmission coefficient and air permeability are slightly decreased. If the same strength properties are maintained, the paper quantity can be reduced by 8 g/m^2. In the process of papermaking, the dewatering capacity of the net part is slightly reduced, the dryness of the wet paper web of the net cloth is reduced by less than 1 percentage point, and the total retention rate is basically unchanged, but the dryness of the paper web is increased.

From the research and development results at home and abroad, NC has a broad application prospect as a papermaking reinforcement material.

Table 3.4 Effect of CNF on the physical strength of paper.

CNF dosage/%	Tensile strength/kN·m^2	Bursting strength/kPa	Folding strength/次	Tear strength/mN
0	6.99	278	23	7.44
2	8.73	331	35	7.40
3	9.14	420	37	7.39

3.3.2 Paper Filling

Filling paper is to add white mineral fillers that are insoluble in water or slightly soluble in water to the fiber suspension of the paper, so that the prepared paper has some properties that are difficult to have without filling. On the one hand, the addition of papermaking filler can replace pulp fiber and reduce the cost of papermaking raw materials. On the other hand, it can improve the water filtering performance of the slurry and improve the opacity, whiteness, gloss, smoothness, and printability of the paper. However, as the amount of filler added increases, the physical strength of the paper will decrease, and the problem of powder and hair loss of the paper will become more and more prominent when the amount of filler is added. At the same time, the low retention rate of fillers is also a major problem plaguing paper workers.

1) Principle of action

NC has a very high specific surface area, a large number of hydroxyl groups on the surface, and a large number of carboxyl and aldehyde groups on the surface of the NC after oxidation, which makes CNF easy to combine with water to the fiber surface to form a low concentration gel. When NC is added to the slurry suspension, the viscosity of the slurry increases and the retention of the filler can be improved. At the same time, the paper strength decline caused by filling can be improved.

However, according to the above retention principle, the application of NC does not have much advantage in terms of retention effect or economy compared with conventional papermaking retention chemicals, so scientists invented the method of preparing NC-filler complex to solve the problem of paper filling. The principle of this method is that the NC is preferentially combined with the filler and then entered into the papermaking system to achieve the double effect of high filler and no loss of strength. The technical paths of preparing NC-filler complex are as follows: (i) the cellulose fiber and filler are co-ground to achieve the purpose of cellulose fibrination; (ii) nano-microfibrillated fibers are wrapped on the surface of the packing through the electrostatic adsorption of cand and ions; and (iii) the NC reacts with the filler at a certain temperature to form a complex.

2) Application example

Example 3.5 FiberLean is at the forefront of the world in preparing NC-filler complexes, and its products are already in use. The CNF–filler complex was obtained by mixing the pulp fiber with the filler and then drying after high pressure homogenization or ultrasonic treatment. When added to the pulp, 1.5% CNF can increase the filling amount by 7.5%, and the performance of the paper does not decrease. This technology can greatly reduce the production cost of paper.

Example 3.6 Su Yanqun et al., China Pulp and Paper Research Institute, disclosed in patent CN 104863008B, a microfibrillated cellulose envelope assisted retaining high-fill papermaking process, with cationic polyacrylamide paper manufacturing,

Table 3.5 Effect of CNF on ash content and tensile index of paper.

Serial number	item	Tensile index/(N·m/g)	Ash content/%
1#	Blank group	11.66	10.97
2#	T-CNF	15.22	14.52
3#	CNF	17.5	10.71
4#	CPAM	8.95	14.08
5#	CPAM+T-CNF	9.79	18.52
6#	CPAM+CNF	13.23	16.81

high-fill paper can be obtained. The system can improve the water filtration performance of the slurry and the retention rate of the filler, the beating degree of the slurry from 44°SR to 28°SR, and the retention rate of the filler from 44% to 69%; the mechanical properties of the paper are greatly improved; and the tensile strength, tear index, and folding resistance are increased by 64%, 47%, and 107%, respectively.

Example 3.7 In the production process of decorative base paper, due to the high price of titanium dioxide, improving the retention rate of titanium dioxide has been a concern of paper mills. The nanocellulose research group of Yao Xianping's team developed two kinds of NC-filler complex products: CNF and T-CNF. The product is applied to the production of decorative base paper, and it is found that the related products can maintain or increase the strength of paper while improving the retention rate of titanium dioxide. The specific research data are shown in Table 3.5. When pure CNF was combined with cationic polyacrylamide (CPAM), the ash content of paper was increased to 16.81%, and the tensile index of paper was increased by 13% on the basis of blank. When T-CNF is used, the ash content of paper is increased by nearly 4%, and the strength of paper is also increased by nearly 30%. When T-CNF is combined with CPAM, the ash content of paper can be increased to 18.52%.

3.3.3 Paper Barrier

Wrapping paper is a kind of packaging material, which has the advantages of light weight, low cost, and sustainability compared with other packaging materials such as glass and plastic. According to statistics, the current global demand for consumer packaging is between US$ 40 and 50 billion and is still growing. Therefore, from the perspective of green and sustainable development, the use of plant-derived paper instead of plastic as packaging materials is the general trend.

Packaging materials need to be able to effectively inhibit the physical, chemical, microbial, and other factors on food, beverages, cosmetics, and other consumer goods damage, but also should have the ability to block oxygen, water vapor, grease and microorganisms, and other factors. Due to the hydrophilic and porous nature of the paper itself, it cannot block water vapor, grease, and other substances, limiting its application field.

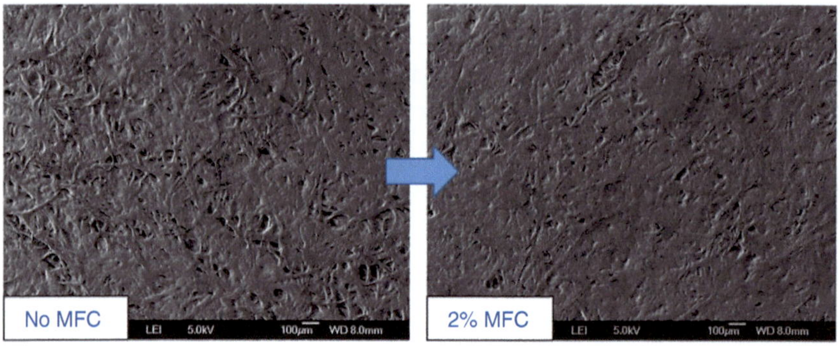

Figure 3.10 Electron microscope image of CNF-coated paper.

NC coating on the surface of the paper can form a dense film, thereby improving the strength and surface properties of the paper, giving the paper good barrier properties. Figure 3.10 shows the microstructure of CNF coating on the surface of the paper. After 2%CNF coating, the voidage of the paper surface is greatly reduced.

1) Blocking principle

NC films have good barrier properties, especially for oxygen and water vapor, which can be attributed to the dense network structure formed by nanofibrils. The study shows that the space in NC films is the main way of molecular infiltration. Data show that NC is composed of a high proportion of crystalline and amorphous regions, and the crystallinity of all kinds of NC ranges from 40% to 90%, among which CNC has the highest crystallinity. Since it is impossible for gas molecules to pass through the crystalline region, the length of the path that gas molecules pass through the film is larger than the thickness of the measured film. The dense, nonporous NC film makes the permeation path of gas molecules more tortuous, as shown in Figure 3.11.

Compared with CNC films, CNF films have better barrier properties, to be exact, better oxygen resistance. This is due to the entanglement of fibril molecules inside NFC, which forms a high-density and more tortuous diffusion path.

2) Application example

Example 3.8 Yao Xianping's team of National Papermaking Chemical Engineering Technology Center conducted a series of studies on the barrier properties of NCs. NC (CNF) was added to polyvinyl alcohol (PVA) to prepare PVA/CNF composite films. The effects of NFC on the oxygen barrier properties, water vapor barrier properties (Table 3.6), light transmittance, and strength properties of the composite films were investigated. The results show that NC can effectively enhance the oxygen and water vapor barrier properties of the composite films, has no obvious effect on the light transmittance of the films, and has obvious improvement on the tensile strength of the films. The prepared PVA/CNF composite film has excellent oxygen barrier properties and water vapor barrier properties, and the oxygen and water vapor permeability can reach $0.072\,cm^3/m^2$ 24h·0.1Mpa and $0.45\,g/m^2$·24h,

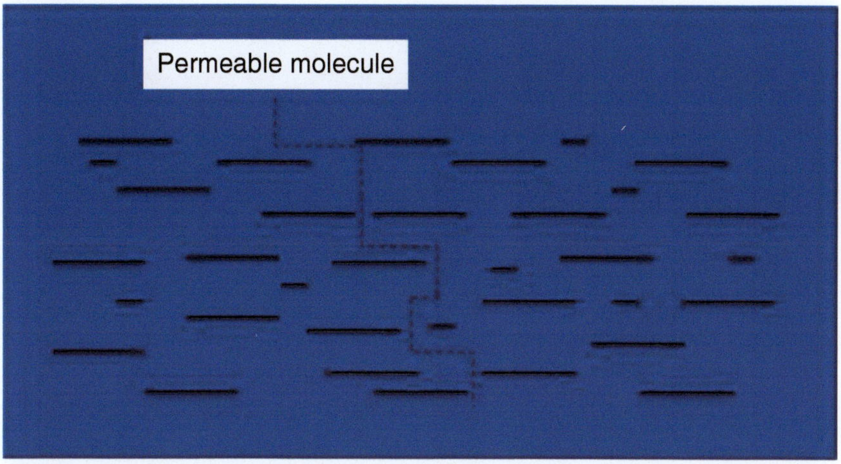

Figure 3.11 Schematic diagram of molecular diffusion paths in NC membranes.

Table 3.6 Oxygen and water vapor barrier properties of different films.

Sample	quantify/ (g/m$^{2)}$	thickness/ μm	Oxygen transmittance/ (cm^3/m^2·24h·0.1MPa)	Water vapor transmittance/ (g/m^2·24h)
Pure PVA film	80	60±3	1.507	850.00
Pure CNF film	30	23±2	0.030	92.13
PVA/CNF film	80	62±2	0.072	0.45

respectively, and the light transmission rate can reach 89%, which is expected to realize the application in the field of green barrier packaging materials.

Example 3.9 Greaseproof paper is a paper that can prevent the absorption and penetration of oil, which is widely used in the packaging of foods such as hamburgers, baked bread, biscuits, and other oil-containing items. At present, the mainstream oil-proof paper mainly uses oil-proof agents containing fluorine to achieve oil-proof effect. However, after the fluorine antioil agent is absorbed into the human body, carcinogens will be produced, thus affecting human life and health. Therefore, the development and application of fluoride-free oil repellent is imminent.

As a natural green product, NC can be coated on the surface of the paper to form a dense film, thus giving the paper excellent resistance to the penetration of oil droplets. However, NC itself has large water content and low viscosity, so it is difficult to directly coat the paper. Yao Xianping's team of researchers carried out a lot of research work on how to coat NC well on paper. By combining NC with other starch, PVA, and other adhesive materials, the team prepared a coating formula with oil-proof properties. At present, the process of the formula paper machine is well

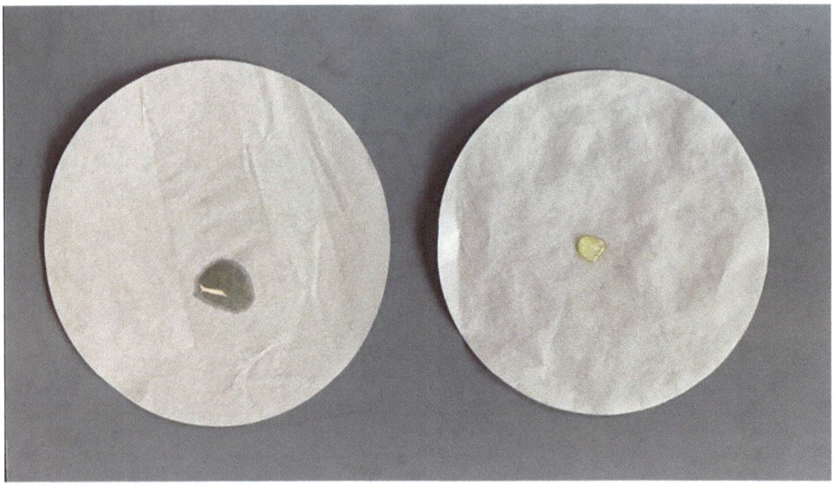

Figure 3.12 Greaseproof paper was prepared by the National Paper Chemical Engineering Technology Center.

adapted, and the oil-proof paper with oil-proof grade 6 can be successfully coated, as shown in Figure 3.12.

Taking into account the characteristics of NC itself, spraying coating on the wet part of the paper machine can better fill the paper gap, and the coating amount is smaller and achieve a higher oil-proof grade. It can be seen from Figure 3.13 that after 10s of spraying time, the permeability of the paper is almost zero, and it can be

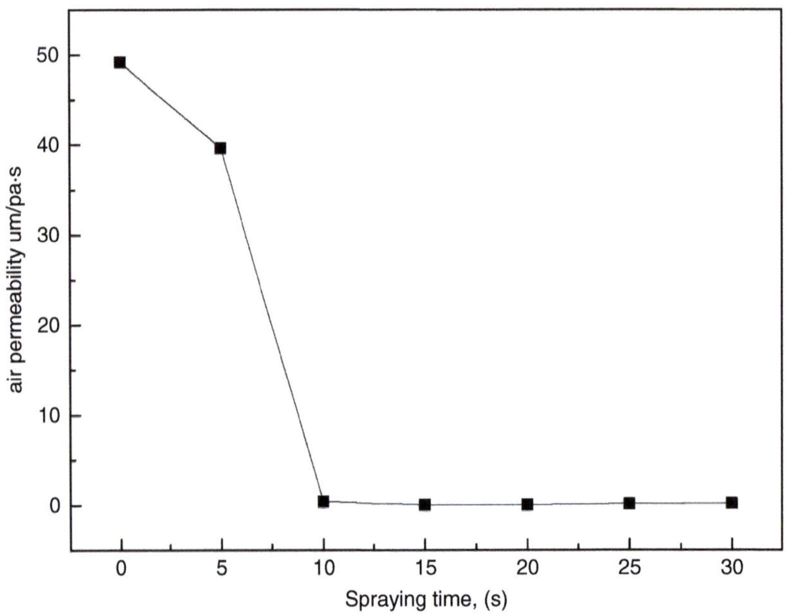

Figure 3.13 Effect of NC spraying time on paper air permeability.

seen that the NC forms a dense film on the surface of the paper. After precoating on the wet part of the paper machine, normal surface sizing or coating can be performed to obtain a greaseproof paper with ideal oil resistance and hydrophobic effect.

Example 3.10 In patent CN 106833139A, Hu Yun designed a preparation and application method of cellulose nanofiber wiki-grease barrier coating. First, the surface of the substrate is treated to improve the surface tension, Then, a coating composed of cellulose nanofiber gel solution, polyvinyl alcohol, modified starch, kaolin, and nano-calcium carbonate is applied on the surface of the substrate. The coating amount is $2 \sim 18$ g/m^2. After drying, the coating forms a barrier coating on the surface of the substrate, especially to prevent oil from penetrating into the interior of the substrate. With a kit oil-proof grade of $\geq$6, it can be widely used in food packaging materials, electronic device surface coating, mechanical parts surface protection, and other fields.

Due to the high water retention of NC, the drying of related products is a major problem. At the same time, many application scenarios will require both oil-proof and waterproof effect. If the hydrophilic problem of NC is solved and the effect of hydrophobic oil repellent is achieved, then the NC oil-repellent agent will have a very great prospect.

3.3.4 Other Applications

1) Applications in tobacco flakes

Tobacco sheet is a paper-like tobacco substrate made by adding some plant fibers and auxiliaries to discarded tobacco raw materials in the process of cigarette production through the principle and process of papermaking.

NC is easy to adsorb negatively charged fillers and fine fibers, such as cigarette dust and tobacco stem, and combine with fine fibers in the slurry to achieve the purpose of retaining fine components of the slurry. In addition, NC was added into the coating solution, and the wood pulp in papermaking tobacco flakes could be partially replaced by its outstanding physical and chemical properties, thus improving the quality of tobacco flakes.

Liu Haoyue et al., the Key Laboratory of Pulp and Paper of Tianjin University of Science and Technology, tried to use cationic NC and guar gum to explore a dual system that does not affect the smoking taste of cigarettes, but also improves the water retention performance of tobacco pulp. It was found that the bicomponent system composed of guar gum and cationic NC could significantly improve the water retention performance of tobacco pulp and effectively reduce the cation demand of filtrate. The effect of the modified NC on the water retention of tobacco pulp increased with the increase of the amount of charge carried by the modified NC. Table 3.7 shows the influence of cationic nanocellulose (C-CNF) with a different charge on the water retention performance of tobacco pulp.

2) Application on battery separator paper

The battery diaphragm is one of the key parts of lithium electronic battery products, which plays a role in preventing the positive and negative contacts in

Table 3.7 Effect of different types of NC on water retention performance of slurry.

	Filtration time/s	Total retention/%	Retention of filler/%	Cation demand/ (μmol/L[1])
Blank group	28.7	76.2	3.1	142
Unmodified CNF	27.2	76.2	3.1	135.8
C-CNF-1	23.7	76.5	4.0	102.2
C-CNF-2	22.4	76.6	4.8	98.8
C-CNF-3	21.7	77.0	14.7	84.6

the battery, blocking the electrons in the circuit when charging and discharging, preventing short circuit, and allowing the lithium ions in the electrolyte to pass freely. Therefore, the performance of the battery diaphragm determines the interface result and internal resistance of the battery, which directly affects the battery capacity and cycle performance.

At present, the market-oriented lithium-ion battery membranes are mainly polyolefin microporous membranes represented by polyethylene (PE) and polypropylene (PP), which have excellent mechanical properties, chemical stability, and relatively cheap characteristics. However, with the development of lithium batteries, the high temperature thermal shrinkage performance of the diaphragm is becoming more and more demanding, and the existing PE and PP diaphragms have not met the requirements. NC itself has good thermal stability, and when it is added to the battery separator produced by papermaking, the product has high porosity, excellent heat resistance, good chemical stability, and so on.

Diaphragm and Celgard 2400PP diaphragm (Figure 3.14). When the temperature of NFC diaphragm is from 140 °C to 200 °C, the thermal shrinkage rate is almost zero. When the temperature of PP diaphragm is 150 °C, the thermal shrinkage rate is 68%, and when it is 160 °C, it reaches 98%, and the diaphragm is almost completely melted and shrunk. Therefore, the NFC diaphragm has good thermal stability and will be a promising choice for the preparation of high temperature–resistant and high-safety lithium-ion batteries.

3) Applications in household paper

Household paper should have a certain dry and wet strength and softness, and wet wipes also need to have a certain moisture. NC, as an excellent reinforcing agent, has a very high water retention value, so it has a great application prospect in household paper.

At present, the application of introducing NC into household paper and toilet paper is still in the development stage, and the main purpose is to improve the strength of paper, reduce the cost, and improve the application performance of paper. The technical routes adopted include: ① adding NC directly into pulp and 2) surface coating on wet web.

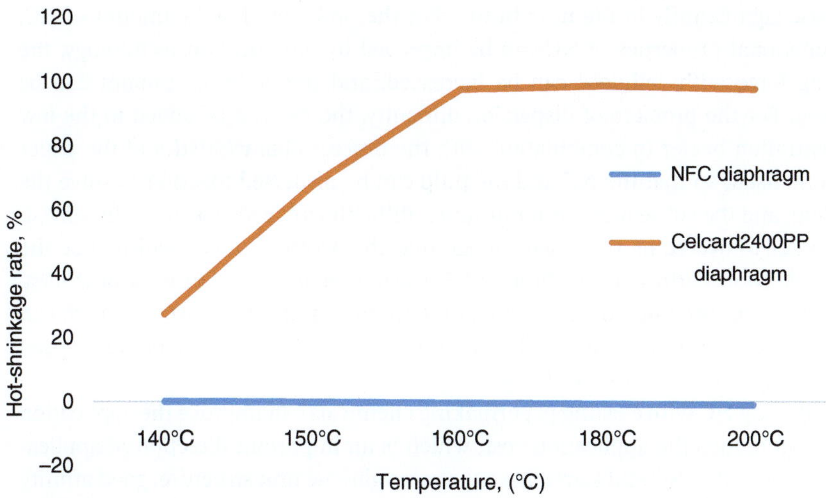

Figure 3.14 Comparison of thermal shrinkage between CNF diaphragm and Celgard2400PP diaphragm.

Mercer International Co., Ltd. in patent CN 201780047631.0 proposed a technical solution, and the technical solution of the paper ingredient consists of 45% to 90% of fine ground softwood pulp fiber mixture and 10% to 55% of hardwood pulp cellulose. The mixture of fine ground softwood pulp fiber consists of 20–89.9% softwood pulp fiber, 0.05–20% NC, and 0.05–5.0% paper auxiliaries. This technology can be used to produce products with good loose thickness, adsorption, softness, high strength, and high durability.

Zhao Qiang et al. from Shandong Sun Household Paper Co., Ltd. pointed out in patent CN 201710091088.6 that the hydrophobic-modified nanocellulose with a particle size of 30–100 nm was coated on the surface of household paper, the coating amount was 0.2–3 g/m^2, and household paper with good wet strength and softness could be obtained eventually.

3.4 Existing Problems and Prospects

3.4.1 Existing Problem

NC development, so far, in the field of papermaking has not been able to achieve large-scale application; the main reason is that the current cost of NC is high and paper mills are more sensitive to cost requirements and cannot afford to use it. On the other hand, the modification and application technology of NC is not mature enough, and some application technology problems have not been effectively solved, such as the amount of NC is large, the dispersion is difficult, and the water filtration performance of the pulp will be reduced.

As for the high cost of NC, with the improvement of preparation technology and the acceleration of large-scale production, it is believed that the price will decrease significantly in the near future. For the problem of large amount of NC, the functional properties of NC can be improved by modification technology, the binding force with cellulose can be increased, and the addition amount can be reduced. For the problem of dispersion difficulty, the NC can be added to the low concentration beater in combination with the process characteristics of the paper machine itself, so that the NC and the pulp can be dispersed together to solve the problem, and the subsequent water filtration difficulties can also be partially solved.

In order to overcome the disadvantage that the addition of NC will reduce the water filtration performance of the pulp, researchers are opening up a new idea, combining NC and filler to form a binary retention and filtration system, which can improve the retention rate of the filler and the strength of the paper, but also speed up the dehydration rate of the slurry.

Combining NC with existing papermaking chemicals can improve the application effect and reduce the application cost, which is an important direction of application. In particular, NC and starch have the same glucose unit structure, good affinity between them, and are natural perfect materials, which need to be further studied.

At present, the application of NC should focus on functional paper, such as battery separator paper, greaseproof paper, and other high-end paper. It is believed that with the continuous efforts of scientific researchers, NC will achieve large-scale application in papermaking.

3.4.2 Prospect

In recent years, the advantages of nanomaterials in papermaking have become more and more significant. NC has both the characteristics of nanomaterials and the biodegradable and environmental protection properties of plant fibers, and its application research has become increasingly hot. The addition of NC in the paper industry can not only be used to replace ordinary pulp, save raw materials, and improve paper performance, but also give paper special properties. For example, Japanese paper has been used in adult diapers by using NC to absorb antibacterial agents; NC can also be used for battery separator, which has the characteristics of many charging times and fast charging speed; it can also be used on the basis of "nano paper" for semiconductor equipment to achieve excellent working performance; and Japan's Prince Holdings and Mitsubishi Chemical are also using "nanopaper" to replace the display screen of tablet computers such as ipads.

There is no doubt that NC is the best choice to improve paper quality, reduce fiber consumption, improve performance quality, and develop functional paper-based functional materials. Once promoted, both the amount and benefits will be huge.

In addition, because of its excellent properties, NC also has very important application value in many other fields. In the past 20 years, domestic and foreign universities, scientific research institutions, and commercial companies have carried out a large number of scientific research and application exploration, and achieved a lot of excellent scientific research results. In the past five years, some commercial products are also being launched, such as Mitsubishi's ballpoint pen and ASics running shoes, but the corresponding large-scale application and commercialization still exist challenges.

At present, the cost of NC is still high, which also limits its commercial application. Based on the survey results of the Japanese Ministry of Economy, Trade, and Industry (METI) on the future prospects and problems of the Japanese paper industry, the METI summarized the relationship between the production scale of CNF material manufacturing and its manufacturing cost. Through the development of chemical pretreatment and fibrillation technology, the overall cost of the first generation of CNF was reduced from $40 to $70 /kg. Through large-scale industrial production, the cost can be reduced to about $10 /kg. By 2030, more advanced preparation processes will improve the production efficiency of CNF, and the cost of second-generation CNF is expected to further decrease to about $5 /kg. New products developed with CNF for green nanomaterials include materials such as automotive parts, information, electronics, packaging, construction, fluid additives, and high-end filters, and the market is expected to reach $10 billion in annual sales.

According to the US Department of Agriculture (USDA), NC has an estimated market potential of 6.4 million tons in the United States and 35 million tons globally. The American Process Corporation (API) believes that the commercialization timeline for NC could be similar to that for plastics and predicts that USDA's global market size estimate for NC will be achieved around 2045.

References

1. Wei, L., Shouxin, L., and Jian, L. (2016). *Fundamentals of nanocellulose preparation and functional applications*, 3. Beijing: Science Press.
2. Haishun, D., Chao, L., Miaomiao, Z. et al. (2018). Preparation and industrialization of nanocellulose. *Progress in Chemistry* 30 (4): 448–462.
3. Yun, H. and Jingang, L. (2013). Preparation and research projects of nanocellulose. *China Pulp and Paper* 34 (6): 33–36.
4. Yao Wenrun, X. and Qinghua. (2014). Research progress in the preparation of nanocellulose. *Paper and Papermaking* 33 (11): 49–55.
5. Qinglin, W., Changtong, M., Jingquan, H. et al. (2018). Current status of nanocellulose preparation technologies and industrialization. *Journal of Forestry Engineering* 3 (1): 1–9.
6. Klemm, D., Kramer, F., Moritz, S. et al. (2011). Nanocelluloses: A new family of nature-Based materials. *Angewandte Chemie International Edition* 50 (24): 5438–5466.
7. TAPPI (2011). Proposed new TAPPI standard: Standard terms and their definition for cellulose nanomaterial. In: *Draft for review*, 1–6.
8. Wenshuai, C., Yu, H., Qing, L. et al. (2014). *Fundamentals of Mechanical Preparation and Application of Nanocellulose*. Beijing: Science Press.
9. Zhang Sihang, F., Runfang, D.L. et al. (2017). Research progress in the preparation of nano cellulose and its composite materials' applications. *China Pulp and Paper* 36 (1): 67–74.
10. Zhu Yachong, W., Dongmei, C.,.Y. et al. (2020). Current research status of nanocellulose preparation methods. *China Pulp and Paper* 39 (9): 74–83.

11 Nechyporchuk, O., Belgacem, M.N., and Bras, J. (2016). Production of cellulose nanofibrils: A review of recent advances. *Industrial Crops and Products* 93: 2–25.
12 Lifen, W., Zheng, L., Shiru, J. et al. (2014). Research progress on the properties and applications of bacterial cellulose. *Microbiology China* 41 (8): 1675–1683.
13 Nelson, K., Retsina, T., Iakovlev, M. et al. (2016). American process: production of low cost nanocell ulose for renewable, advanced materials applications. In: *Materials Research for Manufacturing* (ed. L. Madsen and E. Svedberg), 267–302. Switzerland: Springer, Cham.
14 Yu, H., Qin, Z., Liang, B. et al. (2013). Facile extraction of thermally stable cellulose nanocrystals with a high yield of 93% through hydrochloric acid hydrolysis under hydrothermal conditions. *Journal of Materials Chemistry A* 1 (12): 3938–3944.
15 Chen, L., Zhu, J.Y., Baez, C. et al. (2016). Highly thermal-stable and functional cellulose nanocrystals and nanofibrils produced using fully recyclable organic acids. *Green Chemistry* 18 (13): 3835–3843.
16 Ahola, S., Myllytie, P., Bsterberg, M. et al. (2008). Effect of polymer adsorption on cellulose nanofibril water binding capacity and aggregation. *Bioresources* 3 (4): 1315–1328.

4

High-performance Fiber Paper-based Functional Materials

4.1 Introduction

As a high-tech product, specialty paper is widely used in multiple fields such as aviation, aerospace, transportation, electronics, finance, healthcare, food, and decoration. It has high technical content, high added value, and a wide range of applications. The manufacturing of specialty paper involves multiple disciplines such as papermaking, materials, and chemistry. The raw materials used include plant fibers, synthetic fibers, inorganic fibers, animal fibers, metal fibers, etc. According to the differences in fiber characteristics, specialty papers with different properties and functions can be produced by compounding or blending fibers.

4.2 Typical High-performance Fiber

High-performance fibers generally refer to a category of special chemical fibers with unique physicochemical structures, which are used in specific fields and possess excellent properties such as high strength, high modulus, outstanding high-temperature resistance, flame retardancy, or chemical stability. [1] In comparison with conventional plant fibers, high-performance fibers exhibit remarkable advantages in aspects such as mechanical strength, insulation performance, and temperature resistance. They can serve as structural materials and functional materials. As important raw materials for advanced paper-based composite materials, high-performance fibers and their paper-based composites have become strategic new materials in fields such as electrical insulation, rail transit, aerospace, marine shipping, petrochemical, and wind power generation.

According to their chemical compositions, high-performance fibers can be divided into two major categories: high-performance organic fibers and high-performance inorganic fibers. [2, 3], as shown in Figure 4.1. Typical representatives of high-performance organic fibers include aromatic polyamide fibers, polyimide fibers, ultra-high-molecular-weight polyethylene (UHMWPE) fibers, etc.; typical representatives of high-performance inorganic fibers include carbon fibers, basalt fibers, glass fibers, etc.

Paper-Based Functional Materials: Preparation and Applications, First Edition.
Xianping Yao, Genrong Chen, Zhen Yao, and Liping Zheng.

Figure 4.1 Classification of high-performance fibers.

4.2.1 High-performance Organic Fibers

4.2.1.1 Aromatic Polyamide Fiber

Among aromatic polyamide fibers (aramid fiber), there are two varieties with the greatest practical value: meta-fully aromatic polyamide fiber (meta-aramid, aramid 1313, poly [m-phenylene isophthalamide] fiber, PMIA) and para-fully aromatic polyamide fiber (para-aramid, aramid 1414, poly [p-phenylene terephthalamide] fiber, PPTA).

In the 1960s, DuPont Company developed meta-fully aromatic polyamide fiber. It was put on the market in 1967 under the trade name Nomex®, and its molecular structural formula is shown in Figure 4.2. In 1971, para-fully aromatic polyamide fiber Kevlar® with higher strength and modulus was commercialized. Its molecular structural formula is shown in Figure 4.3. The appearance of Kevlar marks a breakthrough in the field of high-strength and high-modulus fibers. In 1974, the United States Council of Commerce named fully aromatic polyamide as aramid, generally referring to a linear polymer in which the amide group is directly connected to two benzene ring groups. The fiber made from it is called aromatic polyamide fiber, that is, aramid fiber. Aramid fiber has the characteristics of high strength, high modulus, high temperature resistance, and low density. It also has good impact resistance, chemical stability, flame retardancy, and insulation. It is the reinforcing fiber second only to carbon fiber in usage in high-performance composites. It is widely used in personal protection, bulletproof armor, rubber products, asbestos substitutes, automotive friction materials, advanced insulating paper, and paper-based honeycomb structural materials. It is an important material in high-end fields such as aerospace, national defense, electronic communication, petrochemical, and ocean development.

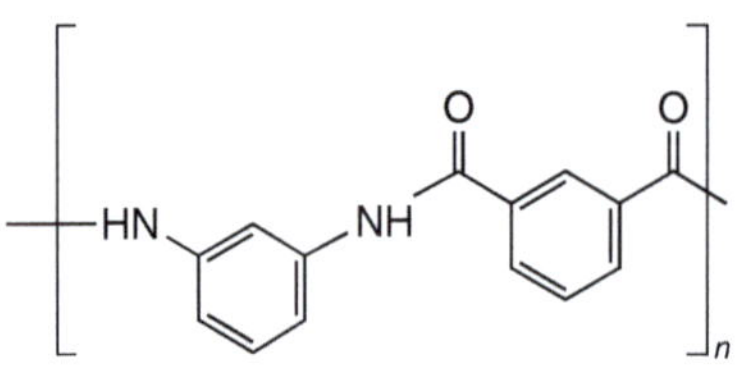

Figure 4.2 Molecular structure formula of Nomex®.

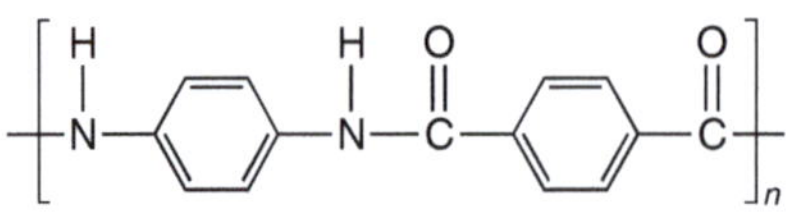

Figure 4.3 Molecular structural formula of Kevlar®.

In China, the research on aramid fiber began in 1972. In 2002, Yantai Spandex Co., Ltd. (now Yantai Tayho Advanced Materials Co., Ltd.) developed high-quality aramid 1313 and realized industrial production, becoming the world's second-largest aramid manufacturer. In 2011, the 1000 t/a para-aramid industrialization project of Yantai Tayho Advanced Materials Co., Ltd. was successfully tested.

Kevlar® is golden yellow due to the existence of conjugate resonance between the amide and aromatic groups, which results in high rigidity. In contrast, this situation does not occur in Nomex fiber with a meta configuration, so it is white. In Nomex® fiber, the phenyl groups and amide units are linked in the meta position, leading to an irregular chain conformation. Compared with Kevlar®, it has lower crystallinity and hence a lower tensile modulus. In Kevlar® fiber, the amide groups appear at regular intervals along the main chain of the linear macromolecule, which is beneficial for the formation of a large number of lateral hydrogen bonds between adjacent chains, triggering effective chain stacking and high crystallinity.

The structure of para-aramid fibers has been analyzed by X-ray diffraction, scanning electron microscopy, and chemical analysis. Many structural models have been proposed [4–6], such as the "radiation-arranged pleated layer structure" model, the "lamellar columnar fibril structure" model, and the "skin-core layer ordered micro-region structure" model. These microstructural models basically reflect the main structural characteristics of para-aramid fibers, that is, there is a fibril structure formed by aggregated extended chain molecules in the fiber, and the cross-section of the fiber has a typical skin-core structure. Due to the use of liquid crystal spinning technology, the outer layer of the fiber cools rapidly during formation, forming a skin structure with uniformly arranged extended chain macromolecules and low crystallinity. The molecules inside the fiber are combined by van der Waals forces and hydrogen bonds. The internal rotation potential energy of para-aramid macromolecules is quite high. The molecular chain segments are planar, rigid, extended chains. During crystallization, extended chain lamellae (extended chain lamellae refer to the fully extended form of long chain molecules in the lamellae) are often formed, making para-aramid fibers easy to form a highly oriented multiple fibril structure. It is the existence of this special structure that enables para-aramid fibers to be fibrillated by beating to prepare aramid pulp.

Due to the smooth surface and lack of active groups of aramid fibers, the adhesion between the fibers and matrices, such as resins, is poor. At present, the application research of aramid fibers in paper-based functional materials mainly focuses on the following four aspects:

1) Surface modification. This includes chemical modification, introducing active groups on the fiber surface to increase the adhesion strength between the fiber and the matrix through chemical bonding or polar interaction. The main methods include surface etching, surface grafting, and modification with coupling agents. Physical modification increases the surface roughness of the fiber through physical action, improves the contact area and wettability between the fiber and the matrix, and improves the interface condition. The main methods include surface coating, plasma modification, gamma-ray radiation technology, and ultrasonic impregnation technology.

2) Preparation of differentiated aramid fibers. DuPont Company and Teijin Company developed differentiated aramid pulps and precipitated fibers through fibrillation or precipitation methods. The prepared differential fibers are more suitable for papermaking, which have a larger specific surface area, stronger interaction between fibers, and better forming quality and paper properties.
3) Enhancement of the intrinsic properties of aramid fibers. Although aramid fibers have excellent properties, there are problems such as intolerance to light, easy aging, and poor ultraviolet resistance. Photodegradation leads to easy degradation, aging, and deterioration of materials, resulting in a gradual decline in the mechanical properties of products. Phenomena such as brittleness, cracking, and yellowing may occur, and even complete destruction may render them unusable. Therefore, the use of aramid fibers under long-term exposure to light has attracted extensive attention. The Toray Fiber Research Institute in Japan found that adding an organic ultraviolet absorber to the spinning solution can improve its light resistance and anti-aging effect [7].

4.2.1.2 Polyimide Fiber

Polyimide fiber, a heterocyclic polymer fiber, is also classified as a typical high-performance fiber. Figure 4.4 presents two common chemical structural formulas of polyimide. The molecular chain of polyimide comprises imide rings, aromatic rings, and heterocycles, exhibiting high rigidity. Additionally, the nitrogen-oxygen double bond on the imide ring possesses a very high bond energy. The conjugate effect generated by the aromatic heterocycles results in a relatively large intermolecular force. Coupled with a high degree of axial orientation during the fiber preparation process, polyimide fiber is endowed with properties such as high strength and modulus, low dielectric constant, resistance to high and low temperatures, radiation resistance, flame retardancy, and low water absorption. Moreover, polyimide fiber also displays good chemical stability and can withstand the corrosion of strong acids. After being irradiated by electrons of a certain intensity, its performance can still remain at approximately 90%, far exceeding that of other fibers. The limiting oxygen index (LOI) of polyimide fiber ranges from 35% to 75%, and its smoke emission rate is relatively low, making it a self-extinguishing material. Compared with aramid fiber, polyimide fiber has higher thermal stability, lower water absorption, higher thermal oxidation stability, higher hydrolysis resistance, and higher radiation resistance. It is a more promising functional material and is expected to be widely applied in fields such as the atomic energy industry, space environment, rescue operations, aerospace, national defense construction, new buildings, high-speed transportation, sports equipment, new energy, environmental industry, and protective equipment. In 2010, Changchun Hipolyking Co., Ltd. collaborated with the Changchun Institute of Applied Chemistry, Chinese Academy of Sciences, to carry out the industrialization of high-temperature resistant polyimide fiber. It has become the only enterprise in China with full-line production capacity and independent research and development ability from raw material synthesis to products.

According to different principles, there are mainly two routes for the preparation of polyimide fibers. One is to use synthesized polyamic acid for spinning, and then

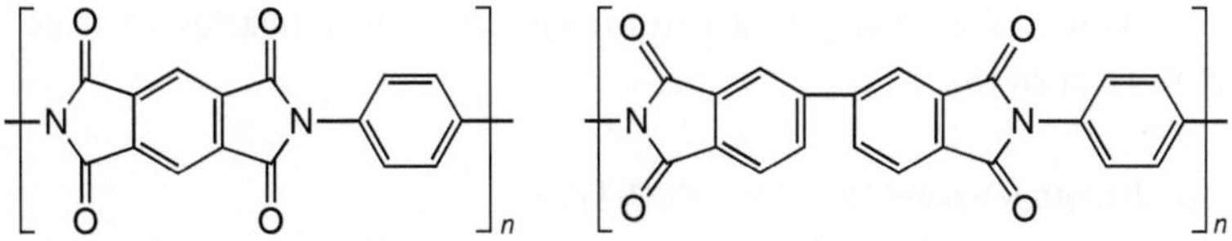

Figure 4.4 Two common polyimide chemical structures.

form a polyimide ring through imidization polymerization. This method is relatively common. The other is to directly use polyimide solution or melt for spinning to obtain polyimide fibers.

The research on polyimide fibers commenced as early as the mid-1960s in the United States and the Soviet Union. Simultaneously, China and Japan also conducted certain research work. Primarily, polyimide fibers with specific mechanical properties were obtained by spinning through polyamic acid solution and then undergoing imidization in a high-temperature or dehydrating agent environment. However, due to the immaturity of spinning technology at that time and the excessively high cost of fiber synthesis, polyimide fibers were not rapidly promoted and applied. Until the 1980s, Kermel Company utilized dimethylformamide (DMF) and phthalate as raw materials to prepare polyimide fibers through wet spinning, with the trade name being Kermel. Prior to 1984, Kermel only supplied the French army and police forces and was employed in the manufacture of high-temperature resistant and fireproof clothing. It was not until the 1990s that it began to supply the global market on a small scale.

Currently, in response to the demand for new fibers in modern industry, an increasing number of scholars have dedicated themselves to the research and development of polyimide fibers, mainly concentrating on fields such as high-temperature separation, high-insulation products, and new biomaterials. Presently, polyimide fibers have been widely utilized in numerous fields such as aerospace, electrical motor insulation, vehicle engineering, electronic communication, and personal protection.

The polyimide fiber paper-based material prepared by wet papermaking using polyimide fibers exhibits excellent mechanical, temperature-resistant, and insulation properties. It can serve as a honeycomb structural material and high-temperature resistant insulation material in fields such as rail transit, national defense and military industry, and aerospace. Nevertheless, polyimide fiber paper by the wet papermaking technology is still challenging due to the lack of differential fiber with highly active, high specific surface area, resulting in poor paper forming quality and low strength, and thus unable to fully exert the excellent performance of polyimide fibers. At present, there is no commercialized polyimide fiber paper-based material internationally. Polyimide fiber paper-based functional material developed by Changchun HiPolyking Co., Ltd. avoids issues such as incomplete imidization and decreased molecular weight of fibers caused by preparing polyimide paper through imidization of polyamic acid fiber or polyamic acid solution under high temperature and high pressure, and significantly enhances the strength performance of paper to meet the growing demand for strength and temperature resistance of lightweight structural materials in industries such as aerospace and rail transit [8].

4.2.1.3 Poly-p-phenylenebenzobisoxazole (PBO) Fiber

PBO fiber, as the super fiber of the 21st century, combines excellent properties such as high strength, high modulus, high heat resistance, and high flame retardancy. In 1991, Dow Chemical Company of the United States and Toyobo Company of Japan jointly developed the PBO fiber spinning technology and realized commercial production at Toyobo Company. The registered trademark is Zylon®. The research on PBO in China started relatively late. In the mid-1980s, East China Institute of Technology (now Nanjing University of Science and Technology) synthesized the earliest domestic PBO polymer. The tensile strength of the prepared fiber is 1.2 GPa, and the modulus is 10 GPa, but its performance is not as good as that of Kevlar fiber. After years of development, many domestic universities and research institutes have achieved certain results in PBO monomer synthesis, PBO polymerization process, and fiber spinning. In 2005, Donghua University prepared PBO fiber with a tensile strength of 4.38 GPa and a thermal degradation temperature higher than 600 °C for the first time, laying a foundation for the localization and large-scale production of PBO fiber [9]. PBO fiber has excellent performance and can be comparable to high-performance fibers such as carbon fiber (T800), para-aramid fiber (Kevlar, Technora®), meta-aramid fiber (Nomex), and polybenzimidazole (PBI) [10]. Therefore, it has broad application prospects in military and civilian fields. However, due to its high price, it is currently mostly used in special fields such as the national defense industry.

In the molecular chain of PBO, the benzene ring and oxazole ring are almost coplanar with the chain axis, and it has a rigid rod-like molecular structure with symmetry. Its molecular structural formula is shown in Figure 4.5. Due to the high degree of conjugation of each structural component in the PBO molecular chain, the molecular bond energy is high, and the stability is good. In addition, the liquid crystal spinning technology enables the rigid rod-like macromolecules in the fiber to obtain a high degree of orientation and regularity, thus endowing the PBO fiber with excellent properties.

PBO fiber has a typical skin-core structure. Under the smooth cortex of approximately less than 0.2 μm is a core layer composed of microfibrils with a diameter of 10 to 50 nm. These microfibrils are arranged by highly oriented PBO rigid rod-like molecules along the fiber axis. There are capillary micropores between microfibrils. This multiple fibril structure enables PBO fiber to be fibrillated through the papermaking beating process to prepare pulp materials with a nanometer size. Researchers have carried out a large amount of research work on the preparation

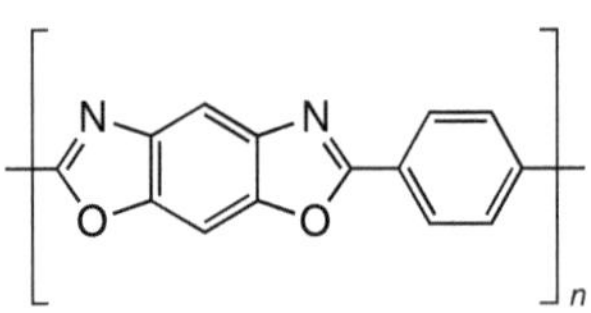

Figure 4.5 The chemical structure formula of the PBO molecule.

process and characterization technology of PBO fibrillated pulp. The functional paper-based composite materials prepared from PBO fibers can be used as friction materials, heat-resistant materials, battery separators, honeycomb sandwich composite materials, etc., and applied in cutting-edge technological fields such as aerospace and new energy.

The molecular structure of PBO fiber determines its excellent thermal and mechanical properties. PBO fiber has outstanding heat resistance. The fiber has no melting point, and the thermal decomposition temperature reaches 650 °C. It can be used for a long time at 300 °C. It is by far the organic fiber with the highest heat resistance. In addition, PBO fiber shows excellent flame retardancy. The LOI is 68%, second only to polytetrafluoroethylene fiber (LOI is 95%) among organic fibers. When PBO fiber burns at 750 °C, the toxic gases such as HCN, NOx, and SOx produced are very few compared with para-aramid fiber.

PBO fiber is an excellent impact-resistant material. When impacted, the fiber can be fibrillated and absorb a large amount of impact energy. The maximum impact load and energy absorption of its composite material are higher than those of aramid and carbon fiber. In addition, PBO fiber has excellent dimensional stability, wear resistance, and bending fatigue resistance. However, the oxazole ring of PBO is prone to ring-opening and fracture under ultraviolet light irradiation, resulting in a decrease in fiber strength. After a 40-hour sunlight resistance test, the tensile breaking strength of aramid can still be stabilized at about 80% of the initial value, while the tensile breaking strength of PBO fiber is only 37% of the original.

4.2.1.4 Ultra-High-Molecular-Weight Polyethylene Fiber

UHMWPE fiber, also known as high-strength and high-modulus polyethylene fiber, is a fiber spun from polyethylene with a molecular weight ranging from 1–5 million. As the fiber with the highest specific strength and specific modulus, UHMWPE fiber, together with aramid and carbon fiber, is known as the world's three major high-tech fibers. The appearance of the UHMWPE fiber is white. It is the fiber with the lowest density among high-performance fibers and also the only fiber that can float on the water surface. It has the advantages of high strength, high modulus, low density, good light resistance, low-temperature resistance, good bending fatigue resistance, and chemical corrosion resistance [11].

UHMWPE fiber has a supramolecular chain structure connected by methylene groups (-CH_2-CH_2-). The fiber has a high degree of axial orientation and high crystallinity, so it has excellent mechanical properties. The tensile strength of UHMWPE fiber is 2.8–4.2 GPa, and the elongation at break is 3%–6%. Its breaking work is higher than that of carbon fiber and aramid fiber. In addition, UHMWPE fiber has outstanding impact resistance and cutting resistance. The tensile strength is 15 times that of steel wire with the same linear density, 40% higher than aramid, and 10 times that of ordinary fiber and high-quality steel fiber. Since the molecular structure of UHMWPE fiber -CH_2-CH_2- does not contain groups such as hydroxyl groups and aromatic rings, it has chemical and optical inertness. As reported, after being immersed in various chemical media such as hydrochloric acid, sodium hydroxide, and seawater for six months, the strength retention rate is still 100%. After being immersed in sodium hypochlorite solution, the strength drops to 91%, while the strength of Kevlar fiber is completely lost [12]. In addition, this fiber is a thermoplastic fiber with a low glass transition temperature. It absorbs energy during the plastic deformation process. The impact resistance of this fiber is higher than that of aramid, carbon fiber, and polyester fiber and is only smaller than that of polyamide fiber (nylon).

The molecular structure of UHMWPE fiber has no polar groups. The intermolecular force is weak, and the molecules are prone to internal rotation. Therefore, the glass transition temperature and melting point are low. The high-temperature resistance and creep resistance are poor, and it is difficult to form chemical bonds with the resin matrix. At the same time, due to the low surface energy and chemical activity of the fiber, it is difficult for the fiber to be wetted. After being combined with the resin matrix to form a composite material, the interface bonding is weak, thus easily leading to interlayer damage and other issues during the material usage process. In order to improve the defects of UHMWPE fiber, methods such as low-temperature plasma, strong oxidants, radiation-induced surface grafting, and corona discharge are used to increase the bonding strength between the fiber and the matrix.

At present, the industrial technology of UHMWPE fiber is the gel spinning-high magnification hot drawing process. China began to study UHMWPE fiber around 1984 and did not achieve industrial production until 1999. It has become the fourth country to produce this fiber after the Netherlands, Japan, and the United States. The global production capacity distribution of this fiber is highly concentrated. The main producers are DSM of the Netherlands, Honeywell of the United States, Toyobo of Japan, and Shandong Aid Polymer Materials Co., Ltd. of China.

In addition to high strength, high modulus, weather resistance, chemical corrosion resistance, seawater resistance, and impact resistance, UHMWPE fiber also has characteristics such as cutting resistance, low-temperature resistance, excellent insulation performance, and ray transmission performance. It is widely used in fields such as protective equipment, aerospace, sports and leisure, ocean engineering, textiles, construction, biomedical, and so on. UHMWPE fiber is mainly used in the production of soft bulletproof vests, stab-resistant clothing, lightweight bulletproof helmets, radar shields, missile covers, bulletproof armor, cables for ships and ocean-going vessels, lightweight high-pressure vessels, aerospace structural parts, fishing nets, racing boats, sailboats, skis, as well as denture materials, medical grafts, and so on.

4.2.2 High-performance Inorganic Fiber

4.2.2.1 Glass Fiber

Glass fiber is an inorganic non-metallic material with excellent performance that was developed earliest. It is an extremely fine fiber formed by stretching, blowing, or centrifugally spinning molten glass with external force. This type of fiber was successfully developed by the United States in 1938 and began to be commercialized in the 1950s. At present, the technology is relatively mature and it is widely used in fields such as petroleum, chemical industry, transportation, smelting, electrical appliances, and electronics. Compared with traditional glass fiber, high-performance glass fiber refers to that some of its service performances are significantly improved, and it has better bearing capacity under external physical actions such as force, heat, light and electricity, and chemical actions such as acid, alkali, and salt. It retains the common characteristics of traditional glass fiber, such as heat resistance, non-flammability, and oxidation resistance, and has excellent properties and special functions that traditional glass fiber does not have.

The main component of glass fiber is silicon dioxide. According to the presence or absence of alkali metal oxides in the composition, it can be divided into four categories: alkali-free glass (E-glass) fiber, medium-alkali glass (C-glass) fiber, high-alkali glass (A-glass) fiber, and special glass fiber. Table 4.1 shows the chemical composition of commonly used glass fibers for paper-based functional materials [13]. Among glass fiber types, E-glass fiber and C-glass fiber were developed earlier. E-glass fiber belongs to aluminoborosilicate glass and has good electrical insulation and mechanical properties. It has a large application volume and a wide range. Low-boron glass fiber is mainly aimed at the strict requirements of clean rooms to reduce pollution caused by the release of boron-containing substances in the fiber. 475 glass fiber can be widely used in many fields, including gas and liquid filtration.

The production methods of glass fiber mainly include the drawing method, centrifugal method, and blowing method. The fibers obtained by drawing have relatively uniform diameters and are widely used in the production of glass filaments. While the fibers obtained by the centrifugal method and blowing method have thinner diameters and are mainly used in the production of glass staple fibers, which are usually called glass wool. The diameter of glass wool is not uniform. In practical applications, beating degree is usually used to reflect the diameter of glass staple

Table 4.1 Chemical composition of glass fibers commonly used in paper-based functional materials.

Chemical composition	E-glass fiber	Low-boron glass fiber	475 glass fiber
SiO_2	50.0~56.0	69.0~71.0	55.0~60.0
Al_2O_3	13.0~16.0	2.5~4.0	4.0~7.0
B_2O_3	5.8~10.0	<0.09	8.0~11.5
Na_2O	<0.60	10.5~12.0	9.5~13.5
K_2O	<0.40	4.5~6.0	1.8~4.0
CaO	15.0~24.0	5.0~7.0	1.0~5.0
MgO	<5.5	2.0~4.0	<2.0
Fe_2O_3	<0.50	<0.20	<0.25
ZnO	<0.02	<2.0	2.0~5.0
BaO	<0.03	–	3.0~6.0

Note: From Lauscha International Fiber Company, Germany.
Source: Adapted from [13].

fibers. A higher beating degree means that the overall diameter of the glass fiber is smaller.

Glass fiber is generally in a smooth cylindrical shape. During the fiber formation process, it contracts into a circular interface under the action of surface tension before the molten glass is drawn and cooled into solid fibers. The surface of glass fiber is smooth, and the bonding area between fibers is small, resulting in low paper strength. Therefore, resin reinforcement treatment must be carried out. Considering environmental protection requirements, water-based resins are mainly used at present. However, the smooth surface of glass fiber has little resistance to the passage of gas and liquid, so it is an ideal material for making filter materials. In addition to glass fibers with circular cross-sections, some companies have also developed glass fibers with special cross-sections, including triangular, flat, dumbbell-shaped, etc. The composite materials prepared from special-shaped glass fibers have great improvements in strength and filtration effect.

The relative density of glass fiber is usually between 2.50 and 2.80 g/cm^3, which is between organic fibers and most metals and close to that of aluminum. In addition, compared with water, due to the relatively high density of glass fiber, it is prone to sedimentation in water. Therefore, the dispersion of fibers is a key technology in the preparation of paper-based functional materials. A better dispersion effect can be obtained by mixing other fibers with glass fibers for papermaking. When papermaking is carried out using pure glass fibers as raw materials, it is generally done under acidic conditions with a slurry pH of 2.5 to 3.0 to achieve good dispersion effects. In addition, in order to ensure good uniformity of the paper, an inclined wire or Fourdrinier paper machine is used, and the forming concentration of the slurry on the wire is generally below 0.1%.

Compared with bulk glass, the strength of glass fiber is increased by more than 10 times and far exceeds that of some natural fibers, synthetic fibers, and alloy materials. The elastic modulus of glass fiber is lower than that of metal alloys but higher than that of most organic fibers. Glass fiber is brittle and easily broken, and its brittleness is proportional to its diameter. In glass fiber filter paper, brittleness is an important factor affecting the material's performance, which may lead to the problem of brittle fracture and shedding of glass fibers during the use of the material.

Glass fiber also has excellent heat resistance, electrical insulation, corrosion resistance, and weather resistance, and can be used as a motor insulation material in high-temperature, high-humidity, and corrosive medium environments. In addition, the sound absorption coefficient of glass fiber is high. When the sound frequency is 1025 Hz, its sound absorption coefficient is 0.5, so it can be used as a good sound absorption material. The porous material prepared from glass wool fiber has an increased sound absorption coefficient in the high-frequency region, and the thicker the material, the greater the sound absorption coefficient, which can meet the use requirements of various sound absorption and noise prevention environments.

4.2.2.2 Carbon Fiber

Carbon fiber is a special fiber mainly composed of carbon elements. Its carbon content is generally above 90%. It has a series of excellent characteristics such as light weight, high strength, high modulus, high temperature resistance, corrosion resistance, X-ray penetrability, and biocompatibility and is known as the fourth category of industrial materials. Carbon fiber is widely used in fields such as aerospace, national defense, transportation, energy, medical devices, and sports and leisure products.

According to different raw material sources, carbon fiber can be divided into polyacrylonitrile (PAN)-based carbon fiber, pitch-based carbon fiber, viscose-based carbon fiber, lignin-based carbon fiber, phenolic-based carbon fiber, and other organic fiber carbon fibers. According to different mechanical properties, carbon fiber is divided into general-purpose and high-performance types. The strength of general-purpose carbon fiber is generally 1000 MPa, and the modulus is 100 GPa. While high-performance carbon fiber includes high-strength carbon fiber, ultra-high-strength carbon fiber, high-modulus carbon fiber, ultra-high-modulus carbon fiber, etc. Among them, the modulus of ultra-high-modulus carbon fiber needs to be greater than 450 GPa, and the strength of ultra-high-strength carbon fiber needs to be greater than 4000 MPa. According to the difference in product forms, carbon fiber can also be divided into small tow fibers of 1–24 k and large tow fibers of 48–480 k (1 k is 1000 filaments). Japan is a major producer of high-performance carbon fiber, and the United States is a major consumer of high-performance carbon fiber. A few enterprises, such as Toray of Japan, SGL of Germany, Mitsubishi Rayon, Formosa Plastics Group, and Teijin of Japan, own the key technologies of carbon fiber research and development and production, have realized large-scale production of carbon fiber manufacturing, and have become the main suppliers of carbon fiber manufacturing and export.

At present, PAN-based carbon fiber and pitch-based carbon fiber are mainly used in the market. About 80%–90% of commercial carbon fibers are PAN-based carbon fibers. The surface morphology of PAN-based carbon fiber is related to the spinning process of the PAN precursor. Most of the existing spinning processes include wet spinning and dry-jet wet spinning. When wet spinning is used, the surface of carbon fiber usually has obvious cracks and grooves of different depths, which increases the specific surface area of carbon fiber and is beneficial to improving the shear strength of the composite interface. When dry-jet wet spinning is used, the fiber structure is uniform, and the strength performance is better. After special processing, carbon fiber can be made into activated carbon fiber with high adsorption. Its specific surface area is significantly increased compared with carbon fiber and can be widely used in the field of filtration and separation.

Carbon fiber has very high specific strength and specific modulus. Compared with metal materials, it has a low density of only 1.5–2.0 g/cm^3. It can be used to prepare lightweight structural composite materials and realize the lightweight of products. Carbon fiber is fatigue-resistant, friction and wear-resistant, has excellent self-lubricating properties, and can be applied to paper-based friction materials. In addition, carbon fiber also has excellent characteristics such as good energy absorption performance and heat resistance, a small coefficient of thermal expansion ($0–1.1 \times 10^{-6}$/K), and high thermal conductivity [10–160 W/(m·K)].

Carbon fiber is generally rarely directly applied. It is mainly compounded with matrices such as resin, metal, and ceramics to make composite materials. It was first applied in aerospace and national defense fields such as large passenger aircraft, military aircraft, missiles, rockets, and artificial satellites. Carbon fiber and its composites, as structural materials and functional materials, are also widely used in industrial fields such as wind power generation, the marine industry, electronic devices, and industrial equipment. In addition, carbon fiber also has a wide range of applications in the field of sports and leisure products, such as bicycles, fishing rods, tennis rackets, golf clubs, and yachts. In the field of specialty paper, carbon fiber is generally mixed with other fiber raw materials to prepare paper-based functional materials such as electromagnetic wave shielding materials, heating elements, fuel cell electrodes, speaker cones, etc. With the improvement of composite technology, the application field of carbon fiber paper will be further broadened.

4.2.2.3 Basalt Fiber

Basalt fiber is an inorganic fiber made by taking natural volcanic ejecta as raw material, crushing it, melting it at 1450–1500 °C, and stretching it through a platinum-rhodium alloy spinneret. It is another high-tech fiber after carbon fiber, aramid fiber, and UHMWPE fiber [14]. The surface of basalt fiber is smooth, and its cross-section is usually circular, as shown in Figure 4.6. The density of basalt fiber is generally 2.6–2.8 g/cm^3, higher than most organic fibers and inorganic fibers. The chemical composition of basalt fiber is very complex, mainly composed of various oxides such as SiO_2, Al_2O_3, Fe_2O_3, CaO, MgO, Na_2O, K_2O, and TiO_2, and the chemical composition varies with different stones and origins. Table 4.2 lists

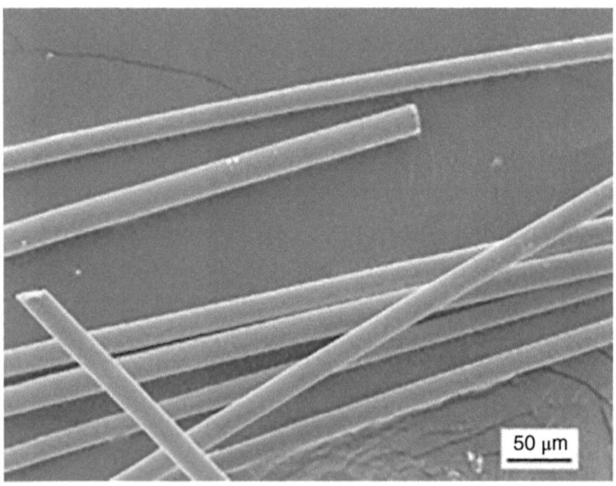

Figure 4.6 Morphology of Basalt fiber.

Table 4.2 Main chemical compositions of basalt ores.

Basalt composition	SiO_2	Al_2O_3	Fe_xO_y	CaO	MgO	$Na_2O + K_2O$	TiO_2
w/%	46–52	10–18	8–16	6–13	7–12	2–10	1.5–2

Source: Adapted from [15].

the main chemical components of basalt ore. SiO_2 is the main component in basalt fiber, accounting for about 50% of the overall mass of the fiber and playing a role as a network former in the silicate network structure. An increase in SiO_2 content is beneficial to improving the thermal stability and chemical stability of the fiber, but it will increase the difficulty of melting and drawing; Al_2O_3 in the fiber composition is beneficial to improving the chemical stability and mechanical properties of the fiber. Generally, when the content of SiO2 and Al_2O_3 is 60%–80%, the fiber has better strength, thermal stability, and chemical stability. The iron oxide in basalt exists in the form of Fe_2O_3 and FeO. When its mass fraction is too high, it can lead to large fluctuations in fiber diameter and increase the number of breakages during the drawing process. At the same time, the color of basalt fiber changes between coffee color and bronze color with different iron contents.

CaO and MgO are alkaline earth metal oxides with a relatively low mass fraction. When their content is low, it is beneficial for the melting of raw materials and the production of fine fibers. The mass fraction of TiO_2 is very low. During melting, it can increase the surface tension and viscosity of the melt, which is beneficial for the formation of long fibers. At the same time, it can also improve the chemical stability and mechanical properties of the fibers. In addition, components such as Na_2O and K_2O can improve the water resistance and corrosion resistance of the fibers.

Table 4.3 gives a comparison of the properties of different fibers. As shown, the tensile strength of basalt fiber is lower than that of carbon fiber, but it is better than that of E-glass fiber. However, the elongation at break of basalt fiber is relatively low, and it belongs to brittle materials. In addition, compared with other fibers, basalt fiber has the widest temperature range. The service temperature is generally −260 to 700 °C. It is an amorphous substance. When working at 400 °C, its breaking strength can be maintained at 85%. If pretreatment is carried out at 720–820 °C, there will be no shrinkage at 860 °C, while glass wool will be completely destroyed [16]. In addition to having good strength properties, basalt fiber also has an extremely low thermal conductivity. The thermal conductivity of thermal insulation materials prepared from ultrafine basalt fiber is 0.03 W/(m·K) at −196 °C. Therefore, thermal insulation materials made of basalt fiber are often used in the production of liquid oxygen. Compared with glass fiber, basalt fiber has higher electrical insulation performance. The volume resistivity is higher than 10^{12} Ω·m. At the same time, it also has excellent sound insulation, wave absorption, flame retardancy, and extremely low moisture absorption.

Currently, the technology and scale of basalt fiber are still in the initial stage. Only a few countries, such as Russia, Ukraine, the United States, Canada, and China, can produce basalt fiber. In China, research on continuous basalt fiber began in the mid-1990s. In recent years, it has developed rapidly. There are more than 10 domestic enterprises engaged in production, and those planning to establish factories. As of 2016, the total production capacity was 20,000 tons. The performance of basalt fiber lies between that of carbon fiber and glass fiber. It offers a relatively high cost-performance and holds significant applications in fields such as the military, aerospace, shipbuilding, highway engineering, safety protection, automobiles, high-temperature filtration, and sports equipment.

Table 4.3 Performance comparison of different fibers.

Fiber type	Tensile modulus/MPa	modulus of elasticity/GPa	Maximum operating Temperature/°C	elongation at break/%
Aramid fiber	2900–3400	70–140	250	2.8–3.6
Carbon fiber	3500–6000	230–600	500	1.5–2.0
E-glass fiber	3100–3800	72.5–75.5	380	4.7
S-glass fiber	4020–4650	83–86	300	5.3
Continuous basalt fiber	3000–4840	79.3–93.1	650	3.1

Source: Adapted from [16].

4.2.2.4 Ceramic Fiber

Ceramic fiber is a fibrous substance made from natural or artificial inorganic substances by different processes. It can also be transformed from organic fibers through high-temperature heat treatment. The fiber diameter is generally 2–5 μm, and the length is 30–250 mm. It has excellent mechanical properties, oxidation resistance, and high-temperature resistance. At the same time, it has characteristics such as low thermal conductivity, low specific heat capacity, and resistance to mechanical vibration. In the structure of ceramic fiber, both the solid phase and the gas phase exist in the form of continuous phases. The solid phase exists in the form of fibers, and the gas phase exists in the gaps of the fiber skeleton, resulting in high porosity, large pore diameter, and large specific surface area of the fibers, thus endowing the ceramic fiber with good thermal insulation performance and low density. According to different compositions, ceramic fibers include aluminum oxide ceramic fibers, silicide ceramic fibers (silicon carbide fibers, silicon nitride fibers), boron, and boride ceramic fibers.

1) Alumina fiber

Alumina fiber, also known as polycrystalline alumina fiber, belongs to high-performance inorganic fiber and is a kind of polycrystalline ceramic fiber. It has various forms such as long fiber, short fiber, and whisker. Alumina fiber takes Al_2O_3 as the main component and contains a small amount of SiO_2, B_2O_3, Zr_2O_3, MgO, etc. The strength of alumina fiber is 1.4 to 2.6 GPa, the modulus is ranged from 190 to 380 GPa, and the long-term service temperature is 1450 to 1600 °C. At the same time, it has small thermal shrinkage, low thermal conductivity, corrosion resistance, and unique electrical properties. Alumina fiber combines well with the resin matrix. The resin-based composite material prepared has greater elasticity than glass fiber and higher compressive strength than carbon fiber. In some fields, it has gradually replaced glass fiber and carbon fiber.

In China, the industrialization of alumina fiber began in the late 1970s. The main international companies supplying alumina fiber are DuPont Company in the United States, 3 M Company in the United States, ICI Company in the United Kingdom, and Sumitomo Company in Japan. At present, the main preparation methods of alumina fiber include sol–gel method, slurry method, prepolymerization method, Brunner method, impregnation method, and melt spinning method. Alumina fiber is mainly used in the form of short fiber and long fiber as thermal insulation refractory materials, high strength materials, aerospace materials, automotive accessory materials, etc. Alumina short fiber has excellent temperature resistance and is widely used in metallurgical furnaces, ceramic sintering and other industrial high-temperature furnaces. In the field of reinforced composite materials, alumina fiber is mainly used for reinforcing metal, ceramic and resin-based composite materials. The most remarkable application is to reinforce metal or ceramic matrix composites. Alumina fiber reinforced metal matrix composites have been applied in aerospace vehicles and automotive piston groove components. Alumina fiber reinforced ceramic matrix composites can be used as nozzle materials for supersonic aircraft and rocket engines, which can greatly reduce the number and quality of nozzle components.

2) Silicon carbide fiber

Silicon carbide fiber (silicon carbide fibers, SCF) is an inorganic fiber with β-silicon carbide structure prepared from organosilicon compounds by spinning, carbonization, or vapor deposition. This fiber has high specific strength and specific modulus. The strength can reach 1960–4410 MPa and the modulus is 176.4–294 GPa. The maximum service temperature of silicon carbide fiber can reach 1200 °C. At this temperature, the strength still remains above 80%. According to different forms, it can be divided into continuous fiber, chopped fiber, and whisker. Continuous fiber is a continuous filament formed by silicon carbide coated on core filaments such as tungsten wire or carbon fiber, or pure silicon carbide filament obtained by spinning and pyrolysis. Chopped fiber is made by cutting a filament with a short-cutting machine. The length is generally in units of mm. Whisker is a single crystal. The diameter of silicon carbide whisker is generally 0.1–2 μm and the length is 20–300 μm. The appearance is powdery.

At present, there are three industrial methods for preparing silicon carbide fiber: the chemical vapor deposition method, precursor conversion method, and activated carbon fiber conversion method. Silicon carbide has good compatibility with substrates such as metals, resins, and ceramics. It can be used as a reinforcing fiber and heat-resistant material for high-performance composite materials and can be used in aerospace, high-performance weapons, high-temperature engineering, and other fields. In addition, with the development of preparation technology, the application of silicon carbide fiber has gradually expanded to civil fields such as advanced sports equipment, civil engineering, medical treatment, and sports equipment.

4.3 Wet-laid Forming Technology of High-performance Fiber Paper

High-performance fiber paper is are functional material with specific properties, which is prepared by modern papermaking technology using high-performance fibers such as aramid fiber, carbon fiber, and polyimide fiber. It has excellent characteristics such as low density, high specific strength, high specific stiffness, and high temperature resistance. It is a structural and functional material with certain strategic significance in fields such as electromechanical equipment, rail transit, and aerospace. High-performance fiber paper-based functional materials are an important field for the papermaking industry to meet the strategic needs of the country. Although significant breakthroughs have been made in the key technologies for preparing such materials at present, there is still a gap in the product performance for high-end fields.

4.3.1 Challenges in Preparing High-performance Fiber Paper

The papermaking process of high-performance fibers (or ordinary synthetic fibers) is roughly the same as the process of ordinary cellulosic fiber papermaking. However, due to the great differences in physical and chemical properties between it and plant fibers, there are distinct characteristics in manufacturing technology, which are as follows [17,18,19]:

1) Hydrophobicity. Most high-performance fibers are hydrophobic and lack dispersion performance in water. Moreover, the fiber length is very long, and the aspect ratio is large. In the process of paper formation, problems such as fiber flocculation and serious cloudiness in the formed paper are easy to occur, resulting in poor uniformity of the formed paper and affecting the comprehensive performance.
2) Low paper strength. Most high-performance fibers cannot be fibrillated, and they cannot form hydrogen bonding between fibers as in plant fibers. Therefore, the paper is not as good as plant fiber paper.
3) Different densities. The densities of most high-performance fibers are very different. For example, the relative density of polyolefin fibers is 0.91 g/cm^3 and that of polyamide fibers is 1.14 g/cm^3. Those with low density are easy to float on the water surface, causing flocculation; while those with high density are easy to precipitate. Therefore, measures should be taken to obtain the good dispersion of fibers according to the density of synthetic fibers. For those with low density, such as polyethylene, they can be mixed with plant fibers for papermaking; for those with high density, a dispersant can be selected according to the fiber type.
4) The forming process is not easy to control. Due to the high freeness of fibers and rapid water filtration, the paper is formed quickly, affecting the uniformity of the formed paper. Besides, strong deformation and hot melting are easy to occur during the drying process.

4.3.2 Key Technologies

The challenges of high-performance fibers in the wet papermaking process include some issues, such as difficulty in fiber dispersion, difficulty in controlling formation, and low strength of formed paper. Therefore, the common technologies for the preparation of high-performance fiber paper-based functional materials mainly include four aspects: fiber preparation, fiber dispersion, paper formation, and paper reinforcement.

4.3.2.1 Differential Fibers

High-performance fibers usually have smooth surfaces, uniform lengths, and strong hydrophobicity. These fibers lack chemically active groups and have a small specific surface area, resulting in a lack of interlacing force between fibers. At the same time, the interfacial bonding with resin is poor, leading to poor wettability and low interfacial bonding strength. In addition, due to the inherent high strength and high-modulus characteristics of traditional high-performance fibers, it is difficult to fibrillate to obtain good paper strength. Therefore, high-performance short fibers cannot form paper alone and lack "adhesive fibers" with matching performance, high surface activity, large specific surface area, easy dispersion, and good compatibility. The use of "adhesive fibers" or application of resin impregnation processing and other methods can solve the above problems and significantly improve the strength of the formed paper. However, this method has some problems. For example, the paper is prone to being brittle after hot pressing, the process is complex, and the performance of some resins or adhesive fibers does not match that of the main fiber, resulting in a decline in comprehensive performance. Therefore, differentiated functional fibers have become an indispensable basic raw material for improving the comprehensive performance of high-performance fiber paper-based functional materials.

At present, the research and application of differentiated functional fibers are mainly focused on aramid fibers, PBO fibers, polyester fibers, etc., while the research in the field of differentiated functional fibers of polyimide fibers and carbon fibers is still in a blank state. The research on the papermaking performance, wet-end chemical characteristics, and the correlation between fiber and paper performance/structure of differentiated functional fibers has been conducted. Yantai Tayho Advanced Materials Co., Ltd. uses its para-aramid filament fiber product Taparan® (with a single filament linear density of 1.5D) as the raw material and prepares Taparan pulp fibers through fibrillation treatment. This fiber has excellent properties such as high temperature resistance, friction resistance, high degree of fibrillation, low static electricity, good dimensional stability, environmental protection, high strength and high modulus, and good dispersion. It is mainly used in fields such as friction materials, sealing materials, composite materials, and aramid paper. The following section will introduce three differentiated functional fibers, namely para-aramid pulp, precipitated fiber, and aramid nanofiber (ANF).

1) Para-aramid pulp

Para-aramid pulp (PPTA-pulp) is a differentiated functional product of para-aramid fiber developed by DuPont [20]. Its appearance and preparation process are shown in Figure 4.7. After the aramid filament is cut to a certain length, stirring and mechanical fibrillation are performed in an aqueous medium. The fiber is longitudinally torn, and micro-fibrillar fuzz is generated on its surface [21]. Compared with chopped

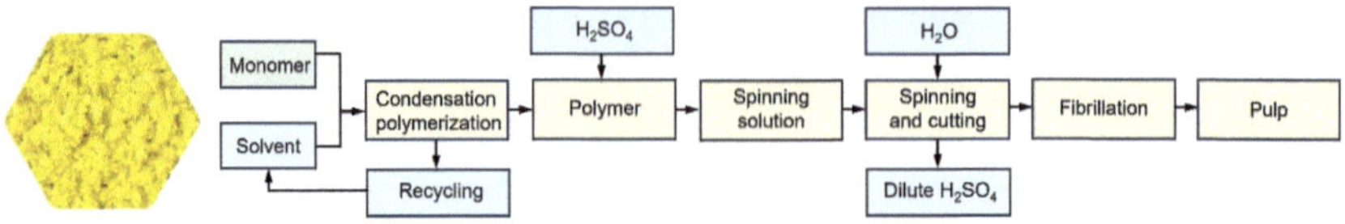

Figure 4.7 Appearance, morphology, and preparation process of para-aramid pulp.

aramid fiber, it looks similar to wood fiber. The fiber surface is highly fibrillated, with an average length of about 1.0 mm and an average specific surface area of more than 7 m^2/g. It retains most of the excellent properties of para-aramid fiber, such as heat resistance, wear resistance, high strength, and good dimensional stability. At the same time, the amino group content on the surface of pulp is higher than that of long fibers. It has high surface activity, better dispersion in water, and paper uniformity. Hence, it is an indispensable bonding material for preparing high-performance fiber paper-based materials such as paper-based friction materials, sealing materials, and honeycomb materials. In the process of preparing aramid paper, aramid pulp, as a filling and bonding material, is softened by heat during the hot pressing process, and endows the paper with mechanical strength through bonding chopped fibers and its own bonding effect.

The morphology of para-aramid pulp fiber is shown in Figure 4.8. Para-aramid pulp fibers are uniformly dispersed in the aqueous suspension. The fibers overlap each other. The appearance is similar to that of woody fibers. The fibers are generally in a fibrillar shape and soft. The obvious fibrillation of the fibers results in a larger specific surface area and higher surface activity compared to chopped fibers.

During the fibrillation treatment, the fibers are torn longitudinally to increase their specific surface area. Although the average length of the fibers after fibrillation treatment will slightly decrease, this action causes the fibers to swell and further fibrillate, generating a larger specific surface area. The uniformity of fiber dispersion and the degree of fibrillation in the slurry are further improved. As shown in Figure 4.9, as the beating degree increases, the degree of fiber fibrillation intensifies, and more feathery fibers are generated on the fiber surface, greatly increasing the specific surface area of pulp fibers. This is conducive to better bonding with chopped

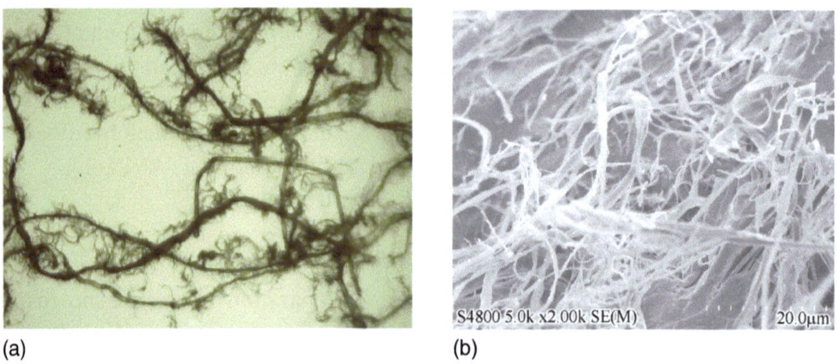

Figure 4.8 Optical microscope (a) and SEM images (b) of para-aramid pulp fiber.

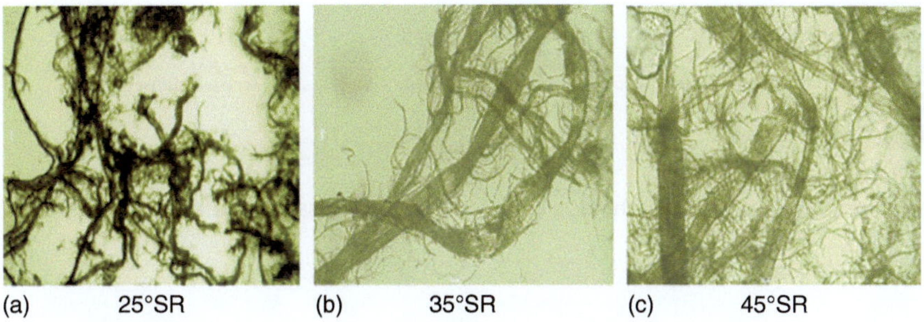

Figure 4.9 Optical images of aramid pulp with different beating degrees.

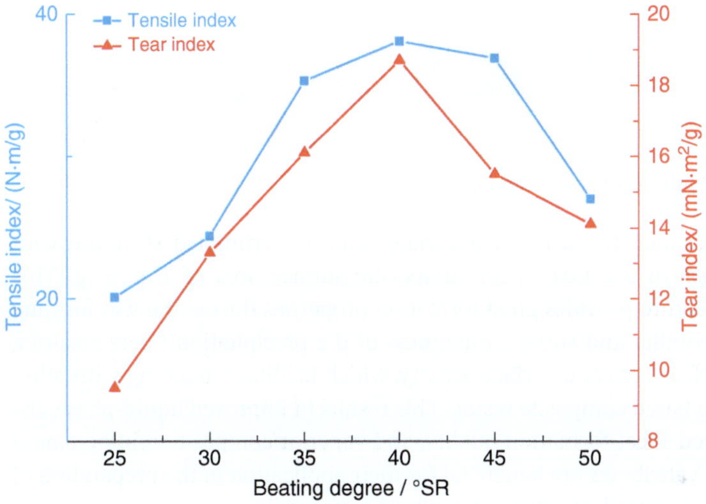

Figure 4.10 Effect of fibrillation degree of para-aramid pulp fiber on the physical strength of paper.

fibers and improving surface smoothness and physical strength. The degree of fibrillation treatment of pulp fibers has a direct impact on its final paper performance. As shown in Figure 4.10, when the degree of fibrillation of aramid pulp fibers is between 40 and 45°SR, the performance of the formed paper reaches the optimum.

2) Aramid fibrid

Aramid fibrid as a differentiated functional fiber is typically used to strengthen the physical properties of aramid paper. It is categorized into para-aramid precipitation fiber and meta-aramid precipitation fiber. The production primarily involves adding a precipitation agent to a low-temperature polycondensation solution of aramid polymers, followed by high-speed centrifugal shearing. The morphology and manufacturing process of the precipitation fibers are illustrated in Figure 4.11.

Aramid fibrid exhibits outstanding thermal stability, with an initial decomposition temperature of 490 °C and a mass loss of 48.6% at 800 °C. The unique manufacturing process endows these fibers with distinct morphological characteristics. As depicted

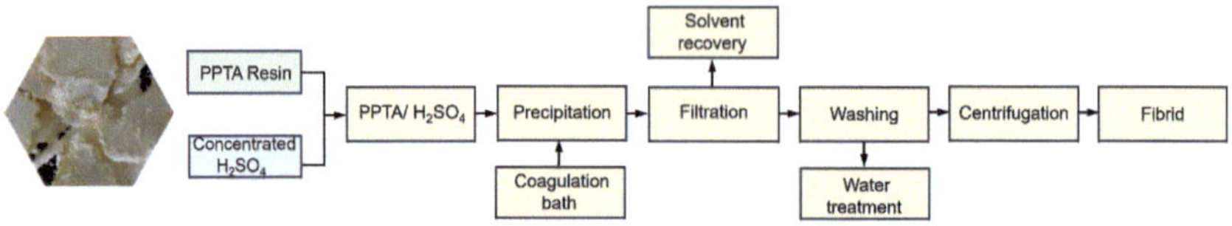

Figure 4.11 Appearance, morphology, and preparation process of meta-aramid fibrils.

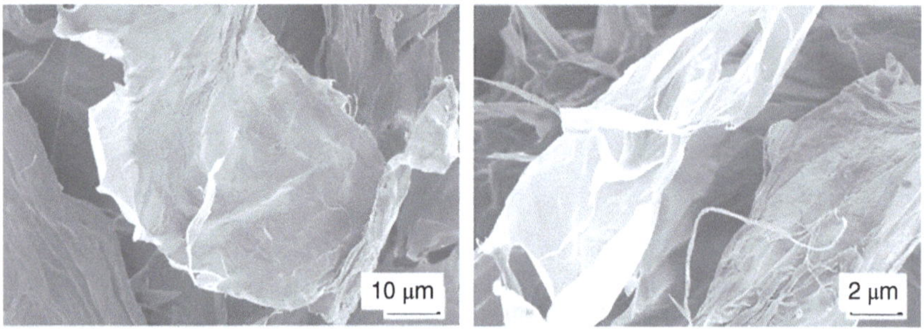

Figure 4.12 SEM images of aramid fibrid.

in Figure 4.12, aramid fibrid has a non-rigid, film-like crumpled structure with an average length of 0.2–1.0 mm and a specific surface area of 7–8 m^2/g. This morphological feature provides good formation properties during the wet forming process. The flexibility and surface roughness of the precipitation fibers ensure a high specific surface area and surface activity, which facilitates bonding with other fibers and amide-based composite resins. This results in improved liquid-phase dispersion, enhanced sheet formation quality, and superior composite reinforcement properties. These attributes are beneficial for their application in the preparation of high-performance aramid paper-based composite materials.

Beating treatment can adequately straighten the crimped fibrid, altering their morphological structure, which in turn enhances their formation quality and bonding properties. During this process, the fibrid undergoes swelling, straightening, and longitudinal tearing, resulting in a finer structural morphology and a larger specific surface area, which facilitates paper formation, mechanical strength, and breakdown strength.

3) ANFs

ANFs are a type of newly developed nanoscale polymer fiber material that retains the excellent mechanical properties and thermal stability of conventional aramid fibers. Additionally, their unique nano-scale structure, high aspect ratio, and large specific surface area endow them with composite reinforcement effects that are unattainable with macroscopic aramid fibers. ANF has emerged as one of the most promising "reinforcing building blocks" for constructing high-performance composite materials, playing a crucial role in interface reinforcement and material toughening [22, 23]. In the field of nanocomposite reinforcement, ANF can significantly improve the issues associated with traditional macroscopic aramid fibers, such as their smooth surface and chemical inertness.

Typically, the preparation techniques for ANFs mainly include electrospinning, polymer dispersion, high-pressure mechanical grinding, and deprotonation methods. Among these, the deprotonation method, reported by Professor Kotov's group at the University of Michigan in 2011, has gained the most widespread application due to its simple preparation process. This technique has successfully overcome the limitations of traditional para-aramid fibers, which are characterized by strong chemical inertness and low reactivity, and has also provided new insights and pathways for the development of high-performance nanoscale building blocks. However, this method has a long preparation cycle of up to seven days, a low reaction concentration (0.2%), and lacks a clear characterization method for the reaction endpoint. In response to these issues, a fibrillation/ultrasonic/proton donor-coupled deprotonation method has been proposed for ANF preparation, which has reduced the ANF preparation cycle from the traditional seven days to four hours [24]. Furthermore, the use of proton donor-coupled deprotonation has enabled the large-scale production of high-concentration ANF with a mass concentration of 2.0%. This approach offers significant advantages in terms of ANF diameter, diameter distribution uniformity, preparation cycle, film strength, and industrial prospects.

The preparation of aramid nanopaper using papermaking techniques involves protonating and reducing ANF to obtain an ANF aqueous dispersion, which is then utilized in modern papermaking technology to fabricate aramid nanopaper. The aramid nanopaper produced through this protonation and reduction, followed by papermaking, not only exhibits high tensile strength and modulus but also demonstrates exceptional toughness. This is primarily due to the vacuum-assisted self-assembly formation method, which results in a multilayered biomimetic structure similar to a natural shell. The interwoven ANF within each layer facilitates stress transfer, thereby ensuring high strength and modulus. The ANF slurry can be converted into aramid nanopaper through processes such as web formation, dewatering, pressing, and drying. During the papermaking process, the thickness, basis weight, fiber aspect ratio, and paper structure of the aramid nanopaper can be controlled by adjusting the stock supply system and papermaking process parameters. This method is characterized by rapid filtration and formation, precise control over paper basis weight and thickness, and roll-to-roll processing capability, making it a promising approach for large-scale industrial production of aramid nanopaper.

4.3.2.2 Fiber Dispersion

Compared to papermaking plant fibers (1–2 mm), high-performance fibers have a greater length (>5 mm), lower surface activity, poorer wettability, and stronger hydrophobicity. The fibers are prone to entanglement, flocculation, and deposition, and the secondary dispersion of flocs is extremely difficult, which severely impacts the forming uniformity and strength of paper. Additionally, the approach flow and mixing of high-performance fiber slurries involve different pipelines, storage facilities, mixing containers, and slurry pumps, making the hydrodynamic characteristics and wet-end chemical environment of the slurry blending system even more complex. Therefore, enhancing the efficient dispersion is a key theoretical and technical issue that urgently needs to be addressed to ensure that high-performance fiber paper-based functional materials possess excellent uniformity and performance.

The dispersion process of fibers in suspension is relatively complex and involves many physical and chemical issues. Fiber flocculation is also a dynamic equilibrium process between physical interlacing, chemical bonding forces, and destructive turbulence and shear forces. Therefore, only by considering multiple aspects comprehensively can improve the dispersion of fibers and improve the uniformity of the formed paper. The dispersion performance of high-performance fiber slurry can be improved from the following aspects:

1) Improving the wetting state of the fiber surface. Since most high-performance fibers are hydrophobic, anionic or nonionic surfactants or their combinations can be added to reduce the surface tension of the slurry suspension, improve the wetting performance of the fiber surface, and improve the dispersion effect of the fibers in the slurry suspension.
2) Increasing the viscosity of the slurry system. The addition of natural or synthetic thickeners as a water-soluble lubricant to the fiber suspension allows the fibers to slide past each other without tangling. Increasing the viscosity of the slurry will limit the degree of freedom of high-performance fibers, reducing the chance of fibers contacting each other, which is beneficial to improving paper formation. In addition, the increase in the viscosity of the suspension is conducive to prolonging the settling time of fibers. Studies have found that when the dosage of polyethylene oxide (PEO) is added from 0.1% to 1% (based on absolutely dry fibers), the increase in dosage is beneficial to improving various properties of aramid paper.
3) Changing the electrokinetic charge on the fiber surface. Most fiber surfaces carry a negative charge. By adding anionic surfactants such as sodium hexametaphosphate, polyphosphate, and pyrophosphate to the system, the ζ potential on the fiber surface can be increased and the repulsion between fibers can be enhanced, thereby reducing the degree of slurry flocculation.
4) Adjusting the crowding factor of the slurry suspension. The crowding factor is commonly used to illustrate the possible degree of fiber flocculation. The crowding factor Ncrowd refers to the number of fibers contained in a spherical volume with a diameter equivalent to the fiber length. This concept was first proposed by Mason [25] and later improved and developed [26]. It is expressed by the formula as:

$$N_{crowd} = (2/3)C_v(L/D)^2 \approx (5C_mL^2)/\delta \tag{4.1}$$

where C_v is the volume concentration of fiber, kg/m^3L is the length of the fiber, m; D is the diameter of fiber, m; C_m is the mass concentration of fiber, %; and δ is fiber coarseness, kg/m.

As can be seen from the above formula, fiber length has a great influence on slurry flocculation. The longer the fiber, the greater the crowding factor. Therefore, the possibility of fiber flocculation is increased. In addition, as the slurry concentration increases, the crowding factor increases and the number of collisions between fibers increases, leading to an increase in the degree of flocculation. Therefore, the simplest way is to improve the uniformity of formed paper by reducing the concentration of fiber suspension, but this method would consume large amounts of water. The coarseness of fiber also has a certain influence on the crowding factor. It affects the degree of fiber flocculation by changing the number of fibers in the fiber suspension. In addition, the coarseness of the fiber also affects the softness of the fiber. The smaller the coarseness of fiber, generally the softer the fiber. During the movement, fibers have more chances of colliding and interlacing with each other, and the probability and degree of flocculation are also greater. Therefore, increasing the coarseness of the fiber can also improve the dispersion of the fiber to a certain extent. However, increasing the coarseness of fiber will also lead to a relatively high porosity of paper, so it is not suitable for some paper-based materials with high density.

Besides, many papermaking equipment are different from traditional papermaking equipment during the flow process of high-performance fibers. For example, for dispersing chopped fibers in water, the beating equipment for plant pulp cannot be used. Generally, hydraulic pulpers and deflakers are used. When transporting long fiber slurry, pumps with small clearances and high shear forces (such as rotary centrifugal pumps) are not suitable and should be replaced by positive displacement pumps (such as reciprocating plunger pumps). In addition, since high-performance fibers have a large length and are easily entangled and deposited, the design of pipelines and slurry tanks should avoid sharp corners and sudden changes in direction. There should be no flow valves and concentration controllers in the pulp supply system. The pipeline material is preferably transparent plexiglass. On the one hand, it can reduce the mixing of metal ions and debris into the pulp and avoid affecting paper properties (such as insulation). On the other hand, the transparent pipeline can clearly see the flow direction of the pulp, and fiber entanglement and deposition can be found and eliminated as soon as possible.

4.3.2.3 Paper Forming

To prevent high-performance fiber from flocculating during wet forming, a large amount of water is required to dilute it. Generally, the forming concentration for high-performance fiber paper needs to be reduced to 0.005%–0.05%. Otherwise, it is prone to cause uneven distribution of fibers, leading to poor uniformity and low quality of paper formation. Moreover, due to the different lengths, densities, and significant differences in slurry drainage properties of various fiber components in mixed slurry, problems such as poor paper formation and uneven Z-direction distribution of fibers arise. At the same time, ordinary fourdrinier forming cannot dewater large amounts of water, resulting in issues such as low retention of functional components and uncontrollable Z-direction distribution of fibers, which negatively affect the properties of paper.

The papermaking machines used for high-performance fibers mainly include the Fourdrinier machine, cylinder machine, and inclined wire machine. The Fourdrinier papermaking machine can be used only when the papermaking performance of synthetic fibers is close to that of plant fibers. The papermaking machine used for non-plant fiber is basically the same as the wire part of the traditional cylinder

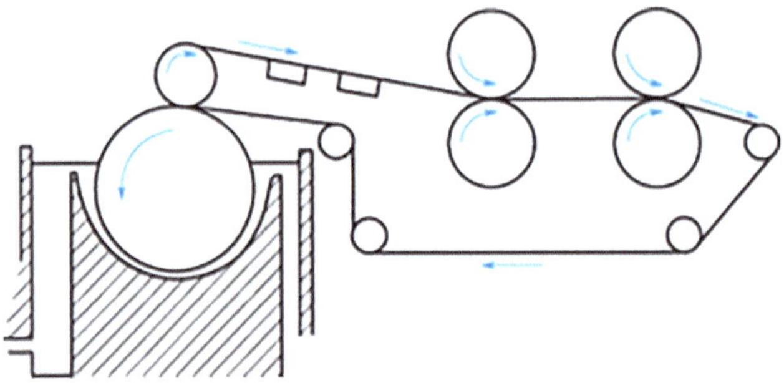

Figure 4.13 Schematic diagram of a cylinder machine.

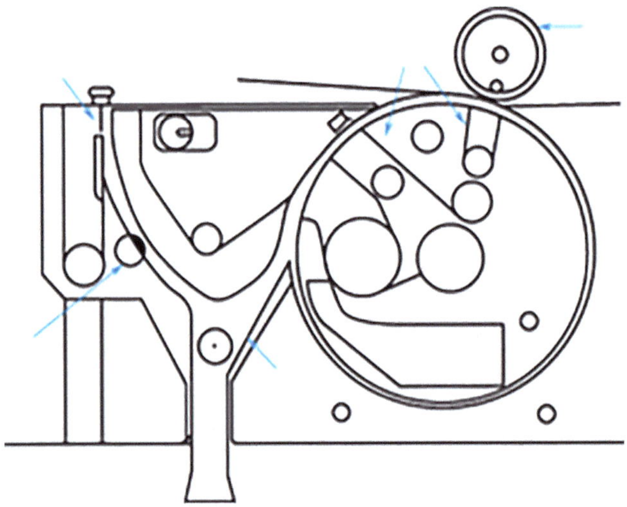

Figure 4.14 Schematic diagram of Rotoformer cylinder machine.

machine. It consists of a wire cage, vat, and couch roll, as shown in Figure 4.13. The Rotoformer cylinder former (Figure 4.14) has an upper slice lip that can be adjusted up and down and back and forth. This structure has a very positive effect on the optimization of slurry speed ratio, dewatering curve, paper uniformity, and structure. Therefore, it has great flexibility and can produce paper with a basis weight range of 15–1000 g/m^2 and can adapt to any fiber raw material formulation from 100% plant fiber to 100% synthetic fiber or any combination in between. The cylinder machine is space-saving, has low investment, and good process flexibility. It is an important equipment for non-plant fiber papermaking. However, it has issues like a large jet-to-wire ratio, poor forming uniformity, and low production capacity.

In order to adapt to the requirements of lower forming concentration, inclined wire formers have been developed. The dewatering capacity of inclined wires is much greater than that of horizontal fourdriniers. The dewatering capacity of an inclined wire with a length of 0.6–0.9 is comparable to that of a 3.5- to 4.5-m-long

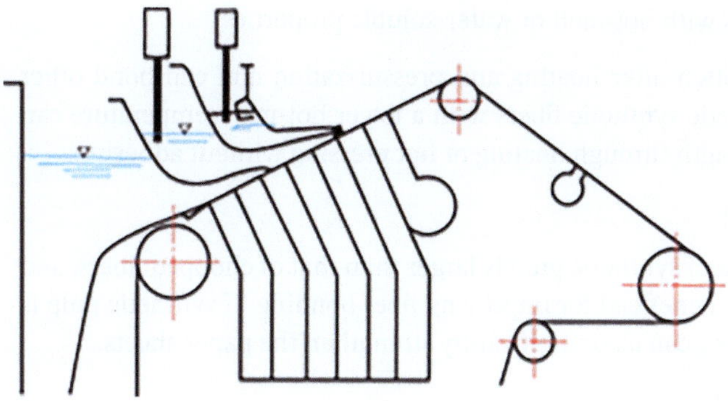

Figure 4.15 Inclined wire former with single-wire and three headboxes.

horizontal wire, thus allowing extremely low furnish concentrations. At the same time, the required length of the wire section is greatly reduced. The inclined wire formers with single-wire double-channel headboxes and single-wire three-channel headboxes after the first generation (open weir pond with adjustable wire table angle) and the second generation (continuously adjustable wire table angle), enabling multilayer paper sheets. Nowadays, inclined wires can adjust the furnish speed, dewatering curve, and the structure and properties of paper, and can achieve multilayer paper, as shown in Figure 4.15.

Yantai Metastar Special Paper Co., Ltd. has carried out computational research on inclined wire formers and designed an ultra-low concentration inclined wire former suitable for manufacturing aramid paper-based materials. Moreover, it has optimized the engineering parameters for forming, such as jet-to-wire ratio and dewatering curve. This enables the equipment to achieve uniform dewatering at a low concentration. Meanwhile, by adjusting the positions of the slice lip, it can regulate the fiber distribution direction and optimize the MD/CD ratio of paper strength to meet the requirements of different grades. With the development and progress of ultra-low-concentration inclined wire forming technology, more long fiber paper-based functional materials have obtained better forming uniformity and quality by adopting this forming method.

4.3.2.4 Paper Strengthen

Most synthetic fibers, including high-performance fibers, cannot be fibrillated and form hydrogen bonds similar to those in plant fiber paper. As a result, synthetic fibers lack dispersibility in water and the ability to fibrillate during beating. This leads to poor interlacing of paper sheets, a loose paper sheet structure, and low strength in both the wet state during papermaking and after drying. Generally, strengthening techniques must be adopted to meet the final usage requirements. Generally, there are the following methods.

1) Adding adhesives

Adding inorganic or organic adhesives (synthetic resin adhesives) through addition in the furnish or by coating, impregnation, spraying, etc. After drying, the base paper can effectively improve paper strength.

2) Using bonding fibers with hot-melt or water-soluble properties

Thermoplastic fibers soften after heating and pressurization and can bond other fibers. Thus, thermoplastic synthetic fibers with a lower hot-melt temperature can form good bonding strength through heating or hot pressing without adhesives.

3) Using synthetic pulp

The specific surface area of synthetic pulp is larger than that of chopped fibers, and its polarity is enhanced, beneficial for improving fiber bonding. If synthetic pulp is hot-meltable, hot pressing can also significantly strengthen the paper sheets.

4) Solvent enhancement

Using a solvent with partial dissolution and swelling effects on fibers can enhance fiber bonding as the fiber surface is dissolved or chain segments are swollen during paper drying.

5) Hot pressing enhancement

During hot pressing, the mechanical properties of aramid paper can be enhanced by adjusting parameters like temperature, pressure, and the number of pressing times. Before hot pressing, the paper-based material is loose with disordered fibers. After hot pressing, it becomes dense. Chopped fibers as reinforcement partially deform due to high temperature and pressure, which is related to fiber rigidity, material uniformity, and hot pressing conditions. Fiber deformation increases the contact area between aramid fibers, facilitating bonding and paper strength. Aramid fibrids, as bonding materials, are evenly distributed around chopped fibers and deform plastically after heating, embedding, and fixing the chopped fibers to form a structure similar to reinforced concrete, ensuring excellent mechanical properties. Hot pressing pressure and temperature have a great impact on the strength of para-aramid paper-based materials. Considering various physical performance, the optimal hot pressing conditions are 240 °C, 0.4 MPa, a roll speed of 1.5 m/min, and two hot pressings times. At this time, the tensile index is 54.27 N·m/g, the tear index is 24.63 mN·m^2/g, and the dielectric strength reaches 23.81 kV/mm.

4.4 Typical High-performance Fiber-based Functional Paper Materials

Paper-based functional materials made from high-performance fibers account for less than 1% of the total paper production. However, their importance is growing, and they are increasingly gaining attention from countries worldwide.

Compared to plant fiber paper, high-performance fiber paper has the following characteristics:

1) Fibers are clean, and the finished paper is almost free of impurities
2) High folding endurance, tear resistance, and tensile strength
3) High chemical resistance, ensuring long paper sheet preservation

4) High corrosion resistance, strong wear resistance, and strong resistance to light-induced aging
5) Special properties such as fire resistance, sound insulation, antibacterial properties, rust prevention, and insect resistance can be imparted through the addition of additives or subsequent processing
6) Excellent electrical properties, with a dielectric strength of up to 40 kV/mm and a tangent of the loss angle that can be less than 0.005, which is significantly superior to cellulose fiber paper.

Therefore, the trend of using high-performance fibers to replace traditional cellulose fibers for improving paper properties is becoming increasingly evident in various industries, such as insulation materials, friction materials, filter materials, battery separator materials, and sealing materials. This has also become one of the research hotspots in this field. This section introduces some typical high-performance fiber-based functional paper materials in recent years.

4.4.1 Aramid Fiber-based Electrical Insulation Paper

Aramid paper is are high-performance sheet-like composite material prepared from chopped aramid fibers and fibrids through the papermaking wet forming process and hot-pressing processes. The preparation process is shown in Figure 4.16. In the structure of aramid paper, chopped aramid fibers act as reinforcing materials, randomly and evenly dispersed in the three-dimensional structure of the paper, playing the main load-bearing role. Aramid fibrids act as matrix materials, distributed and filling the spaces between the chopped fibers. During the hot-pressing process, the fibers undergo plastic deformation, firmly embedding the chopped fibers and enabling them to transfer stress [27].

Electrical insulation products, such as those used in motors and transformers, benefit significantly from the use of high-performance fiber-based electrical insulation papers. These papers offer substantial advantages over natural fiber-based products in terms of thermal stability, electrical properties, and mechanical strength, which can notably enhance the service life and safety of electrical equipment. Since the 1970s, aramid paper has been utilized in dry-type transformers due to its exceptional mechanical properties, insulation capabilities, and resistance to high temperatures, thereby upgrading the temperature class of these transformers.

The application of aramid paper and its composites in the field of insulation has been expanding ever since. Presently, aramid insulation materials are extensively employed in various applications, including rail transit and wind power generation equipment, power transformers, and as materials for transformer winding coils,

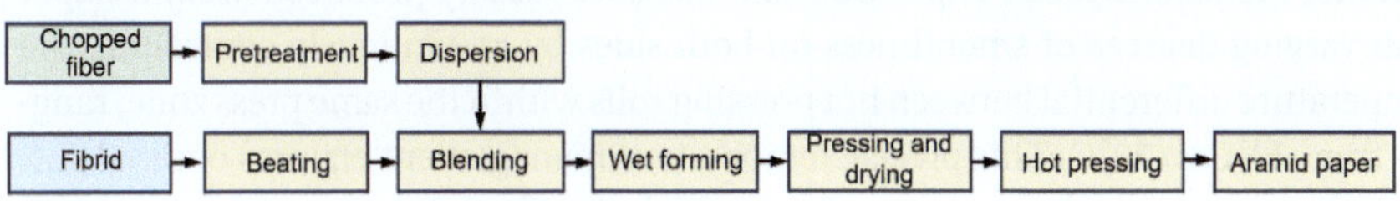

Figure 4.16 Preparation process of aramid paper.

phase insulation, and turn insulation in traction motors. They are used in the form of paper, paperboard, soft composite materials, and aramid mica paper. These materials are particularly suitable for use as upgrade materials for insulation products in harsh environments characterized by high temperatures and humidity. They are especially adept at meeting the stringent insulation performance requirements necessary for the acceleration and braking processes of high-speed rail and subway systems.

Currently, research on aramid paper is primarily concentrated on two main aspects.

1) Improvements in corona resistance and aging resistance

Although aramid fibers exhibit excellent properties that meet the needs of most applications, they exhibit poor resistance to corona discharge and aging. In the context of insulation materials subjected to high electric fields, resistance to corona is more critical than dielectric strength. Although the dielectric strength of various insulation materials may not differ greatly, their resistance to corona can vary significantly. Insufficient corona resistance can result in polymer cracking, leading to brittle and fissured insulation layers, which in turn compromise the safety and lifespan of the insulation products. To address this, researchers have undertaken extensive studies aimed at enhancing the lifespan and corona resistance of aramid paper, thereby improving the reliability of electrical equipment operations. The incorporation of inorganic nanoparticles has been shown to not only bolster the mechanical properties and thermal stability of polymers but also to enhance their dielectric characteristics. Specifically, the addition of nanoscale aluminum nitride (AlN) and silicon dioxide has been found to effectively reduce the conduction of electric current through the insulation paper and to increase its alternating current breakdown field strength. Moreover, these nanoparticles play a crucial role in mitigating the injection, migration, and accumulation of space charges within the dielectric material. This results in a more uniform distribution of space charges within the insulation paper, which is essential for maintaining the integrity and performance of the insulation system [28, 29].

2) Functionalization and Structural Design

Research institutes and manufacturers should integrate critical technologies in high-performance fiber preparation, dispersion, formation, and hot pressing. Through raw material composition, process optimization, and system control, they can achieve responsive regulation of fiber morphology, paper structure, and material properties. This enables the production of high-performance paper-based functional materials that meet the strength, insulation, and temperature resistance requirements of various application fields.

Yantai Metastar Special Paper Co., Ltd. has successfully produced aramid paper with varying degrees of smoothness on both sides by meticulously controlling the temperature differential between hot pressing rolls within the same press zone, ranging from 30°C to 70°C. This precise temperature management ensures optimal surface glue adhesion and composite effects, which significantly improve the material's processability. Institutions such as the South China University of Technology and the Shaanxi University of Science and Technology have pioneered the development of multiratio layered forming technologies. These advancements involve a strategic design of the structural and compositional ratios within the paper, enabling each layer to serve a specific function. This innovative approach allows for the precise control of the apparent density, which can be adjusted within a range of 0.5–1.2 g/cm^3, and the structural porosity, which can be varied between 18.0% and 60.0%. These capabilities meet the diverse requirements of various application fields [30–32].

When aramid paper is subjected to external forces and breaks, only a portion of the fibers actually fracture, while the majority are pulled out. This phenomenon, from a material failure perspective, demonstrates that the binding strength between fibers in traditional aramid paper is relatively weak. Zhang and colleagues [33] have developed aramid nanopaper through wet forming and sol–gel processes, achieving tensile strength, modulus, and toughness of up to 165 MPa, 6.4 GPa, and 9.7 MJ/m^3, respectively. These properties exhibit the characteristics of being "strong and tough," showcasing excellent mechanical performance. The aramid nanopaper, which is only 20 μm in thickness, has a strength performance that is 2.3 times that of conventional aramid paper with a thickness of 55 μm. This implies that when aramid nanopaper is applied in the field of insulation, it enables the "thinning" of insulation layers, which is conducive to the miniaturization and integration of insulating equipment. Besides, ANF can be adopted to enhance the electrical insulation strength of aramid paper. With a 6% addition, the dielectric strength of aramid paper increased by 44% compared to the control sample (14.5 kV/mm) [34]. The utilization of ANF, as opposed to the macroscopic aramid fibers typically used in common aramid paper, offers a substantial advantage. The nanofibers contribute a greater number of bonding sites and surface areas, which not only strengthen the paper strength but also minimize the paper porosity, thus significantly boosting the dielectric strength of the aramid paper.

Furthermore, Yang et al. [35] have developed a CNF@ANF nano-composite insulation paper through layer-by-layer self-assembly techniques. This composite boasts mechanical strength and dielectric strength, along with superior water resistance, aging resistance, and commendable temperature tolerance. The dielectric strength has been notably enhanced, more than doubling to reach 29.6 kV/mm. This material holds promise for application in transformers that are rated Class F and above, where higher demands are placed on temperature resistance, dielectric strength, and mechanical properties. It can be a candidate as an upgrade and a new generation replacement for cellulose-based insulation paper.

4.4.2 Aramid/ Mica Insulation Paper

With the rapid advancement of power electronics technology, an increasing number of AC variable frequency speed drives and motors have been widely implemented. The proliferation of electrical equipment operating in various complex engineering environments has set new standards for traditional insulating materials. These include enhanced mechanical strength, temperature resistance, moisture-proof capabilities, corona resistance, and service life. Traditional aramid paper is known for its excellent dielectric properties and good chemical stability, but it falls short in terms of corona resistance. Mica paper relies on van der Waals forces between its flakes and typically requires a glass fabric substrate, along with adhesives or synthetic resins for reinforcement. However, the temperature resistance and insulating performance of these adhesives and resins often fail to meet the demands of motor applications.

Aramid/mica paper is a high-performance paper-based composite insulating material made from aramid fibers and mica. It is produced using papermaking techniques combined with composite material processing methods. This material offers a combination of high-temperature resistance, excellent corona resistance, and superior strength, making it suitable for use in high-voltage and variable frequency motors, as well as traction motors, representing a premium class of electrical insulating materials.

Aramid mica paper features high mechanical strength, temperature resistance, flexibility, moisture resistance, and dielectric strength and is used in high-voltage motor insulation, high-voltage transformer insulation, V-rings, coil winding insulation, and interturn insulation. There are mainly two types of products: one with low aramid fiber content (fiber content <10%) and another with high aramid fiber content (about 50%). Currently, Nomex T418 and Nomex T419 aramid mica papers, which are widely used as insulation materials. Table 4.4 shows typical representative products.

Aramid/mica paper retains excellent mechanical properties at high temperatures, effectively enhancing the material's aging resistance and extending its lifespan and safety. Compared to Nomex T418, aramid/mica paper with low fiber content maintains higher strength at elevated temperatures. Additionally, aramid/mica paper features a very high LOI, which can exceed 60%. With an LOI value over 20%, the material is non-flammable in air, and the high LOI of aramid mica paper provides excellent flame retardancy, enhancing the safety of the material during use. Tables 4.5 and 4.6 demonstrate the mechanical and insulation properties of aramid/ mica paper.

Table 4.4 Aramid/mica paper grades and performances.

Performance	KMF-1-200	KMF-2-200	KMF-3-200	Nomex T418	Nomex T419
Fiber content /%	2.5	5.6	7.7	53.5	50.69
Thickness/mm	0.14	0.17	0.17	0.08	0.20
Density/(g/m^3)	1.42	1.20	1.19	1.13	0.45
Grammage/(g/m^2)	199.4	204.5	202.0	89.2	91.5
Tensile strength/(N/cm)	4.8	7.2	8.1	29.0	18
Dielectric strength/(kV/mm)	18.6	16.5	15.3	30.3	15.6
Air permeability/(s/100 mL)	415	580	664	–	–
Permeation time/s	83.9	70.4	76.2	–	–

Table 4.5 Mechanical performance of aramid/mica paper.

	Aramid mica paper (low fiber content)			NomexT418		
Performance	**1**	**2**	**3**	**1**	**2**	**3**
Thickness/mm	0.076	0.114	0.185	0.080	0.130	0.210
Grammage/ (g/m^2)	90.6	149.1	238.7	89.2	148.4	236.8
Tensile strength/(N/cm)	33.65	54.46	62.09	29.00	52.00	87.00
Elongation/%	1.0	1.5	2.0	2.4	2.9	3.7
Tear strength/N	0.5	0.9	1.2	1.1	2.2	3.6

Table 4.6 Electrical insulation performance of aramid/mica paper.

	Aramid mica paper (low fiber content)			Nomex®T418		
Performance	**1**	**2**	**3**	**1**	**2**	**3**
Thickness/mm	0.076	0.114	0.185	0.080	0.130	0.210
Dielectric strength/(kV/mm)	33.84	35.61	37.95	30.30	35.00	40.20
Dielectric constant (60 Hz)	3.1	3.0	3.0	2.3	2.5	2.5
Dielectric loss factor (60 Hz)/$\times 10^{-3}$	11	10	10	6	6	6
Volume resistivity/ Ω·cm	$(10)^{14}$	$(10)^{15}$	$(10)^{15}$	$(10)^{16}$	$(10)^{16}$	$(10)^{16}$
Surface Resistivity/Ω	$(10)^{14}$	$(10)^{15}$	$(10)^{15}$	$(10)^{14}$	$(10)^{15}$	$(10)^{15}$

4.4.3 Lightweight and High-strength Para-aramid Honeycomb Core Material

In recent years, with the advancement of transportation technologies, composite materials have played a crucial role in promoting the lightweight and high-performance characteristics of aircraft and spacecraft. Aramid paper-based honeycomb structural materials, manufactured through processes such as resin impregnation, honeycomb lamination, stretching, bonding, and composite processing, offer excellent properties including light weight, high strength, impact resistance, corrosion resistance, sound and heat insulation, and ease of large-area integral forming. These materials are key to weight reduction in the structural design of aircraft, high-speed trains, ships, and automobiles. Kevlar paper-based honeycomb is manufactured using para-aramid fiber paper and phenolic resin, offering lower moisture absorption and superior mechanical performance compared to Nomex paper-based honeycomb.

1) Aerospace Sector

Aramid paper-based honeycomb structural materials, known for their lightweight and high strength, efficient flame retardancy, good design flexibility, adaptable processing, and capability for monolithic forming, are utilized in both military and civilian aircraft for load-bearing and secondary load-bearing components (such as wings, fairings, floors, engine nacelles, and cabin doors) as well as interior filling materials (such as sidewalls, ceilings, partitions, and luggage compartments). This application facilitates the lightweighting of aircraft, contributing to energy conservation and emission reduction. The use of high-performance aramid honeycomb structural materials has brought significant advantages and technological transformations to aircraft manufacturing enterprises, such as environmental friendliness, cost-effectiveness, and efficiency. Moreover, due to the excellent dielectric properties and high wave-transparency of aramid paper-based honeycomb materials, these materials, with their light weight, sufficient compressive strength, shear strength, and good fatigue resistance, as well as good adhesion and assembly compatibility with composite materials, have widely replaced glass cloth honeycomb cores as the core material for radar domes, used in airborne, shipborne, and satellite radar antenna covers.

2) Rail Transit Sector

As para-aramid paper-based honeycomb structural materials gradually become more cost-effective, they are also being applied in floors, sidewalls, ceilings, and doors of high-speed trains, as well as yachts and racing boats. Therefore, they have become key materials for achieving vehicle light-weighting in high-speed trains and maritime vessels. Not only do they possess high mechanical strength and lightness, but they also exhibit good bending resistance. Additionally, they have excellent self-extinguishing properties, emit low heat, and can form a fire-resistant layer, minimizing the release of smoke and toxic gases. New high-speed trains such as America's Acela and Italy's ETR500 use aramid paper-based honeycomb structural materials. Companies like Hexcel supply products to high-speed train manufacturers such as Bombardier, Siemens, and Alstom.

4.4.4 High-temperature-resistant Polyimide Fiber Paper-based Functional Materials

The demands for high-temperature-resistant lightweight materials that reduce the weight of structures in aerospace and high-speed rail transit are continuously increasing. Polyimide fiber paper-based composite materials have higher strength and temperature resistance compared to aramid papers, making them a more promising type of functional paper-based material. However, many technical challenges existed during wet forming of polyimide fiber, resulting in poor paper forming quality and low strength. Therefore, the excellent performance of polyimide fibers cannot be fully utilized. Currently, there is no commercialized polyimide fiber paper, but only polyimide films are available in the field of electronic electrical insulation. However, the dense structure of polyimide film leads to poor wetting of insulation products during subsequent processing and weak interfacial composite performance.

Polyimide fibers have inherent chemical inertness. Besides, there is no fibrillation phenomenon during the beating/refining process, and there are no hydrogen bonds between the fibers during the drying process of the paper sheet, resulting in lower strength of the original paper obtained by the wet forming process. Selecting appropriately adhesive fibers to enhance the binding between fibers is key to developing high-performance polyimide fiber paper. Currently, there are three main methods for manufacturing polyimide fiber paper: a) utilizing polyamic acid fibers that are easily dispersed in water to prepare polyamic acid paper, which is then subjected to imidization treatment to obtain polyimide paper; b) using polyimide fibers to produce primary polyimide paper, and then impregnating and reinforcing it with polyimide resin or epoxy resin to obtain polyimide paper; c) employing aramid pulp or fibrids as adhesive fibers to produce polyimide fiber paper by wet forming process [36, 37].

Due to the inherent susceptibility of polyamic acid fibers to decomposition and the tendency for incomplete imidization during the imidization process, the performance of polyimide fibers is greatly affected. Additionally, paper made from polyimide fibers has extremely low strength due to the weak bonding between fibers, which makes it prone to breaking. To enhance the performance of polyimide paper, researchers mainly focus on the pretreatment of polyimide fibers, the dispersibility of the fiber furnish, and optimization of hot pressing processes. In the future, it is necessary to prepare polyimide fibrid with a high specific surface area and surface activity, which can be used as a self-assembling adhesive material for papermaking. Besides, efforts should be made to further reduce costs and expand its applications in the field of national defense, aerospace, and rail transit.

References

1 Committee on High-Performance Structural Fibers for Advanced Polymer Matrix Composites National Research Council (2005). *High-Performance Structural Fibers for Advanced Polymer Matrix Composites*, 1–58. Washington: The National Academies Press.

2 Afshari, M., Sikkema, D.J., Lee, K. et al. (2008). High-performance fiber based on rigid and flexible polymers. *Polymer Reviews* 48 (2): 230–274.

3 Bhat, G. (2016). *Structure and Properties of High-Performance Fibers*, 1–417. Cambridge: Woodhead Publishing Ltd.

4 Dobb, M.G., Johnson, D.J., and Saville, B.P. (1977). Supramolecular structure of a high-modulus polyaromatic fiber (Kevlar 49). *Journal of Polymer Science Polymer Physics Edition* 15 (12): 2201–2211.

5 Rebouillat, S., Peng, J.C.M., and Donnet, J.B. (1999). Surface structure of Kevlar Fibers studied by atomic force microscopy and inverse gas chromatography. *Polymer* 40 (26): 7341–7350.

6 Rao, Y., Waddon, A.J., and Farris, R.J. (2001). The evolution of structure and properties in poly (p-phenylene terephthalamide) fibers. *Polymer* 43 (13): 5925–5935.
7 Xu, Z.A. and Shen, L. Para-aramid fiber. CN201410238057.5. 2017-02-15.
8 Lu, Z.Q., Hua, L., Ding, M.X. et al. A Method for modifying polyimide fiber paper: CN201210264309.2. 2014–11-19.
9 Wang, J.M. (2009). Development survey and application prospect of poly-P-phenylene benzobisoxazazole fiber. *Hi-Tech Fiber & Application* 34 (2): 42–47.
10 Chae, H.G. and Kumar, S. (2006). Rigid-rod polymeric fiber. *Journal of Applied Polymer Science* 100 (1): 791–802.
11 Dingenen V, Jan L J. 3-Gel-spun high-performance polyethylene fibres [M] //HearleJ W S. *High-Performance Fibers*. Cambridge: Woodhead Publishing Ltd, 2001: 62–92.
12 Wang, J.L., Liang, G.Z., Lv, S.H. et al. (2003). Application of UHMWPE fibre on bulletproof materials. *New Chemical materials* 31 (1): 21–23.
13 Zhang, M.Y. (2019). *Converted Paper and Specialty Paper*, Fourthe, 284–285. Beijing: China Light Industry Press.
14 Cao, H.L., Yan, Y.W., Yue, L.P. et al. (2017). *Basalt fiber*, 41–43. Beijing: National Defense Industry Press.
15 Torop, L.V., Vasyuk, G.G., Kornyush, V.L. et al. (1995). New cloth from basalt fibers. *Fibre Chemistry* 27 (1): 67–68.
16 Hu, X.Q. and Shen, T.N. (2005). The applications of the CBF in war industry & civil fields. *Hi-tech Fiber & Application* 30 (6): 7–13.
17 Liu, J.A., Chen, K.F., Lei, Y.C. et al. (2005). Wet-forming papermaking of synthetic fiber. *China Pulp and Paper* 21 (2): 59–61.
18 Dang, Y.H., Wang, Z.J., Wang, L.X. et al. (2006). Application of water soluble PVA synthetic fibre in light weighted paper. *China Pulp & Paper Industry* 27 (12): 42–45.
19 Casey, J.R. *Pulp and Paper Chemistry and Chemical Technology* (Wang J H, Zhang C L, Zhang Y F, et al. Trans) [M]. Third Edition, Beijing: China Light Industry Press,1998: 400–431.
20 Long, X.L. and Liu, Z.F. (2003). Development on technology of PPTA-pulp preparation. *Polymer Materials Science & Engineering* 19 (3): 45–48.
21 Liao, Z.L. (2008). The market demand of aramid fiber and application of aramid fiber pulp. *Hi-tech Fiber & Application* 33 (3): 36–39.
22 Yang, B., Zhang, M., Lu, Z. et al. (2018). Comparative study of aramid nanofiber (ANF) and cellulose nanofiber (CNF). *Carbohydrate Polymers* 208: 372–381.
23 Zhang, M.Y., Luo, J.J., Yang, B. et al. (2020). Research progress of preparation of application of aramid nanofibers. *Materials Reports* 34 (5): 5158–5166.
24 Yang, B., Wang, L., Zhang, M. et al. (2019). Timesaving, high-efficiency approaches to fabricate aramid nanofibers. *ACS Nano* 13 (7): 7886–7897.
25 Mason, S.G. (1954). Fiber motions and flocculation. *Tappi Journal* 37 (11): 494–501.

26 Kerekes, R.J. and Schell, C.J. (1992). Characterization of Fiber flocculation regimes by a crowding factor. *Journal of Pulp and Paper Science* 18 (1): 32–38.

27 De Lange, P.J., Mäder, E., Mai, K. et al. (2001). Characterization and micromechanical testing of the interphase of aramid- reinforced epoxy composites. *Composites Part A: Applied Science and Manufacturing* 32 (3): 331–342.

28 Liao, R.J., Liu, H.B., Bai, G. et al. (2016). Space charge characteristics and dielectric properties of Nano-SiO_2/aramid paper composite. *Transactions of China Electrotechnical Society* 32 (12): 40–48.

29 Bai, G., Liao, R.J., Liu, N. et al. (2015). Influence of Nano-AlN modification on the dielectric properties of meta-aramid paper. *High Voltage Engineering* 41 (2): 461–467.

30 Zhang, S.F. and Zhang, M.Y. Muti-layered para-aramid paper with high interfacial bond strength. CN200710017244.0. 2009-10-28.

31 Sun, M.J., Huang, J.M., Wang, D.X. et al. Preparation method of meta-aramid paper with different smoothness on both sides. CN200910216912.1. 2012–05-30.

32 Hu, J., Wang, Y., Sun, Y. et al. A method for preparing aramid paper and aramid paper. CN201410166497.4. 2014-07-30.

33 Yang, B. (2019). *High-Efficiency Preparation of Aramid Nanofiber and its Application in Paper-Based Insulating Materials*. Xi'an: Northwestern Polytechnical University.

34 Yang, B., Zhang, M.Y., Lu, Z.Q. et al. (2018). Toward improved performances of Para-aramid (PPTA) paper-based nanomaterials via aramid nanofibers (ANFs) and ANFs-film. *Composite Part B: Engineering* 154: 166–174.

35 Yang, B., Zhang, M.Y., Lu, Z.Q. et al. (2019). Comparative study of aramid nanofiber (ANF) and cellulose nanofiber (CNF). *Carbohydrate Polymers* 208: 372–381.

36 Ding, M. X., Tan, H. Y., Lv, X. Y., et al. Preparation method of polyimide paper. CN201210334745. 2. 2015-04-08.

37 Lu, Z.Q., Hua, L., Ding, M.X. et al. Preparation method of polyimide paper strengthened by meta-aramid fibrid. CN201210434470.X. 2015-11-12.

5

Long-fiber Filter Paper-based Functional Materials

5.1 Long Fibers for Filter Paper-based Functional Materials

5.1.1 Overview

Paper-based materials are generally made using wood and herbaceous plant fibers as the main raw materials. The average fiber length varies depending on the type of raw material, production process, and product category, but typically does not exceed 3 mm. However, some paper-based materials made from plant fibers, synthetic fibers, and other raw materials often have an average fiber length exceeding 3 mm, with fibers of varying forms, some even reaching several tens of millimeters. These types of paper-based materials are referred to as long-fiber specialty paper bases. It is generally accepted that fibers with a length greater than 3 mm can be classified as long fibers. These fibers tend to flocculate and bond together, making them difficult to use with conventional papermaking machines to produce uniform paper sheets.

5.1.2 Classification and Characteristics of Long Fibers

Long fibers are classified into plant fibers and non-plant fibers. Non-plant fibers include synthetic fibers, inorganic fibers, and animal fibers.

5.1.2.1 Long Fibers from Plants

Plant-based long fibers are mainly bast fibers and cotton fibers. Among them, bast fibers include hemp, mulberry bark, kozo, and others. The following are descriptions of several typical plant-based long fibers:

1) Abaca Fiber

Manila hemp is one of the most commonly used long-fiber raw materials in the world. It is a type of Abacá, named after the Philippines, where it is primarily produced. More than 80% of the world's Manila hemp production comes from the Philippines, with the remaining production coming mostly from Ecuador.

Paper-Based Functional Materials: Preparation and Applications, First Edition.
Xianping Yao, Genrong Chen, Zhen Yao, and Liping Zheng.

The main applications for Manila hemp fiber are ropes, handicrafts, and long-fiber paper bases. In the Philippines, there are four major pulp mills producing Manila hemp pulp.

Manila hemp fibers are uniform in thickness, with thin walls, blunt ends, and wide, clearly visible lumens that are continuous. The fiber walls have few transverse striations, and the cross-sectional shape of the fibers is typically irregularly oval or polygonal with rounded corners, often containing vessel elements and parenchyma cells. The fiber length generally ranges from 2.0 to 6.0 mm, with a width between 6.6 and 22 μm. The average fiber length is 4 mm, and the average width is 16.8 μm. As a result, Manila hemp fibers are known for their high strength and good toughness [1].

Manila hemp fibers have a wide range of applications. Table 5.1 provides reference on applications for different grades of Manila hemp fibers produced by a certain supplier, according to research by Hangzhou Xinhua Group Co., Ltd. [2].

2) Mulberry Fiber

Mulberry bark fiber is one of the most typical raw materials used to produce long-fiber paper in China, derived from the bark of the mulberry tree. The mulberry tree is a perennial woody plant, and the bark is harvested from the tender stems or branches of the tree's phloem. The mulberry tree is one of the oldest and most economically valuable horticultural plants in China.

Mulberry bark fiber is slightly coarser than Manila hemp, and it has a cylindrical shape with a central channel that is sometimes very prominent and sometimes

Table 5.1 Applications for different grades of Manila hemp fibers for various products.

Order Number	Basic Use	Model	Advice Application Field
1	High Air Permeability, Low Strength	IPP-25	High breathable paper base, filter rod forming paper base
2	High Air Permeability, Medium Strength	IPP-01	Tea filtration and sausage paper base
3	Medium Air Permeability, High Strength	IPP-45, IPP-42	Electrolytic capacitor, securities paper, tea filter paper base
4	Medium Air Permeability and Strength, High Tear Degree	IPP-40, IPP-37	Industrial filtration, impregnated substrate, handmade paper, etc.
5	Low Air Permeability, High Strength	IPP-17, IPP-16	Roof and wall cloth substrate, securities paper, handmade paper, etc.
6	Low Air Permeability, Medium Strength	IPP-53	Filter materials and medium-and low-grade tea filtration
7	To Replace or Fill with Other Fibers	IPP-16	Filter substrate, cigarette paper base, industrial filtration,roof and wall covering

only a narrow line. The fibers often have parallel longitudinal cracks, with fewer but prominent transverse striations. The arrangement of the fibers is relatively regular, and sometimes the middle of the fiber has a protruding node, which is darker in color. Additionally, there are membrane-like thin cells that overlap irregularly, forming various longitudinal and transverse fine striations [1]. Mulberry bark fiber also possesses the characteristics of high strength and good toughness.

For thousands of years, mulberry bark fiber has been an excellent traditional raw material for long-fiber papermaking. Historically, it has been used in the production of famous book and painting papers, window paper, Zhejiang's leather paper, and umbrella paper, as well as in modern products like typewriter carbon paper and tea bag filter paper, among others. Before many foreign manufacturers started using Manila hemp in bulk, mulberry bark had already been widely used in long-fiber papermaking in China, where it is still the best-utilized raw material today.

However, unlike Manila hemp, which follows an integrated industrial model involving planting, pulping, and sales, mulberry bark fiber production is more decentralized. The procurement of raw materials and commercial pulp is more challenging and less convenient.

3) Kozo Bark Fiber

Kozo bark, also known as "mulberry bark," has long fibers that closely resemble mulberry bark. The average fiber length is between that of flax and jute. The fiber walls have distinct transverse striations, and the lumen is clearly visible, with some fibers having large lumens and others smaller ones. The fiber ends are sharp and often branch, with the ends sometimes forming small spherical tips. The outer wall of the fibers is often covered by a transparent layer known as the "gum coat".

One notable characteristic of kozo pulp, compared to mulberry pulp, is the presence of a large amount of calcium oxalate crystals. These crystals are usually prism- or square-shaped and are commonly found in thin cell cavities or between cells. Beating and washing the pulp can cause the crystals to be removed, reducing their presence.

Kozo has been an excellent traditional papermaking material for centuries. The most used part for papermaking is the tender bark from the branches of the shrub, which is softer and contains fewer impurities than the older bark of the main trunk. While the trunk bark becomes thicker with age, it also contains a higher percentage of black, coarse bark. This black coarse bark is free of fibers and mainly consists of cork material and ash content, which, if not completely removed, can affect the quality of the pulp.

4) Cotton Fiber

Cotton fiber is the fiber derived from cotton plants. It is long and soft, which makes it easier to interlace and bond together during papermaking. Cotton fibers also have a high degree of orientation, with strong intermolecular bonding, resulting in good wear resistance and fold resistance. Cotton fibers are rich in cellulose and low in lignin, giving the resulting paper products higher brightness and durability.

Cotton fibers contain 98% α-cellulose. The cellulose macromolecules contain numerous hydrophilic groups, enhancing the moisture absorption properties of cotton fibers. Additionally, substances such as pectin, proteins, hemicellulose, and inorganic salts present in the fiber are also hydrophilic, further improving moisture absorption. The fats and waxes contained in cotton fibers are hydrophobic substances, which reduce moisture absorption but offer protective properties to the fibers.

The outer macromolecular arrangement of a single cotton fiber is relatively loose, with low orientation and crystallinity, providing the fiber with good flexibility and elasticity. The inner macromolecular arrangement is denser, with high orientation and crystallinity, giving the fiber good rigidity. High-maturity cotton fibers have higher polymerization, better orientation, higher crystallinity, and stronger intermolecular bonding. Cotton fibers have good bending and twisting resistance and are not easily broken. During growth, the helical deposition of cellulose leads to the natural twist of the fiber, which gives cotton fibers unique friction and bonding properties, as well as excellent mechanical characteristics.

When cotton fibers are highly beaten, they become longer and more uniform in length. Cotton fibers are primarily used in high-end filter papers, medical papers, test papers, security papers, and other products.

5) Softwood Fiber

The wood used in papermaking can be classified into softwood and hardwood, with softwood named for its needle-like leaves. The softwoods primarily used in papermaking are conifers, such as pine and fir trees. Softwood fibers contain tracheids, ray cells, and ray tracheids. The fiber length of softwood is approximately 3–3.5 mm, with an average width of about 30–75 μm, making it a high-quality fiber for papermaking. Additionally, due to its dense structural organization, impurities in chemical pulps are washed away during the pulping process, resulting in better-quality pulp. Softwood fibers are uniform, relatively long, and ideal for producing high-quality paper pulps. They are commonly used to make high-grade specialty papers, cultural papers, and are often selected as fibers for filtration paper bases.

5.1.2.2 Long Fibers from Non-Plant Sources

Non-plant fiber long fibers can be divided into synthetic fibers, inorganic fibers, and animal fibers.

6) Synthetic Fibers

The synthetic fibers primarily used for the production of long-fiber specialty papers include Vylon fiber, polypropylene fiber, viscose fiber, polyester fiber, and lyocell fiber.

a) Vylon Fiber

Vylon fiber, scientifically known as polyvinyl alcohol aldehyde fiber, commonly referred to as polyvinyl alcohol fiber, is primarily used as a raw material for long-fiber specialty paper in two types: high-melting-point fiber, which is mainly used to enhance paper strength and the air permeability of specialty papers, such as for automotive filters, medical tape substrates, and seedling cloths. The other is low-melting-point fiber, which, when blended with other high-melting-point fibers, forms a web via a wet process. The web is then passed through a drying cylinder where the surface temperature causes the low-melting fibers to melt, creating a bonding effect.

Vylon has excellent moisture absorption properties and a strong affinity with both plant fibers and water. It can form webs without the need for any dispersants. After dewatering, it retains good wet strength, which not only improves the strength of the product but also increases the output per unit. Vylon can be made into fibers with different melting-point temperatures. Currently, these fibers are available in melting points of 35 °C, 60 °C, 90 °C, 120 °C, etc., giving users the flexibility to choose according to their needs. Vylon is a chemically processed industrial fiber, primarily used in the textile industry but also suitable for papermaking.

b) Polypropylene Fiber

Polypropylene fiber, also known as polypropene fiber, has seen rapid development in the field of sanitary materials in recent years. This fiber has a low relative density, good strength, and superior resistance to acids, alkalis, and chemical solvents compared to most other synthetic fibers. Currently, polypropylene fiber is mainly used in various fields such as filtration materials, medical materials, household decoration materials, the food industry, and composite packaging materials.

c) Viscose Fiber

Viscose fiber, commonly known as "rayon," has a chemical composition very similar to cotton fiber, with properties also resembling cotton fiber. However, it has superior moisture absorption, softness, and dispersibility compared to cotton fiber. Its length, fineness, and crimp can be controlled. Compared to natural cotton fiber, viscose fiber offers consistency in both length and width, making it an excellent raw material for specialty paper bases.

d) Polyester Fiber

Polyester fiber, commonly known as "polyester" or "PET fiber," is a synthetic fiber made by the polycondensation of organic diacid and diol. It is a high polymer compound and is the largest category of synthetic fibers available on the market today.

Polyester fiber is used in industrial weaving, medical and health applications, filtration materials, engineering fibers, and nonwoven fabric industries. As a specialty raw material for paper bases, polyester fiber not only has the typical characteristics of large-fiber fineness, high strength, and easy dispersion seen in regular polymer fibers, but also exhibits outstanding high-temperature resistance. It has high fiber stiffness, excellent strength, a small-fiber diameter, and good chemical resistance, making it the preferred material for specialty filtration products.

e) Lyocell Fiber

Lyocell fiber is a type of regenerated cellulose fiber made from wood pulp as the raw material, with N-methylmorpholine-N-oxide (NMMO)-H_2O as the solvent, using a dry-wet spinning method. The process involves dissolving wood pulp in an amine oxide solvent and directly spinning it, which is completed entirely under physical action. The amine oxide solvent is recycled, with a recovery rate exceeding 99%, making this a green and environmentally friendly spinning technology. Lyocell fiber is biodegradable, and products made from it are disposable and do not pollute the environment.

Lyocell fiber is highly resilient in both dry and wet states, with its wet strength far surpassing that of cotton or viscose fibers. Due to its good orientation, the crystallization degree of the macrofibrils in the fiber is high, and they tend to align along the fiber axis. When subjected to external factors like friction or vibration, these macrofibrils easily separate from the fiber surface, a phenomenon known as fibrillation. By controlling the extent of fibrillation, different product structures can be produced. Lyocell fiber also has good water and moisture absorption properties, and its high wet modulus gives it minimal deformation under moderate loads, providing the product with excellent dimensional stability and wrinkle resistance. These properties make it suitable for producing high-performance medical paper.

Compared to natural fibers, synthetic fibers have better consistency in length and fineness and can be processed into finer fibers. Additionally, synthetic fibers generally have superior strength, elongation, and wear resistance compared to natural fibers.

7) Inorganic Fibers

The inorganic fibers used in the production of long-fiber special paper mainly include glass fibers, asbestos fibers, carbon fibers, and activated carbon fibers.

a) Glass Fibers

In the production of long-fiber specialty paper, glass fibers have been widely applied. In addition to their characteristics of insulation, wear resistance, dimensional stability, high strength, high-temperature resistance, and corrosion resistance, glass fibers can also be made into ultrafine fibers. These fibers are commonly used as substrates or base fabrics in waterproof materials, insulating materials, and reinforcement materials. They are also a primary raw material for specialized filtration materials.

b) Asbestos Fibers

Asbestos fibers are made from rock that is centrifugally melted. These fibers are used to manufacture materials with good heat resistance and chemical stability. Asbestos fibers typically have varying lengths and fineness, and can be used directly for the production of specialty paper bases without pre-treatment.

c) Carbon and Activated Carbon Fibers

Carbon fibers are brittle materials with small elongation before breaking and limited flexibility. However, they have high tensile strength, conductivity, heat conductivity, high-temperature resistance, and chemical corrosion resistance. They are used to make conductive materials, static dissipation materials, high-voltage electrical protection layers, and fireproof clothing, among other products. Activated carbon fiber is a special type of carbon fiber with many micropores ranging from 0.5 to 50 nm in diameter. These pores, which are deeper than the diameter of the fiber itself, can adsorb a significant number of harmful gases. Therefore, wet-processed activated-carbon-fiber products are commonly used as gas filtration materials and are widely applied in power plants, chemical factories, metal smelting plants, and in vehicles for exhaust gas treatment. After a certain period of use, the fibers can be regenerated and recycled through desorption treatment.

8) Animal Fibers

Animal fibers are rarely used in long-fiber specialty papers and are typically utilized only when specific properties are required. Below is a brief introduction to wool fibers, silk fibers, and chitosan fibers.

a) Wool Fibers

Wool fibers are primarily composed of sulfur-containing proteins, and their structure can generally be divided into the surface layer, outer layer, and inner core. The surface layer consists of irregular, scale-like structures, which are the main feature of wool. The outer layer is the main body of the wool fiber, and the inner core, also known as the medulla, has a honeycomb-like structure. Wool fibers vary greatly in thickness, typically ranging from 10 to 70 μm, with differences in thickness even within a single fiber. Wool fibers are generally not used in papermaking, but they can be added to nonwoven fabrics or paper to enhance esthetics or provide anti-counterfeiting capabilities.

b) Silk Fibers

Silk fibers are also animal protein fibers. They are formed when silkworms secrete fibroin to create cocoons. The length of silk fibers can reach several thousand meters, with diameters ranging from 10 to 20 μm. After reeling, multiple threads are combined and can be used directly for weaving. In papermaking, silk is generally not used as a raw material. Instead, waste silk is degummed and cut into suitable lengths for use as a substrate in sanitary materials or cosmetics.

c) Chitin Fibers

Chitosan (chitin) is a natural polymer similar in structure to fiber. It is widely found in the exoskeletons of insects and crustaceans, as well as in the cell walls of certain fungi, and is the second-most abundant renewable resource in nature after cellulose. It is also the only known naturally occurring positively charged biopolymer. Chitosan contains numerous hydroxyl and amino groups, giving it high chemical reactivity and bioactivity. It can activate biological processes without toxic side effects, is easily broken down and absorbed by enzymes in the human body, and is transported through the bloodstream to various body parts. Chitosan is non-repellent, can penetrate skin, muscle, bone, and cells, and aids in tissue repair.

It has immune-boosting, blood-purifying, and circulatory benefits, and it degrades naturally in the environment.

Chitosan fibers made from crustacean exoskeletons, when used in papermaking and textile materials, not only offer excellent physical and chemical properties, but also exhibit biological activity, biocompatibility, and biodegradability. They have a wide range of applications in medical dressings due to their hemostatic, antibacterial, anti-inflammatory, wound-healing, and tissue repair properties.

This section focuses on fibers related to filtration and separation, commonly used in filtering paper-based materials.

5.2 Mechanisms of Filtration

Filtration is a unit operation used to separate solid particles suspended in a gas or liquid. It involves passing the gas or liquid in the suspension (filter slurry) through a porous material (filter medium), which retains the solid particles (filter cake) that are left behind. Based on the type of filter medium, filtration is generally classified into gas filtration and liquid filtration.

5.2.1 Air Filtration

The mechanisms of gas filtration are mainly as follows: interception (or contact, trapping) effect, inertial effect, diffusion effect, gravitational effect, and electrostatic effect [3].

5.2.1.1 Interception Effect

In the fiber layer, the fibers are arranged in a complex pattern, creating numerous meshes. When a micro-particle of a specific size moves along the streamline and gets close enough to the fiber surface (the distance from the particle's center line to the fiber surface is equal to or smaller than the particle's radius), the particle is intercepted and deposited on the fiber surface. This is known as the interception effect, as shown in Figure 5.1.

Interception collection efficiency is denoted as η'_{R}. The sieve effect is also considered a form of interception effect, sometimes referred to as the filtration effect, as shown in Figure 5.2.

5.2.1.2 Inertia Effect

Due to the complex arrangement of fibers, the gas flow undergoes sharp turns as it passes through the fiber layer. When the particles are relatively large or have high velocity (which can be considered as the gas flow velocity), their inertia prevents them from following the flow streamline as it curves. As a result, the particles deviate from the streamline, move closer to the fibers, and collide with them, leading to deposition. This is known as the inertial effect, as shown in Figure 5.3. Inertial collection efficiency is denoted as η'_{St}.

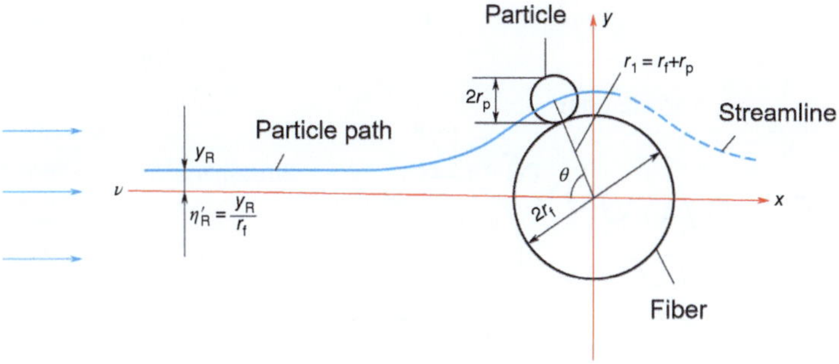

Figure 5.1 Interception effect diagram.

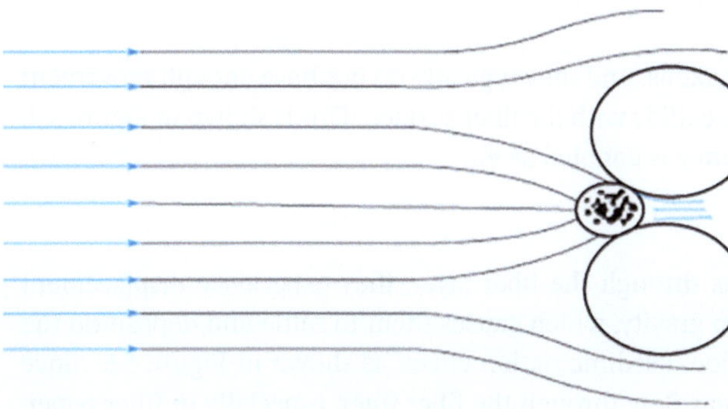

Figure 5.2 Diagram of sieve effect or filter effect.

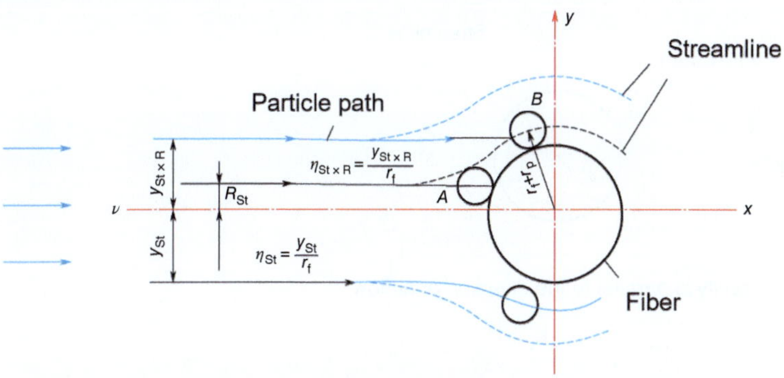

Figure 5.3 Schematic diagram of inertial effect.

5.2.1.3 Diffusion Effect

The diffusion effect refers to the Brownian motion of particles, caused by the collision of gas molecules with the particles. The smaller the particle, the more significant the diffusion effect is. At room temperature, a particle with a diameter of 0.1 μm can diffuse up to 17 μm per second, which is several to dozens of times greater than the distance between fibers. This increases the likelihood of the particle reaching the fiber surface and depositing. However, for particles larger than 0.3 μm in diameter,

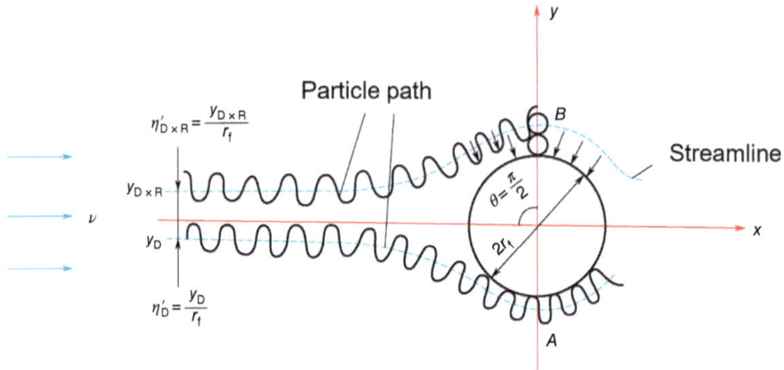

Figure 5.4 Diagram of the diffusion effect.

their Brownian motion weakens, and they typically do not have enough movement to leave the streamline and collide with the fiber surface. This is shown in Figure 5.4. Diffusion collection efficiency is denoted as η'_D.

5.2.1.4 Gravity Effect

When micro-particles pass through the fiber layer, they experience displacement from the streamline due to gravity, which causes them to settle and deposit on the fibers. This is the gravitational sedimentation effect, as shown in Figure 5.5. Since the residence time of the gas flow through the fiber filter, especially in filter paper,

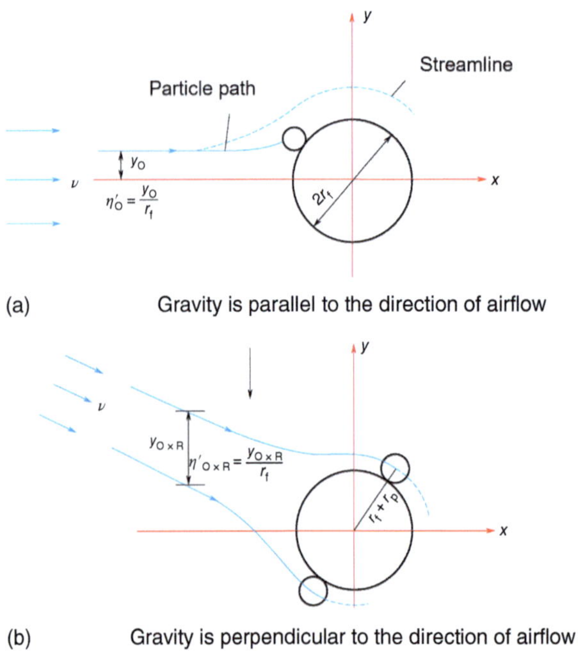

Figure 5.5 Schematic diagram of gravity effect.

is much shorter than one second for particles with a diameter smaller than 0.5 μm, if they have not settled onto the fibers, they have already passed through the fiber layer. Therefore, gravitational sedimentation can generally be neglected for such small particles. Gravitational collection efficiency is denoted as η'_O.

5.2.1.5 Electrostatic Effect

For various reasons, both fibers and particles may acquire charges, which can generate electrostatic attraction. However, unless fibers or particles are deliberately charged, any charge accumulated during fiber processing due to friction or particle-induced polarization will not last long. The electric-field strength will be weak, and the resulting attraction will be minimal, so it can be completely ignored. The electrostatic effect is illustrated in Figure 5.6. Electrostatic collection efficiency is denoted as η'_E.

In addition to the five main filtration mechanisms mentioned above, there are also chemical filtration mechanisms and membrane filtration mechanisms. Chemical filtration can be further divided into physical adsorption and chemical adsorption, with the filtration action of activated carbon belonging to physical adsorption [4]. Membrane filtration differs from traditional fiber filtration in that it mainly operates through surface filtration. The filtration process takes place solely on the membrane surface, and the dust particles accumulated on the membrane surface are easily removed. After backflushing, the resistance can return to its initial state, giving membrane filters a longer lifespan. In the current air filter market, fine glass fiber filters dominate, but membrane filters have also emerged and are increasingly replacing fine glass fiber filters in high-efficiency air filtration and dust collection applications.

5.2.2 Liquid Filtration

5.2.2.1 Principle of Liquid Filtration

Under the action of external forces (gravity, pressure, centrifugal force), the liquid in a suspension passes through the pores of a porous medium, while solid particles are retained, thus achieving solid-liquid separation. The suspension being processed in the filtration operation is referred to as the filter slurry (or filter material), and the porous substance used is called the filtration medium, also known as filter paper. The liquid passing through the pores of the medium is called the filtrate, while the

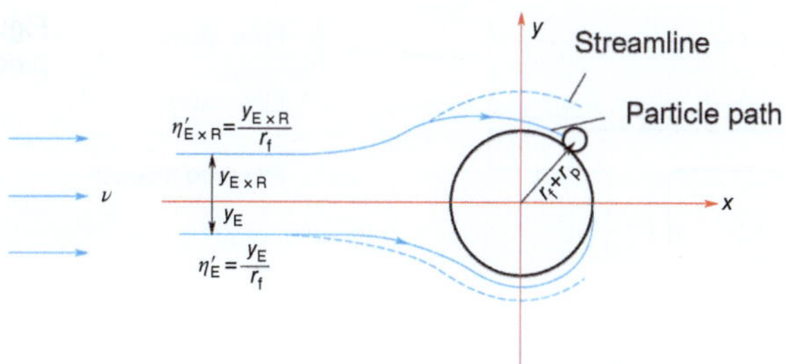

Figure 5.6 Schematic diagram of electrostatic effect.

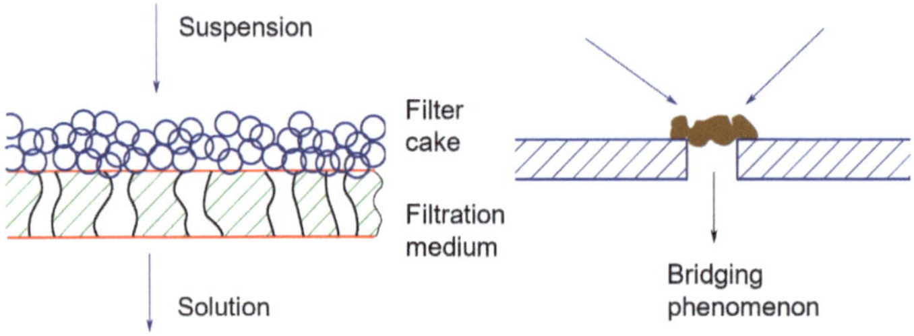

Figure 5.7 Schematic diagram of liquid filtration.

retained material is referred to as the filter cake or filter residue. A simple schematic of liquid filtration is shown in Figure 5.7.

5.2.2.2 Filtration Types

Liquid filtration is typically divided into cake filtration and depth filtration.

1) Cake Filtration

In cake filtration, most of the particles are larger than the pores of the filtration medium. Initially, some fine particles may pass through the medium, resulting in turbid filtrate. However, these particles rapidly form a "bridge" within the pores, which allows even smaller particles to be trapped. Once the filter cake begins to form, the filtrate becomes clear, and effective filtration can occur. In cake filtration, the filter cake, not the filtration medium itself, plays the primary role. This is often referred to as cake filtration or filter cake formation in filtration processes. A schematic diagram of cake filtration is shown in Figure 5.8.

2) Depth Filtration

In depth filtration, the particles are much smaller than the pores of the medium, but the pores are tortuous and elongated. Once the particles enter, they are easily retained due to the complex path within the medium. The compression and collision effects caused by fluid flow further enhance particle retention, causing the particles to adhere tightly to the pore walls. This method is particularly suited for clarification processes or for removing fine suspended particles from liquids, such as in drinking water purification or solid-liquid separation [5]. A schematic diagram of depth filtration is shown in Figure 5.9.

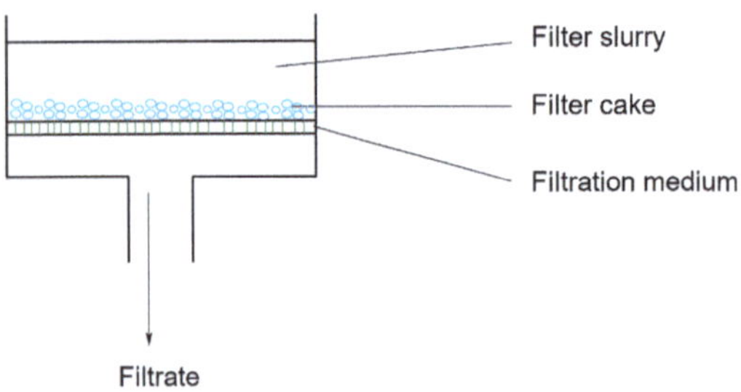

Figure 5.8 Schematic diagram of pied layer filtration.

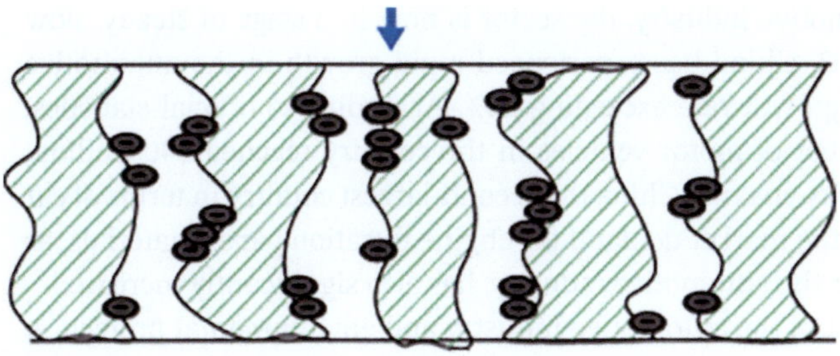

Figure 5.9 Schematic diagram of deep filtration.

5.3 Applications of Filter Paper-based Functional Materials

The development of long-fiber specialty paper substrates has a history of more than half a century. Over this time, advancements in fiber manufacturing technology, equipment innovation, automation, and integration have progressed rapidly, and applications have become increasingly diverse. The range of products continues to expand as well. This section mainly focuses on three representative applications of long-fiber paper substrates in the field of filtration: automotive engines, analytical testing, and tea filtration.

5.3.1 Paper-based Functional Materials for Engine Filtration

5.3.1.1 Overview

Automobile engine filtration paper, also known as internal combustion engine industrial filter paper, hereafter referred to as engine filtration paper, is used as a filtering medium. It is primarily applied in automotive air filters, lubricating oil filters, heavy-duty air filters, hydraulic systems, and other special applications. The main function of this material is to filter out impurities from the air, engine oil, and fuel needed by automobile engines or other types of internal combustion engines during operation, thus protecting the engine and improving its operational efficiency.

Depending on the processing techniques used, engine filtration paper can be divided into two main types: cured filtration paper and uncured filtration paper (also known as solvent-based filtration paper and water-based filtration paper). Due to the use of different types of resin impregnation, cured and uncured filtration papers have different filtration properties. These two types of products have their own advantages and disadvantages in different application fields, and users typically select the appropriate type based on the specific purpose of use.

5.3.1.2 Development of Engine Filtration Paper

The automotive industry is a very mature sector in Europe, the United States, and Japan. Currently, its production and market share have reached saturation, and the growth rate has slowed down. As Europe, the United States, and Japan play a leading role in the global automotive industry, the sector is now in a stage of steady, slow development. Since 2003, China has experienced rapid growth in the automotive sector, with an annual growth rate exceeding 20%. According to official statistics, by June 2019, the number of motor vehicles in the country reached 340 million, including 250 million cars, making China the second-largest country in terms of car ownership. As a result, the market demand for engine filtration paper materials, as a supporting product for the automotive industry, has also significantly increased.

With the development of the automotive industry and environmental protection requirements, engine filtration paper is evolving toward being lightweight, low-resistance, environmentally friendly, efficient, high-dust-holding, and long-lasting. Currently, the main suppliers of engine filtration paper internationally include Ahlstrom and H&V in the United States, Donaldson Filters and FiberMark-Gessner in Germany, Awa Co. and Azumi Co. in Japan, and Unitop Filter Paper in South Korea. The paper-based filter materials from Europe and the United States are of high-end quality, finely classified, and highly specialized, primarily due to their excellent raw materials and advanced equipment. Some of these high-end products are still irreplaceable domestically.

In the mid-1990s, Hangzhou Xinhua Paper Co., Ltd. in China has introduced a non-cured engine filtration paper production line from Germany (with an annual production capacity of 3000 tons). The product quality reached international standards, which changed the previously poor reputation of China's engine filtration paper quality. This development made it possible to replace the internationally leading brands of engine filtration paper. As a result, China's engine filtration paper became a key filtering material for original equipment manufacturer (OEM) air filters in brand-name cars, marking a new milestone for the development of automotive engine filtration paper in China.

At the same time, foreign companies have paid significant attention to China's engine filtration paper market, especially internationally influential enterprises. Many have acquired or invested in Chinese companies, competing in the market with superior product quality, excellent services, and relatively low prices. Notably, in Binzhou, Shandong, Ahlstrom acquired and invested in a filtration paper production line, with an annual capacity of 20,000 tons of paper-based filtration materials. Another example is H&V (Suzhou) Special Materials Co., Ltd., a wholly owned subsidiary of the US H&V Company in Suzhou, China, which primarily produces non-cured engine filtration papers and high-end fiberglass paper-based filtration materials, with an annual output of approximately 20,000 tons. As a result, the competition in China's automotive engine filtration paper market has become highly intense.

Currently, the demand for automotive engine filtration paper in China is approximately 90,000 to 100,000 tons per year.

5.3.1.3 Technology of Engine Filtration Paper Materials

Engine filtration paper materials are primarily used in the filtration of air, engine oil, fuel, and other fluids in automotive engines, generator sets, and other types of internal combustion engines, as well as gas turbines and other related fields.

The production of these materials involves selecting appropriate fiber raw materials and chemical additives, which are processed into base paper through specific papermaking processes. This base paper needs to have adequate strength and be suitable for filtration. It is then impregnated or coated via an impregnation and coating system to enhance the paper's strength and other filtration properties. Finally, the paper is slit and packaged according to customer specifications.

The key technical requirements for the product are as follows: As a filtration medium in engines, the material must have excellent filtration performance. For the paper itself, it must be loose and porous, possess sufficient strength and stiffness, have high dimensional stability, and exhibit certain resistance to water, oil, and other factors. Specific technical requirements may vary depending on the type of air filter or filtration system used.

5.3.1.4 Types of Engine Filtration Paper Materials

Engine filtration paper materials can be mainly classified into two types: cured paper-based filter materials and uncured paper-based filter materials. The primary differences between these two types are as follows:

1) Different Production Processes for Filtration Paper

While the base paper production process for both cured and uncured paper-based filters is similar, there are significant differences in the impregnation and coating processes. In the impregnation process, if solvent-based resins (such as phenolic resin) are used as the impregnation solution, the resulting product is a cured paper-based filter. In contrast, if water-based emulsions (such as acrylic acid) are used, the result is an uncured paper-based filter.

2) Different Processing Methods for Filter Manufacturers

Due to the use of different chemical raw materials in the impregnation process, the characteristics of the filter paper also vary. Cured paper-based filters require high-temperature curing in the filter production line to achieve the necessary strength and stiffness for final use. During the curing process, certain chemicals (mainly phenol) volatilize, which can have environmental and health impacts. On the other hand, uncured paper-based filters do not require high-temperature curing and are more environment-friendly, making them a green product.

3) Different Areas of Application

Cured paper-based filters generally offer better high-temperature resistance, water resistance, and dimensional stability than uncured filters. As a result, cured paper-based filters are more suitable for use as oil and fuel filters. Additionally, due to their higher stiffness, cured paper-based filters are more dimensionally stable, making them better suited for heavy-duty air filters. In contrast, uncured paper-based filters have lower stiffness and are more prone to deformation, making them more appropriate for use in smaller, lighter vehicle air filters. If the suitability of cured and uncured paper-based filters is similar, users typically make their selection based on preferences, cost-effectiveness, and environmental requirements.

5.3.1.5 Classification of Engine Filtration Paper Materials

1) Air Filtration Paper
Air filtration paper is a key material for the air filter element in automotive engine air filtration systems, often referred to as the "lungs" of the engine. The intake system of a car engine consists of the filter element and the housing. The filter element, made from air filter paper, primarily serves the purpose of filtering the air [6], while the housing provides the necessary external protection. The main function of the air filter element is to provide clean air to the engine, preventing the intake of air containing impurities that could cause abrasion and damage to the cylinders, pistons, and valves during engine operation. If excessive wear occurs in the engine cylinders, it can reduce the sealing capability, leading to a decrease in engine power, increased fuel consumption, and incomplete combustion of the fuel in the cylinders. This results in higher emissions of harmful substances through the exhaust, which is detrimental to the environment. Additionally, some of the combustible mixture may seep into the oil pan through gaps between the cylinders and pistons, causing oil degradation, impacting engine lubrication, increasing wear, and shortening engine life. The performance of the air filter directly determines the cleanliness of the air entering the cylinders, affecting the engine's reliability and lifespan, as well as the quality of vehicle emissions. For this reason, air filter quality is a mandatory inspection item for determining the quality of internal combustion engines in many countries.

2) Engine Oil Filtration Paper

The engine oil filter is a crucial component of the engine's lubrication system, and its key material is the engine oil filtration paper. The oil filter plays an important role as it acts as a "cleaner" for the engine. The engine oil filtration paper removes metal shavings, carbon particles, and other impurities that mix with the engine oil during combustion, ensuring that only clean oil reaches critical engine parts such as the crankshaft, connecting rods, camshaft, turbochargers, and piston rings. This clean oil is essential for maintaining normal engine operation and performs additional functions such as lubrication, cooling, and cleaning, significantly extending the lifespan of these components and, in turn, prolonging the overall lifespan of the vehicle [7].

3) Fuel Filtration Paper

Fuel filtration paper is the main filtering material used in fuel filters and is classified into gasoline filter paper and diesel filter paper. During storage, transportation, and refueling processes [8], fuel inevitably becomes contaminated with impurities and water. The quality of the fuel entering the fuel injection system is a key factor affecting the overall performance and lifespan of internal combustion engines. The fuel filtration paper helps protect the engine by preventing the entry of harmful particles and water, reducing erosion and corrosion, optimizing fuel consumption, and minimizing atmospheric emissions. This ensures the safe and efficient operation of the engine, thereby improving the engine's performance and longevity.

Table 5.2 Key technical specifications for engine filtration paper products.

Model	K230	J400	R45
Basis Weight/(g/m^2)	123.0 ± 6.0	135.0 ± 7.0	140.0 ± 7.0
Air-resistor/mbar	≤ 3.8	≤ 1.3	≤ 22
Thickness/mm	0.38–0.48	0.50–0.62	0.40–0.5
Corrugated Depth/mm	0.18–0.30	0.18–0.30	0.18–0.30
Maximum Pore Size/μm	≤ 80	≤ 110	≤ 65
Mean Pore Size/μm	≤ 68	≤ 92	≤ 50
Burst Strength/kPa	≥ 230	≥ 230	≥ 230
Stiffness/mg	≥2000	≥2000	≥2000
Resistance to Humidity	1 min impermeability	1 min impermeability	1 min impermeability
Moisture Content/%	5 ± 2	5 ± 2	5 ± 2
Use	Air filter paper	Oil filter paper	Fuel filter paper

5.3.1.6 Technical Specifications for Engine Filtration Paper

Engine filtration paper, also known as internal combustion engine filter paper, is governed by the existing technical standard QC/T794–2020 "Industrial Filter Paper for Internal Combustion Engines". The key technical specifications are outlined in Table 5.2.

5.3.1.7 Production Processes for Engine Filtration Paper

At present, there are two commonly used production methods and processes for engine filtration paper at the international level: the online coating and impregnation method and the offline coating and impregnation method.

1) Online Coating and Impregnation Method

In this process, the base paper is formed using a wet-laid machine, followed by online impregnation and coating, and then drying. The final product is produced by completing the process steps. The process flow is shown in Figure 5.10. This method is simple and economical but lacks flexibility. Non-curing engine filter paper typically adopts this process.

2) Offline Coating and Impregnation Method

In the offline method, the base paper is first formed by a wet-laid machine (see Figure 5.11), and then impregnated or coated using a separate coating and impregnation line (see Figure 5.12). The final product is completed by finishing the process steps. This method offers greater flexibility, produces higher-quality products, and is more cost-effective. Most curing-type engine filtration papers use this type of production line.

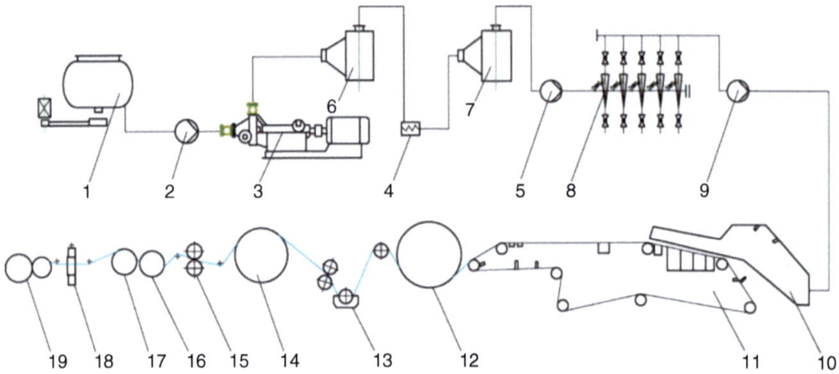

Figure 5.10 Schematic diagram of the process flow for online coating and impregnation of engine filter paper base.

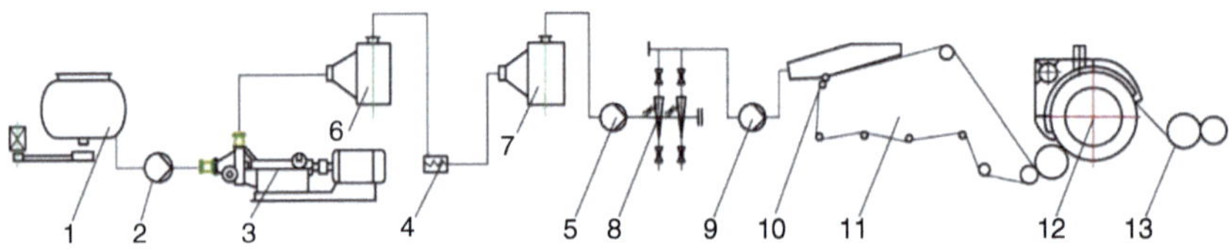

Figure 5.11 Schematic diagram of the process flow of the offline base paper production line for solidified type engine filter paper.

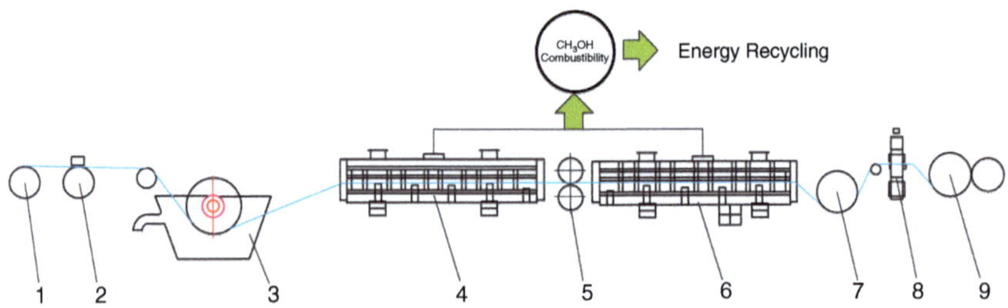

Figure 5.12 Schematic diagram of the process flow for impregnation coating of cured type engine filter paper.

5.3.1.8 Prospects for Engine Filtration Paper

Engine filtration papers with characteristics such as high efficiency, low resistance, light weight, environmental friendliness, high dust capacity (high dirt-holding capacity), and long service life are the ideal products sought after in the industry. These qualities also represent the future direction of development for engine filtration papers. The key approach to achieving these ideals is through the composite of different materials, particularly the combination with ultrafine fiber materials, which is seen as the main method to achieve these goals in the future.

Currently, mainstream automotive air filters or oil filters primarily use single-layer filter elements [9]. These filters trap impurity particles through the pore structure of the fiber network, using methods such as interception, adsorption, and the formation of a cake layer, thereby achieving purification and filtration. However, due to the limitations of the single-density fiber layer structure, single-layer filter elements struggle to achieve excellent comprehensive filtration performance. They can only balance filtration efficiency, accuracy, dust-holding capacity, and service life to meet different functional requirements. To improve filtration accuracy, the filter element's pores must be reduced, and the filtration path lengthened, but this leads to a decrease in both filtration efficiency and dust-holding capacity. On the other hand, reducing filtration resistance and increasing dust-holding capacity generally involves increasing the filter element's pore size, which results in a loss of filtration accuracy. Therefore, single-layer traditional filter elements face challenges in balancing filtration efficiency, accuracy, and dust-holding capacity, reaching a bottleneck in terms of overall performance improvement.

As a result, the development of new composite filtration materials has become a key area of focus for the automotive air filtration industry.

Composite filter elements with two or more layers can effectively combine high filtration accuracy with a high dust-holding capacity and low filtration resistance, leveraging the advantages of different materials to complement each other. By using multi-layer filter materials, density-gradient filtration is achieved, where each layer of the filter material has a different density and filtration efficiency. As the air flows through the filter element, each layer efficiently filters out particulate matter of progressively smaller sizes in the direction of the airflow [10].

In general, the composite materials of filter media have smaller pore sizes and higher porosity compared to the layers beneath them. This structure enables the filter material to achieve high efficiency and dust-holding capacity. Due to the small diameter of ultrafine fibers, they can significantly improve the filtration efficiency of the material and increase its dust-holding capacity. As a result, the application of ultrafine fibers in composite filter materials has received increasing attention, and ultrafine fiber composite air filtration paper has become a major development trend in both domestic and international research aimed at improving filter material performance.

There is no unified international standard for the definition of ultrafine fibers, but fibers with diameters below 5 μm are generally classified as ultrafine fibers. Industrialized methods for producing ultrafine fibers include direct spinning, composite spinning, blending spinning, electrospinning, melt-blowing, and flash-spinning techniques.

The main preparation methods for ultrafine fiber composite air filtration paper currently include electrospinning, wet paper-forming multi-layer composites, melt-blown nonwoven fabric composites, and membrane-coated composites.

Electrospinning is one of the primary methods for producing nanometer-scale ultrafine fibers with fiber diameters of around 20 nm. The dimensions and morphology of electrospun fibers can be controlled, and they have the advantages of a high length-to-diameter ratio and a large specific surface area, making them highly suitable for use as filtration materials. The process involves applying high voltage to a polymer solution to create a strong electric field, which draws the solution from the nozzle to form a jet. Under the influence of electrostatic repulsion, the jet splits continuously, and the diameter decreases while the solvent evaporates, ultimately solidifying into very fine fibers that deposit onto conventional filter paper. This dry composite technique enhances the filtration efficiency of the paper without significantly increasing resistance. Donaldson, an American company, was one of the pioneers in successfully commercializing electrospun fiber-layer composite air filtration papers.

Wet multi-layer composite preparation of ultrafine fiber composite filter papers is based on the papermaking process, using water as a carrier. A suspension of ultrafine fibers mixed with other fibers is formed and dewatered on a double- or multi-layer inclined screen, followed by a series of treatments such as drying to form the composite filter paper.

Electrospinning and wet multi-layer composite papermaking are two of the primary methods for producing ultrafine fiber composite air filtration papers. These methods result in significant structural differences in the composite filter papers they produce. Table 5.3 and Figure 5.13 present comparative data on the performance of these two composite materials, as evaluated by a university [11].

As shown in Table 5.3 and Figure 5.13, electrospun composite materials have clear advantages over wet papermaking multi-layer composite materials in terms of overall filtration performance.

The development of composite filtration media is a key trend in the evolution of filtration materials. Leading global filtration material companies, such as Donaldson (USA), Ahlstrom (USA), H&V (USA), and Gessner (Germany), all produce composite filter paper-based materials to meet the high-end filtration needs of various industries. For example, the electrospun composite filter paper used in gas turbine air filters by Donaldson has remained at the forefront and is irreplaceable.

The combination of multiple materials and the use of ultra-fine fibers will enable the achievement of highly efficient, low-resistance, environmentally friendly filters with high dust-holding capacity and other special functional requirements [12].

Table 5.3 Comparison of the performance of electrospun composite filter paper and wet papermaking multi-layer composite filter paper.

Order Number	Paper Sample Name	Quantify/ (g/m^2)	Thickness/ mm	Air Permeability/ (mm/s)	Mean Pore Size/μm	Maximum Pore Size/ μm	Filter Efficiency/%
1	Electrospun Composite Filter Paper	106	0.493	243	7.5	37.3	60.4
2	Papermaking Wet Process Multilayer Composite Filter Paper	129.8	0.657	128	7.9	23.3	64.2

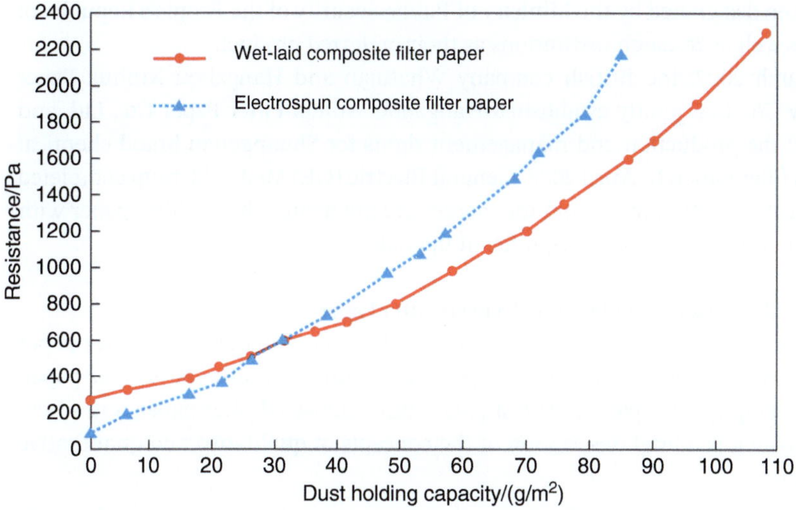

Figure 5.13 Comparative performance of electrospun composites and wet-laid multi-layer composites in papermaking.

These characteristics represent the essential performance that future engine filter papers must possess.

5.3.2 Filter Paper-based Functional Materials for Analysis and Testing

The primary type of filter paper-based functional material used for analysis and testing is chemical analysis filter paper. This material has gained widespread application across various sectors, including laboratories, research institutes, medical technology, pharmaceuticals, biotechnology, academia, forensics, and diagnostics. The rigorous demands of these industries ensure that the filter paper products meet specific standards in terms of quality and performance.

5.3.2.1 Overview

In China, the "Shuangquan" brand filter paper is widely used, while globally, Whatman filter paper is recognized as a leading brand. This is the positioning and evaluation of chemical analysis filter paper brands by users over the years.

Chemical analysis filter paper differs from ordinary filter paper as it is a specialized filtering medium. Its functions include separation, purification, concentration, decolorization, deodorization, and recovery, all of which are of significant importance for environmental protection, human health, equipment maintenance, and resource conservation [13].

The "Shuangquan" brand chemical analysis filter paper was developed independently by the Hangzhou Xinhua Paper Factory in the late 1950s. With its exceptional quality, it gained wide application in industry standard analyses and industrial processes in China. The "Shuangquan" name has become synonymous with chemical analysis filter paper and has won the national product silver award.

It has been designated by the Ministry of Public Security of the People's Republic of China as well as research institutions as their preferred product.

In March 2007, the British company Whatman and Hangzhou Xinhua Paper Industry Co., Ltd. jointly established Hangzhou Wohua Filter Paper Co., Ltd. and acquired the production and management rights for Shuangquan brand chemical analysis filter paper. In April 2008, General Electric (GE) Medical Group completed its acquisition of Whatman, and the merger of Shuangquan brand filter paper with Whatman products marked a significant upgrade.

5.3.2.2 Classification of Chemical Analysis Filter Paper

Chemical analysis filter paper can mainly be classified into qualitative filter paper and quantitative filter paper. Other types include diffusion-dispersive filter paper, chromatography filter paper, pH test paper base, industrial filter paper, and so on. The following is a brief description of the concepts of qualitative and quantitative filter papers:

9) Qualitative Filter Paper

Qualitative filter paper is used for identifying the properties of substances in qualitative analysis techniques. In analytical chemistry applications, when inorganic compounds are filtered and the precipitate is collected on the filter paper, the residual material can be used to calculate the loss rate during the experimental process. After filtration, qualitative filter paper generates a large amount of cotton fiber, making it suitable only for qualitative analysis.

10) Quantitative Filter Paper

Quantitative filter paper refers to filter paper that undergoes treatment with hydrochloric acid and hydrofluoric acid during production, followed by washing with distilled water. After high-temperature ashing, the amount of residue (ash) produced should not exceed 0.0009%. After treatment with hydrochloric acid and hydrofluoric acid, most impurities in the paper fiber are removed, so the residual ash after ashing is minimal, which has almost no effect on the analytical results, making it suitable for precise quantitative analysis. There are two types of quantitative filter paper: low-ash quantitative filter paper and ash-free quantitative filter paper. Ash-free quantitative filter paper undergoes special treatment, making it more resistant to chemical reactions, thus generating fewer impurities and being suitable for high-precision quantitative analysis.

11) Differences Between Qualitative and Quantitative Filter Papers

The primary difference between quantitative and qualitative filter papers lies in the amount of ash produced after ashing. For qualitative filter paper, the ash content is no more than 0.13%, while for quantitative filter paper, it is no more than 0.0009%. Ash-free filter paper, as a type of quantitative filter paper, produces less than 0.1 mg of ash after ashing, a mass that can be ignored in analytical balances. The concepts of quantitative and qualitative filter paper are specific to cellulose filter papers and do not apply to other types of filter papers, such as glass microfiber filter paper.

5.3.2.3 Technical Specifications and Processes of Chemical Analysis Filter Paper

1) Qualitative Filter Paper

Qualitative filter paper is made from high-quality cotton fibers with an α-cellulose content of 98%. It has high strength, good purity, and is uniformly clean, making it suitable for use in qualitative analysis techniques for material determination and identification. It is widely applied in most laboratories.

Qualitative filter paper is generally used for gravity filtration, where it separates solid-liquid phases by trapping solid particles. It is typically used in single or multiple layers stacked together. Qualitative filter paper has natural mechanical strength and some toughness. If faster filtration is needed, the paper can be folded multiple times (e.g., folding two layers into a fan shape or crimped pattern). During vacuum filtration, two or three layers of filter paper can be stacked in the funnel, or a dense filter cloth can be placed at the bottom with filter paper on top. Ordinary filter paper should not be used for filtering highly corrosive concentrated nitric acid and sulfuric acid solutions; specialized filter papers, such as Whatman #1573, should be used instead. Qualitative filter paper is also commonly used for cleaning experimental tables and wiping various laboratory equipment and tools. Some filtration processes have higher requirements for wet strength, and regular qualitative filter paper may not have sufficient wet strength. As a result, reinforced qualitative filter paper has been developed. This product has significantly improved wet burst resistance, and due to its high wet strength and excellent retention ability, it performs well in vacuum or positive pressure filtration while also offering good chemical resistance.

In China, the basic grades of qualitative filter paper are typically categorized as fast, medium, and slow based on pore size. Fast filter paper generally has a pore size of 80–120 μm, medium-speed filter paper has a pore size of 30–50 μm, and slow filter paper has a pore size of 1–3 μm. In China, the filtering performance of the paper base is often differentiated by filtration speed. Table 5.4 provides the performance specifications for different qualitative filter papers. The key technology used to produce these three different types of filter papers lies in the pulping process. By controlling the current and the gap of the knives during pulping, a pulp with different levels of pulping can be produced. After forming and drying, fast, medium, and slow filter paper bases can be obtained. The production process for qualitative filter paper is shown in Figure 5.14.

2) Quantitative Filter Paper

Quantitative filter paper is known for its smooth surface, low background, and excellent filtration effect. It is mainly used in precise quantitative analysis experiments and applications that require a clean environment with minimal interference, such as devices, electronic components, and laboratory conditions. Quantitative filter paper is generally available in three different flow rates to meet various applications: fast, medium, and slow. The physical and chemical properties of different grades of filter paper vary to balance filtration speed and particle retention ability, and

Table 5.4 Performance indexes of different qualitative filter papers.

Rank	Quantify/ (g/m^2)	Separating Property	Water Filter Time (t)/s	Wet Breaking Resistance/ mmH_2O	Ash Content/ %	Apply
Fast	80.0 ± 4.0	Ferric Hydroxide	≤ 35	≥ 120	≤ 0.13	With a high load, suitable for the filtration of coarse particles, culture medium, and colloidal sediment
Intermediate Speed		Lead Sulfate	$35 \sim < t \leq 70$	≥ 140		Moderate separation performance, moderate filtration speed, covers most of the laboratory applications of liquid filtration; industrial filter is also often used in medium-speed filter paper
Low Speed		Barium Sulfate (Heat)	$70 \sim < t \leq 140$	≥ 180		Strong retention ability, obvious adsorption effect, can intercept particles to a large extent, takes a long time

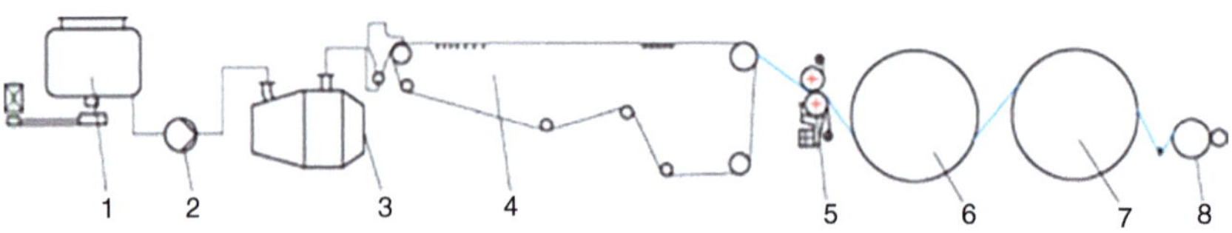

Figure 5.14 Schematic diagram of the production process of qualitative filter paper.

are used in applications such as food analysis, soil analysis, construction, mineral analysis, and inorganic analysis in the steel and mining industries. During the manufacturing process, strict physical controls and cleaning procedures remove most impurities from the filter paper. After calcination, the residual ash content is minimal, ensuring accurate quantitative analysis. The performance specifications for different types of quantitative filter paper are shown in Table 5.5, and the production process flow is shown in Figure 5.15.

Table 5.5 Performance indexes of different quantitative filter papers.

Rank	Quantify/ (g/m^2)	Separating property	Water filter time (t)/s	Wet breaking resistance/ mmH$_2$O	Ash content/%	Apply
Fast	80.0 ± 4.0	Ferric Hydroxide	≤35	≥120	≤0.01	Filtering of large particles or gel-like precipitate
Intermediate Speed		Lead Sulfate	35 ~ <t ≤ 70	≥140		Scientific research, can be used as the main filter paper for the separation of solid material from aqueous extract in soil analysis, conventional filtration in industry, and density analysis of cement soil and metal composition
Low Speed		Barium Sulfate (Heat)	70 ~ <t ≤ 140	≥180		Particle collection in food analysis, soil analysis, and air pollution monitoring; is also used for inorganic analysis in the construction, mineral, and steel industries

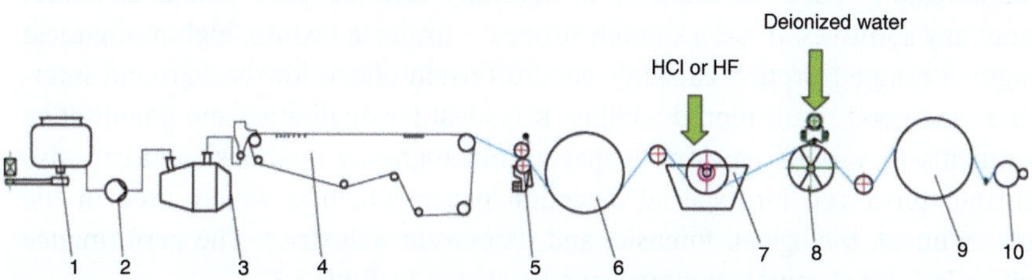

Figure 5.15 Schematic diagram of quantitative filter paper production process flow.

3) Diffusion-Type Filter Paper and Dispersion-Type Filter Paper

Diffusion-type and dispersion-type filter papers have excellent solution diffusion and dispersion properties, low background interference, and are the standard filter papers used for analytical testing in the dye industry. When evaluating the quality of dyes, the test solution is dropped onto the surface of the diffusion-type and

Table 5.6 Performance indexes of diffusion-type filter paper and dispersion-type filter paper.

Type	Separating property	Water filter time (t)/s	Ash content/ %	Wet breaking resistance/ mmH_2O	Quantify/ g/m^2	Breaking length/ km	Water Extraction Solution pH	Luminance/ %
Diffuse Filter Paper	Ferric Hydroxide	≤25	≤0.13	≥120	80.0 ± 4.0	≥1.50	6.0 ~ 8.0	≥85.0
Dispersed Filter Paper	Lead Sulfate	35 ~ < t ≤ 45		≥140		≥1.90		

dispersion-type filter papers. The shape of the droplet on the filter paper is then observed to assess the quality. Additionally, these filter papers can also be used in other special analytical testing industries. The performance specifications for diffusion-type and dispersion-type filter papers are shown in Table 5.6.

4) Base Paper of pH Test Functional Material

The base paper of pH test functional material is made from high-purity cotton cellulose raw materials, which have strong adsorption capacity, good diffusion effect, low background interference, and good reproducibility of results. It is specifically used to manufacture different types of test papers, such as pH test paper, lead acetate test paper, potassium iodide test paper, and ammonia nitrogen test paper. The performance specifications for base paper of pH test functional material are shown in Table 5.7.

5) Chromatography Paper

Chromatography paper is made from carefully selected pure cotton cellulose, without any additives. It has a smooth surface, a uniform texture, high mechanical strength, strong adsorption capacity, good diffusion effect, low background interference, and good result reproducibility. It is ideal for qualitative and quantitative experiments in various types of "paper chromatography methods" and can also be further processed into special detection paper, which is widely used in the pharmaceutical, biological, forensic, and diagnostic industries. The performance specifications for chromatography paper are shown in Table 5.8.

Table 5.7 Performance index of base paper of pH test functional material.

Iron Content/%	Breaking Length/km	Wet Breaking Resistance/ mm H_2O	Quantify/g/m^2	Water Extraction Solution pH	Luminance/%
≤0.005	≥1.80	≥150	80.0 ± 4.0	6 0.5 ~ 7.5	≥85.0

Table 5.8 Performance specifications for chromatography paper.

Class	Iron Content/ %	Breaking Length/ km	Ash Content/ %	Hygroscopicity/ mm/10 min	Quantify/ g/m^2	Aqueous Extract Solution pH	Luminance/%
Xinhua 1 #	≤0.003	≥1.80	≤0.10	91 ~ 120	95.0 ± 5.0	6.5 ~ 7.5	≥85.0
Xinhua 3 #	≤0.003	≥1.80	≤0.10	91 ~ 120	180.0 ± 9.0	6.5 ~ 7.5	≥85.0

5.3.3 Paper-based Functional Materials for Tea Filtration

Since the 21st century, with the rapid development of the equipment manufacturing and materials industries, the tea bag filter paper industry has basically formed a complete industrial chain, including upstream planting and harvesting, midstream production and processing, downstream sales markets, and further market extensions.

5.3.3.1 Principle of Tea Bag Filter Paper

Raw tea leaves, packaging materials, and tea bag packaging machines are the three key elements in the production of tea bags, with packaging materials being the fundamental requirement. The soluble and insoluble components in tea leaves vary in molecular size, state, and properties. The purpose of drinking tea is to utilize the water-soluble components while discarding the insoluble ones. Therefore, products such as tea bags rely on diffusion and osmosis principles, utilizing the filtration properties of special long-fiber filter paper. This allows the water-soluble substances inside the tea bag to continuously diffuse and permeate outward, achieving the extraction process.

The insoluble tea dust, with relatively large particles, is unable to pass through the filter paper bag's pores and is thus retained inside the bag, serving the filtration function. The tea processing products, which involve wrapping, folding, or pressing various tea leaves and other ingredients into the paper bag, offer advantages such as accurate dosing, speed, convenience, hygiene, environmental friendliness, and saving time.

5.3.3.2 Preparation Process of Tea Bag Filter Paper

Tea bag filter paper is classified into non-heat-seal tea bag filter paper and heat-seal tea bag filter paper.

1) Preparation Process of Non-Heat-Seal Tea Bag Filter Paper

   a) Materials

Currently, non-heat-seal tea bag filter paper is generally made by blending bleached wood pulp and hemp pulp in a specific proportion, forming and initially drying the mixture, followed by coating with modified natural polymer materials and drying. The bleached wood pulp consists of bleached softwood pulp fibers and bleached hardwood pulp fibers, which are dispersed and defibrated using a hydropulper. The hemp pulp can be made from a variety of fibers, including sisal,

Manila hemp, hemp, and flax, with Manila hemp pulp being the most commonly used fiber. The hemp pulp is defibrated and processed through a pulping machine, with the pulping degree controlled between 30 and 50 °SR. The modified natural polymer materials include carboxymethyl cellulose (CMC), hydroxycellulose, oxidized starch, cationic starch, and anionic starch, with CMC often being the preferred choice for coating and reinforcement. The bleached wood pulp, which has not been heavily pulped, helps increase the thickness and stiffness of the filter paper. The higher surface temperature of the drying cylinder also aids in rapid drying, further improving the thickness of the paper. The longer fibers of the hemp pulp act as reinforcing materials, and through pulping, the fibers become defibrated, enhancing the strength of the paper without significantly reducing its thickness. The use of CMC for sizing further strengthens the filter paper.

b) Process Equipment and Workflow

Hangzhou Xinhua Paper Industry Co., Ltd. (formerly Hangzhou Xinhua Paper Mill) was one of the first manufacturers in China to develop non-heat-seal tea bag filter paper. The production equipment used is a side-laying long-wire paper machine, which was independently designed and manufactured by the company. This equipment features several side-by-side and symmetrically distributed headboxes (pulp distributors), with an additional white water headbox on the drive side. During operation, the headbox rotates, filling with pulp from the wire trough. As it reaches the top position, the pulp is transferred onto the moving long-wire forming fabric, and the process is repeated several times until a well-distributed wet paper sheet is formed. The last headbox serves as a white-water headbox, further refining the wet paper sheet.

This pulp feeding method is particularly beneficial for producing long-fiber papers, such as those made from hemp pulp or mulberry bark pulp, which are low grammage and thin papers. However, side-laying long-wire paper machines have some inherent limitations, such as low machine speed (typically under 100 m/min) and narrow web widths (typically under 2 m). With significant advancements in papermaking equipment technology, the use of inclined-wire papermaking machines for producing long-fiber, low-grammage specialty papers has become increasingly mature. Figure 5.16 illustrates the process flow of a domestic company's long-fiber inclined-wire production line. This process can produce high-transparency paper, non-heat-seal tea bag filter paper, and other products, with options for additional operations like calendaring and moisture control based on product performance requirements.

2) Preparation Process of Hot-Sealed Tea Bag filter Paper

a) Materials

Heat-seal tea bag filter paperis made by mixing bleached wood pulp or unbleached natural wood pulp with heat-sensitive fibers (some products may also include a small amount of hemp pulp depending on performance and style requirements), in a specific proportion, to form the paper through a pulp-making process. After drying, the paper is heat-treated to complete the production. During processing, the sealing

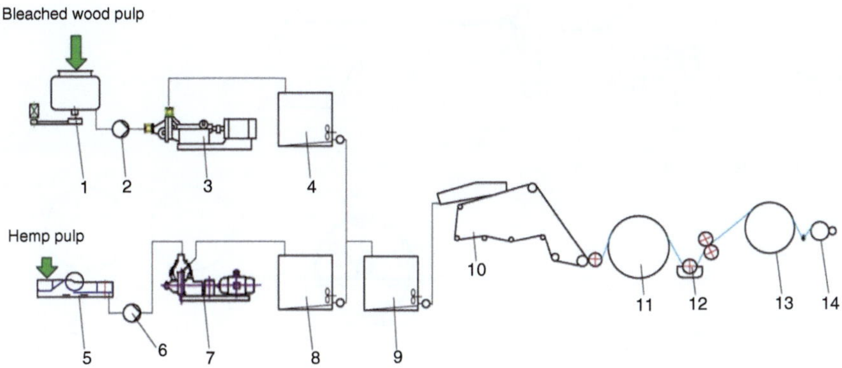

Figure 5.16 Schematic diagram of the process flow for non-heat seal type tea bag filter paper base.

equipment is heated to the melting point of the heat-sensitive fibers, allowing for quick heat-seal bonding of the tea bag [14].

The bleached wood pulp is composed of bleached softwood and hardwood fibers, with the softwood pulp being primarily dispersed using a hydraulic pulper. The pulp's beating degree can also be adjusted with refining equipment based on product performance requirements. If higher strength is needed for certain heat-seal tea bag filter paper products, a small amount of Manila hemp pulp fibers may be added to increase the dry and wet tensile strength of the paper. The heat-sensitive fibers are mainly polypropylene fibers.

b) Process Equipment and Flow

Xin Hua Paper Industry Co., Ltd. (formerly Hangzhou Xin Hua Paper Mill) was one of the earliest companies in China to introduce production equipment and technology for heat-seal tea bag filter paper. The production process uses the Japanese Da Chang paper machine. The process involves forming a double-layer composite of round and long net structures, followed by the use of heat processing equipment to bond the synthetic fibers, preventing shedding and improving both dry and wet strength. This process also achieves high-temperature sterilization. Domestic papermaking equipment technology has been continually updated, but this dual-layer or multi-layer composite process remains the foundation. The production process flow for heat-seal tea bag filter paper is shown in Figure 5.17.

5.3.3.3 Technical Parameters of Tea Bag Filter Paper

The main technical parameters for non-heat-seal tea bag filter paper are outlined in the current national standard "Non-Heat-Seal Tea Filter Paper," while the technical parameters for heat-seal tea bag filter paper are specified in the current national standard "Heat-Seal Tea Filter Paper."

Additionally, the filter bags should be made of non-heat-seal or heat-seal tea bag filter paper, and the main technical parameters of the filter paper must meet the specifications outlined in Table 5.9.

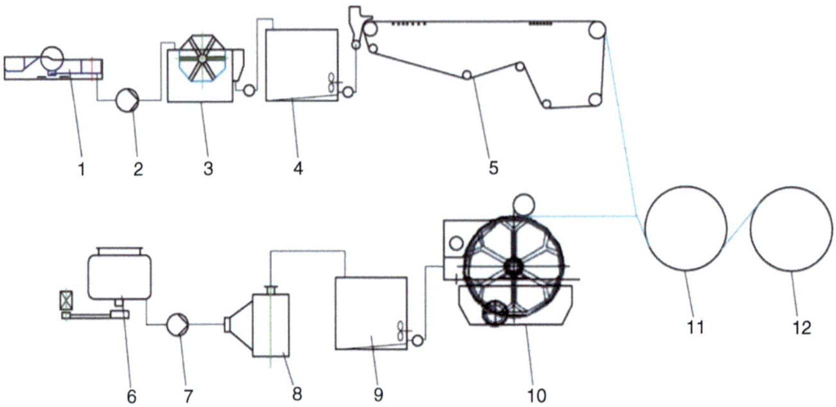

Figure 5.17 Schematic diagram of the production process of heat-sealable tea bag filter paper.

Table 5.9 Main technical parameters of tea bag filter paper.

Metric	Non-Heat-Sealed Tea Bag Filter Paper	Hot-Sealed Tea Bag Filter Paper
Basis Weight/(g/m^2)	13±1	17±1
M Tensile Strength/(N/mm)	≥0.6	≥0.52
D Tensile Strength/(N/mm)	≥0.2	≥0.12
Filter Velocity/s	≤3	≤5
Moisture Content/%	≤7	≤7
Rare Delicacy	No	No
Wet Strength (Boiling 10 min)	No Break	No Break

5.3.3.4 Prospects for Paper-based Functional Materials for Tea Filtration

With the rapid development of China's economy and the continuous improvement in people's living standards, bagged tea and tea products are becoming increasingly popular among consumers. Tea filter paper substrates are used as the inner packaging filter paper for tea bags. In addition to being applied in the packaging of tea bags, they can also be used to make filter bags for traditional Chinese medicine and health drinks. Filter bags made from tea filter paper substrates do not tear when exposed to hot water and can quickly release the tea essence without allowing tea dust to escape. Tea bag filter paper is made from plant fibers, with uniform texture, moderate softness, high dry and wet strength, and fast filtration speed. It is non-toxic and odorless, making it a green and environmentally friendly product that fully complies with food hygiene standards. The future of tea filter paper substrates will focus on the development of products with higher health benefits, greater strength, faster filtration, quicker tea extraction, and lower basis weight.

References

1 Wang, J. (1999). *Chinese Paper Raw Materials and Microscopic Atlas*, 43–53. Beijing: China Light Industry Press 163–185.

2 Yao, X. and Wang, L. (2015). Application of slanted mesh forming technology in long-fiber specialty papers. *Donghua Paper Industry* 46 (5): 30–36.

3 Durst, M., Klein, G.M., and Moser, N. (2002). *Filtration in Vehicles*, 9–11. Germany: Verlag Moderne Industrie.

4 Xu, Z. (2003). *Principles of Air Purification Technology*, 3e, 85–87. Beijing: Science Press.

5 Zhao, Y., Ji, Z., and Wang, Z. (2010). Current status and development trends of liquid filtration technology in China. *Journal of Hefei University of Technology (Natural Science Edition)* 33 (6): 812–816.

6 Loesecke, D. and Murphey, B. (2008). An overview of engine filtration. *Filtration and Separation* 45 (7): 17–22.

7 Yu, T., Jiang, Y., and Hu, J. (2013). Performance and development of engine oil filter paper. *Paper Science and Technology* 32 (6): 58–61.

8 Li, H., Wu, Q., and Pan, Z. (2018). Structure and performance analysis of high-precision fuel filtration materials. *Modern Textile Technology* 6: 10–15.

9 Shao, L., Zhang, Y. Preparation Method for a Fiber Composite Filter Element Material for Engine Air Filters: CN201610334363.2. 2016-10-12.

10 Liang, Y. and Hu, J. (2005). Development of gradient structure filter paper for automotive applications. *China Paper* 24 (8): 18–21.

11 Yang, J. and Liang, Y. (2017). Structure and performance of two types of ultrafine Fiber composite air filter papers. *Paper Science and Technology* 36 (2): 8–14.

12 Xie, X. (2007). The impact of inlet treatment quality on gas turbines. *Turbine Technology* 49 (3): 226–227.

13 Liu, R. (2004). The difference between filter paper and chemical analysis filter paper. *Tianjin Paper* 26 (1): 43–45.

14 Schoeller & Hoesch Papierfab. Heat Sealing Filter Materials: DE 03800077.6. 2005-03-30.

6

Hydrophobic/Oleophobic/Hydrophilic/Lipophilic Paper-based Functional Materials

6.1 Introduction to the Mechanism of Surface Wettability of Paper-based Materials

6.1.1 Wetting Model of the Surface of the Object

The wettability of an object is mainly characterized by the contact angle, and Figure 6.1 is a schematic representation of the main models describing the contact angle of an object, including Young's model, Wenzel model, and Cassie–Baxter model [1].

Young's model describes the wettability of a liquid on a smooth, homogeneous solid surface under ideal conditions: it is hydrophilic when the contact angle is less than 90°, hydrophobic when the contact angle is $90° \leq \vartheta \leq 150°$, and superhydrophobic when it is greater than 150° [2].

Young argues that when a liquid is resting on a solid surface, the contact angle (ϑ) of the liquid on the solid surface is caused by the interfacial tension between the solid, the gas, and the liquid and satisfies Young's model:

$$cos\,\vartheta = (\gamma_{SA} - \gamma_{SL})/\gamma_{LA} \tag{6.1}$$

where γ_{SA} is the tensions between solid–gas interfaces; γSL is the tensions between solid–liquid interfaces; and γLA is the tensions between liquid–gas interfaces:

$$cos\,\vartheta = r\,(\gamma_{SA} - \gamma_{SL})/\gamma_{LA} = r\,cos\,\vartheta_0 \tag{6.2}$$

People refer to Eq. (6.2) as the Wenzel model, where ϑ_0 is the value of the contact angle of the smooth surface and r is the surface roughness factor, which is the ratio of the solid–liquid contact area (ASL) of the rough surface to the solid–liquid contact area (AF) of the smooth surface:

$$r = A_{SL}/A_F \tag{6.3}$$

Since $r > 1$, it is obtained from Eq. (6.2) that the roughness of the hydrophobic surface ($90° \leq \vartheta_0 \leq 150°$) increases, resulting in an increase in ϑ and a more hydrophobic surface, while conversely, the roughness of the hydrophilic surface ($\vartheta_0 < 90°$) decreases, resulting in a decrease in ϑ and a more hydrophilic surface. The Wenzel model is used for chemically homogeneous solid surfaces and is not suitable for

Paper-Based Functional Materials: Preparation and Applications, First Edition.
Xianping Yao, Genrong Chen, Zhen Yao, and Liping Zheng.

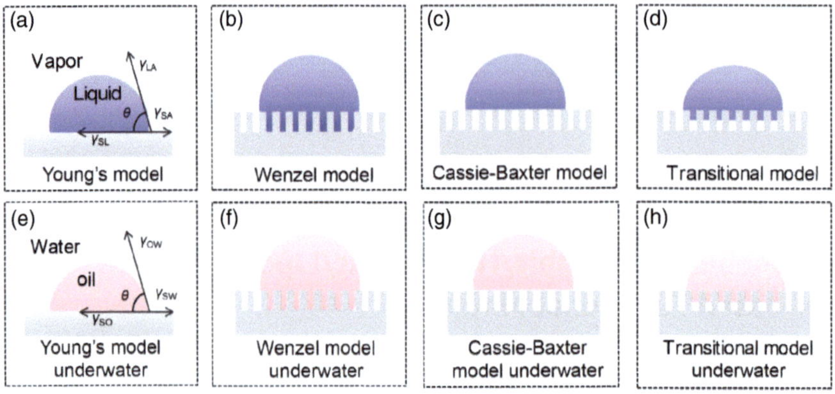

Figure 6.1 Typical wetting states of liquid droplets on solid surfaces with different roughnesses in air (a), (b), (c), (d); wetting states of oil droplets on different solid surfaces under water (e), (f), (g), (h).

chemically inhomogeneous rough surfaces. Therefore, in order to get closer to the real surface, Cassie and Baxter propose that when the liquid comes into contact with a solid surface, this rough solid surface with an uneven chemical composition is a solid–gas composite surface, and the contact angle of this composite surface can be given by the Cassie–Baxter model:

$$\cos \vartheta = f_1 \cos \vartheta_1 + f_2 \cos \vartheta_2 \tag{6.4}$$

where ϑ_1 and ϑ_2 are the contact angles of the liquid on the surface of these two media; f_1 and f_2 are the percentages of the liquid–solid interface and the liquid–gas interface to the total surface, respectively, and $f_1 + f_2 = 1$.

When the liquid comes into contact with the composite surface, the void of the rough structure is completely filled with air, and the droplets come into contact with the solid not only to form a liquid–solid interface ($f_1 = f_{SL}$, $\vartheta_1 = \vartheta_0$), but also to form a liquid–gas interface ($f_2 = f_{LA} = 1\text{-}f_{SL}$, $\vartheta_2 = 180°$) in contact with the air, as shown in Figure 6.1c. So from Eq. (6.2) and (6.4), the following equations can be obtained:

$$\cos \vartheta = f_{SL}(r \cos \vartheta_0 + 1) - 1 \tag{6.5}$$

From Eq. (6.5), it can be seen that for the hydrophobic surface ($\vartheta_0 > 90°$), the hydrophobic surface becomes more hydrophobic with the increase of f_{SL}, ϑ increases, and for the hydrophilic surface ($\vartheta_0 < 90°$), ϑ decreases with the increase of f_{SL}, and the hydrophilic surface becomes more hydrophilic.

The Cassie–Baxter model can be applied to rough surfaces with uneven chemical composition, where the voids of the rough structure are filled with air when the droplets are placed on the surface. However, under the action of external force, the droplets are immersed in the voids of the rough structure by downward pressure, and the model of solid–liquid contact is changed from the Cassie–Baxter model to the Wenzel model. That is, there is a new transitional state model in the Cassie–Baxter model and the Wenzel model, as shown in Figure 6.1d [3].

6.1.2 Definition of Different Wetting Interfaces

6.1.2.1 Definition of a Hydrophobic and Oleophobic Interface

Biphobic surfaces are both hydrophobic and oleophobic, with large contact angles (>90° for both aqueous and oil) and small sliding angles (SA), and droplets do not stay stably on the surface, thus having the ability to repel droplets [4]. Roll angle hysteresis and contact angle hysteresis (CAH) are often used to describe the wetting behavior of droplets when placed on an inclined solid surface. Generally speaking, when the water contact angle and oil contact angle of the interface are both greater than 90°, it can be called a biphobic interface, and when the water contact angle and oil contact angle of the interface are both greater than 150°, and the rolling angle to the water and oil phases is less than 10° and there is a low CAH, the interface can be called a super-biphobic interface.

6.1.2.2 Definition of a Hydrophobic and Lipophilic Interface

Hydrophobic lipophilicity is a special phenomenon of surface wettability of materials, which exhibit both hydrophobicity and lipophilicity. When the water contact angle of the material interface is greater than 90° and the oil contact angle is less than 90°, the interface is called a hydrophobic lipophilic interface. Similarly, if the water contact angle of the hydrophobic lipophilic interface is greater than 150° and the oil contact angle is close to 0°, it is called a superhydrophobic superlipophilic interface. If the surface free energy of the solid is between water and oil, the smooth solid surface should be hydrophobic and lipophilic [5], and a superhydrophobic superlipophilic surface can be prepared by combining the Wenzel model and the Cassie model to construct a certain roughness [6].

6.1.2.3 Definition of Hydrophilic and Lipophilic Interfaces

The hydrophilic and lipophilic (amphiphilic) interface refers to the phenomenon that the contact angle between the water and oil phases at the material interface is less than 90°, and it can be wetted by both water and oil. In contrast to the superhydrophobic interface, when both the aqueous and oil phases of the interface have contact angles close to 0°, the liquid at the material interface can be completely absorbed, exhibiting extreme wetting behavior, which is called a superhydrophilic interface. The superhydrophilic interface, a surface where both water and oil can diffuse completely, has been extensively studied in recent years. In general, there are two main ways to obtain super-friendly surfaces. One is a photocatalytic coating with UV irradiation, and the other is a very rough surface [7]. At present, super-amphiphilic surfaces are usually prepared by various methods such as electrospinning, nanopolymer coating, plasma etching, solid surface etching and oxidation, and impregnation.

Among them, the basic principle of superhydrophilic binary synergistic interface material technology (hereinafter referred to as superamphiphilic nanomaterials) is that under light irradiation, the surface of the material can cause the formation of a binary synergistic nanointerface structure with hydrophilic and lipophilic phases coexisting in the nano region, so that a wonderful superamphiphilic will be formed on the surface of the material. The new material made by using this principle can be modified on the surface of building materials such as glass and ceramic tiles to make them self-cleaning and antifogging. Among them, antifog nano-coating materials can be widely used in bathroom mirrors, various glasses, shop windows, agricultural greenhouse films, military gas masks, tanks, ships, vehicle windows, etc., to make them have an antifog effect. The super double personal cleaning nano-coating material can be modified to the surface of buildings, military and civilian tents, to make it have a self-cleaning function.

6.1.2.4 Definition of Hydrophilic and Oleophobic Interfaces

The hydrophilic and oleophobic interface refers to the water contact angle of the interface is less than 90° and the oil phase contact angle is greater than 90°, when the water contact angle of the hydrophilic and oleophobic interface is close to 0° and the oil contact angle is greater than 150°, it is called a superhydrophilic superoleophobic interface. Superhydrophilic superoleophobicity is a special phenomenon of solid surface wettability, which is mainly related to the roughness and chemical composition of the surface. Superhydrophilic superoleophobic materials, which are completely nonoily but completely hydrophilic, have important application prospects in many fields, such as greaseproof paper.

Illustrates the relationship between the four basic surface superwetting/wetting properties (in the blue box) and other special surface superwetting/wetting features (in the yellow box) obtained by combining two of the basic properties, as shown in Figure 6.2.

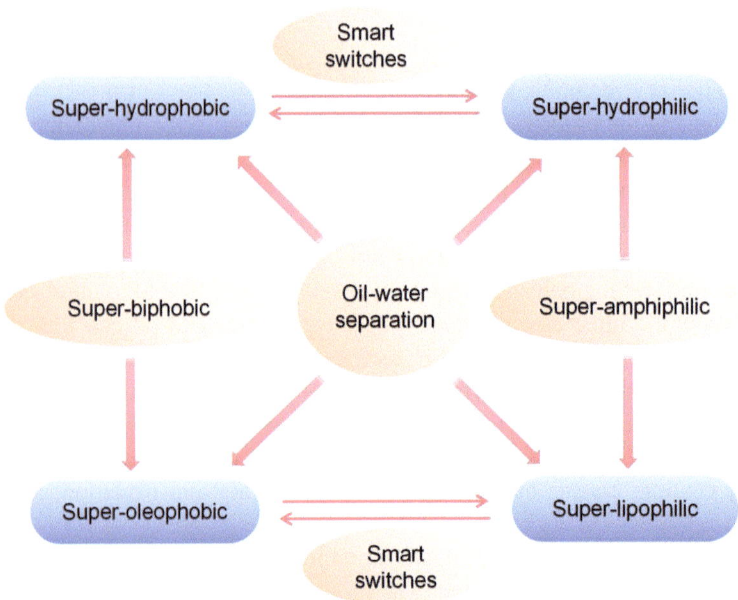

Figure 6.2 An explanation of the relationship between the four basic surface superwetting/moisture protection properties (blue box) and other special surface superwetting/moisture protection features (yellow box).

Since water and oil have different surface tensions, it is possible to achieve high-efficiency and high-flux oil–water separation at atmospheric pressure by constructing a special wetting coating with different wettability for water and oil, and applying it to a mesh substrate. At present, there are two types of special wettability surfaces used for oil–water separation: superhydrophobic, superlipophilic surfaces and superhydrophilic superoleophobic surfaces. The adhesion between the superhydrophobic superlipophilic coating and the oil is large, and the oil droplets and impurities in the oil are easy to adsorb on the surface of the omentum during the separation process, causing pollution to the omentum, also causing rapid attenuation of the permeation flux, and seriously affecting the performance and working life of the omentum. At the same time, because the density of oil is generally smaller than that of water, in the process of oil–water separation, water will be in the lower layer of the mixture, and the oil floats on the water surface, while on the superhydrophobic superlipophilic separation membrane, the water cannot pass through the separation membrane, and it hinders the oil that can pass through the separation membrane from contacting the omentum, forming an obstacle.

In contrast, superhydrophilic and superoleophobic surfaces can make up for the above shortcomings and have broader application prospects. According to the classical theory of surface free energy, it is difficult to prepare "anomalous" wettable surfaces that are both superhydrophilic and superhydrophobic because superoleophobic surfaces are often superhydrophobic at the same time. At present, most of the indirect methods such as the preparation of superhydrophilic and underwater superoleophobic coatings, are used to realize the superhydrophilic and superoleophobic function. This type of coating generally exhibits superhydrophilic and superlipophilic (or lipophilic) properties in air. When in contact with water, due to the superhydrophilic nature of the coating surface, it is very easy to be wetted by water to form a stable hydration layer, and because oil and water are incompatible, it forms an oleophobic or superoleophobic surface. Although the oil–water separation web film constructed by superhydrophilic and underwater superoleophobic coating has a relatively wide application prospect, its application is limited because the oleophobicity of this type of web coating depends on the formation and retention of the surface water film.

6.2 Hydrophobic and Oleophobic Paper-based Functional Materials

6.2.1 Preparation

Hydrophobic surfaces based on the "lotus leaf effect" can be prepared in two ways: one is to use hydrophobic materials to construct rough surface structures, and the other is to modify low surface energy species on rough surfaces. Because the interfacial tension of oil is less than that of water (72 mN/m), especially some oils (such as n-hexadecane and dodecane) have a relatively low interfacial tension (20 ~ 30 mN/m), which is much smaller than that of water. Therefore, in order to repel such oil droplets, the surface roughness and surface free energy must be more tightly controlled [8].

The chemical composition determines the surface energy of the material and plays a vital role in the wettability of the material. In order to achieve superoleophobicity, the surface energy of solid surfaces must be as low as a few millinewtons/m (mN/m), which can only be achieved by special molecular functional groups such as – CF3. For example, 1H,1H,2H,2H-perfluorophosphoric acid, 1H,1H,2H,2H-perfluorooctanoic acid (PFOA), 1H,1H,2H,2H-decafluorooctanoic acid trichlorosilane, 1H,1H,2H,2H-perfluoroundecyltrichlorosilane, 1H,1H,2H, 2H-perfluorodecane-1-mercaptan, and 1H,1H,2H,2H-perfluoromethacrylate. In addition, the superdiphobicity of the material can be achieved by spraying or spinning a fluorinated polymer (fluoropolymer) solution, and fluoromonomers can also be applied to grafted fluorinated layers by vapor or liquid phase deposition. The stringent requirements for specific fluorine-containing functional groups are a major bottleneck in the fabrication of superoleophobic surfaces, which limits the selectivity of superbiphobic materials.

Another important factor influencing the surface wettability of a material is the topology of the material surface. When the liquid comes into contact, the air can be trapped in rough areas, greatly reducing the contact between the surface and the droplets. However, superoleophobicity is only suitable for determining rough structures with several concave corner shapes, such as cantilevered structures, negative slope structures, anti-trapezoidal structures, and mushroom-like structures. Tuteja et al. first proposed the concept of concave geometry, and in their work, a synthetic polyhedral oligosiloxane (POSS) molecule was used to create a fiber pad that exhibited superhydrophobic properties due to its concave geometry, despite the contact angle of each component being less than 90° [9]. This means that even if a low-contact object is placed on a flat surface, the contact angle on the actual rough surface is still high enough in this particular case. If the sum of the geometric angles and contact angles of the concave corners of the texture is greater than 90° on a plane, this geometry can support a good shape at the liquid–gas interface with its surface tension upward. Due to the low contact angle of the airbag and liquid–solid interface in this structure, the surface cannot be wetted by liquids. On the other hand, although their planar contact angles are below 90°, various liquids can also show high contact angles on surfaces with concave angular structures, inducing a negative pressure difference at the gas–liquid interface, which transforms the gas–liquid interface from concave to convex and prevents oil wetting [10–12]. Therefore, liquids with various contact angles can wet the reentry with different curvatures to different degrees to obtain the ideal liquid–gas interface shape, and the multiple concave structures can effectively maintain the oil-gas interface, which is of great significance for the formation of a stable superoleophobic effect.

It can be seen that the key to constructing a biphobic surface is to construct the micro-nano rough structure of the surface, and second, to modify the micro-nano rough structure with low surface energy substances for hydrophobic oleophobic treatment. Compared with the single hydrophobic and oleophobic interface, the biphobic surface has higher requirements for micro and nano structures, and the low surface energy substances required are superior, so it is more difficult to construct the biphobic surface. Methods for preparing biphobic surfaces are mainly divided into bottom-up method and top-down method: the bottom-up method refers to the construction of micro-nano microstructures from the bottom by using the accumulation of atoms, molecules, or clusters, including the sol–gel method and electrospinning method, and the top-down method refers to the crushing or destruction of the substrate material from the top to obtain the micro-nano structure, including etching technology and template method [13]. The preparation of hydrophobic and oleophobic paper-based functional materials is mainly introduced from the aspects of physicochemical modification method, surface coating method, nanoparticle deposition method, and chemical modification method.

6.2.1.1 Physicochemical Modification Method

The physicochemical modification method mainly includes the impregnation method, stencil method, plasma method, and papermaking filler method. Among them, it is simple and convenient to prepare biphobic paper-based functional materials by the impregnation method. Jiang et al. used commercially available degumming agents to effectively remove fines and modify interfiber hydrogen bonds to control the size of the fiber network, then used oxygen plasma to generate nanoscale roughness on micron-sized fibers and remove residual fine fractions that clogged the pores between fibers, and finally impregnated the paper into a fluorosilane solution to obtain an ultra-thin, low-surface-energy coating biphobic paper-based material [14].

The template method prepares superhydrophobic surfaces by forming nano-rough structures after depositing soot on the surface. Zhang et al. prepared superhydrophobic paper with high hydrophobicity and low adhesion by depositing a layer of ash on the surface of the paper by vapor deposition of SiO_2 coating. If the paper is wet-stretched, the adhesion of the paper surface can be significantly improved, and the surface can be changed from a roll-off type to a sticky type [15].

Plasma is composed of positively and negatively charged ions and electrons, and after etching cellulose paper, microcrystalline bundles can be formed in the amorphous region, but at the same time, the interfiber adhesion is also reduced, which has an impact on the internal fiber structure and the surface morphology of the coating. Li used plasma to etch the surface of the paper, and then deposited low surface energy fluoropolymers on the surface of the paper to prepare a biphobic paper-based material, and the oil contact angle of the paper sample reached $(149 \pm 3)°$. The exchange of water in the pulp solution with sec-butanol further controls the fiber spacing, resulting in a superbiphobic paper-based material with a contact angle of more than 150° between water, ethylene glycol, oil, and n-hexadecane [16]. Jiang et al. produced nanoscale roughness and increased the surface porosity of paper-based materials by thinning the fibers and removing paper scraps by oxygen plasma, and combined these properties with naturally occurring concave fiber structures and micron-sized fiber sizes to form layered structures on paper. Then, in order to obtain the necessary surface chemistry, a thin layer of fluorosilane was formed on the surface by vapor deposition, and finally, a superbiphobic paper-based material was successfully prepared [17].

When using papermaking fillers such as precipitated calcium carbonate (PCC) and kaolin to prepare superhydrophobic paper, it is essential not only to ensure that the mineral particles can be uniformly dispersed but also to modify the hydrophobic nature of the mineral particles or improve the surface roughness of the mineral particles. PCCs are typically attached to the surface of the paper using cross-linkers such as latex to form a bi-scale rough structure, and then impregnated in a solution of potassium stearate or the paper sizing agent alkyl enone dimer (AKD) to obtain a superhydrophobic and oleophobic paper-based material [18]. The use of fibrillated cellulose (MFC) as a PCC linker to prepare superhydrophobic paper can improve both PCC retention and the rough structure of the paper surface. After MFC is coated on the surface of the paper, it can form a layered nano-rough structure, and the fluorination modification can impart superhydrophobic properties to the paper [19]. Mirvakili et al. studied the amount of papermaking filler in the plasma preparation of superhydrophobic paper, and found that the amount of calcium carbonate and talc in superhydrophobic paper can reach up to 45%, while the maximum amount of kaolin is only 35%. The application example of adding modified inorganic particles to the pulp is not common due to the large amount of addition and the difficulty of process adjustment.

Wang Haisong's team from Dalian Polytechnic University mixed 10% polyolefin (ES) fibers (polyethylene with an outer layer with a melting point of 110 ~ 130 °C; polypropylene with an inner layer of polypropylene, melting point 160 ~ 170 °C) and pulp fibers to prepare a superhydrophobic paper with a contact angle of up to 153°, while the contact angle of the paper without the addition of ES fibers under the same conditions was only 25° [20]. This is because the ES fibers not only cover the hydroxyl groups on the surface of the plant fibers after melting, but also fill the holes between the fibers and the fibers, so the hydrophobicity of the paper surface is greatly improved, and the mechanism of obtaining the superhydrophobic surface is shown in Figure 6.3.

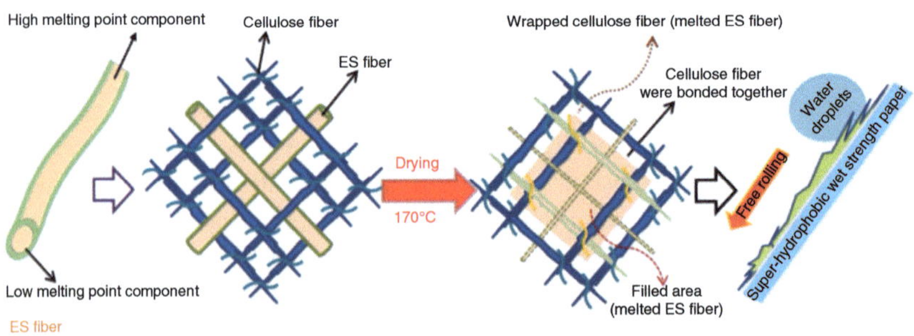

Figure 6.3 Schematic illustration of the fabrication and structure of the proposed cellulose-based hydrophobic composite.

6.2.1.2 Surface Coating Method

The sol–gel method often uses surface coating to achieve the covalent bonding of SiO_2 to the hydroxyl group on the fiber, forming a micro/nanorough structure. The coating methods include spin coating, brush coating, etc., but the preparation process of the gel is relatively long, and the hydrogen bond between silica alcohol and cellulose hydroxyl group is low, the adhesion performance is poor, and the water resistance is low. Therefore, this method requires the use of cross-linkers or siloxane precursors to improve the bonding properties between inorganics and cellulose and form a stable network structure on the surface. Qu et al. successfully prepared a highly stable superbiphobic paper-based material that can be triggered by ethanol by a simple coating method using zinc oxide (ZnO) particles and organosilane (PFDS) with switchable surface wettability, with a water contact angle of 160° and glycerol, ethylene glycol, and sunflower oil with an oil contact angle greater than 150°[21]. Li et al. fabricated translucent super-biphobic paper by spraying a fluorinated suspension of SiO nanoparticles on the surface of A4 paper. Firstly, fluorinated SiO_2 nanoparticles are prepared by injecting perfluorodecyltrichloro groups into the surface of SiO_2 nanoparticles, and secondly, the aggregation protrusions, the binding of pores, and the high perfluorocarbon content obtained by the spraying process can maintain the concave curved shape and have low surface energy, thus forming a strong composite interface with the low surface tension liquid. This ultra-biphobic paper is sensitive to a wide range of liquids, from water to organic liquids, even rapeseed oil and cetane have contact angles of more than 150° and are prone to slipping at low slip angles. In addition, the obtained superhydrophobic SiO_2 paper was found to be translucent, and the visibility of the characters on the paper barely changed after spraying fluorinated SiO nanoparticles [22]. Tang et al. studied the effect of alkyl chain length on the hydrophobicity and oleophobicity of cellulose-based paper prepared by alkyl trichlorosilane (methyl, butyl, dodecyl, and octadecyltrichlorosilane) using alkyl trichlorosilane as a model, and found that the superhydrophobic oleophobic performance of methyl trichlorosilane coated paper was the best, and the water contact angle could reach 152° [23].

In addition, polycarboxylic acids can form stable network structures through ester bonds between fibers and silicon nanoparticles and can also be used as catalysts for sol formation [24, 25]. For example, citric acid, catalyzed by sodium hypophosphite, creates a covalent bond between cellulose and silica gel. Cyanate-based functional group precursors are also used to link fibers to SiO_2 nanoparticles. Cunha et al. grafted propyl 3-isocyanate trimethoxysilane onto cellulose and copolymerized with silane polymers to improve the stability of the superhydrophobic and oleophobic paper-based surface [26]. Hydrophobic and oleophobic paper can also be prepared by directly dispersing SiO_2 nanoparticles in siloxane solution. Liu et al. prepared hexane suspensions of trichloro (1H,1H,2H,2H tridecafluo-n-octyl) silane, tetraethyl orthosilicate, SiO_2, and TiO_2 nanoparticles and obtained ultra-diphobic filtered paper separating polar and nonpolar liquids, such as water, glycerol, 1,4-butanediol, soybean oil, and 1-octadecene, by immersing the filter paper for 10 minutes at contact angles of 168°, 158°, 154°, 145°, and 121°, respectively [27]. Zhang et al. used polystyrene-polymethyl methacrylate copolymer mixed with SiO_2 and coated it on the surface of the paper to produce a hydrophobic paper with high load-bearing capacity and low weight from biomimetic water strider [28], and superhydrophobic and oleophobic paper could be prepared by heating and coating a waxy emulsion on the surface of the paper [29]. Styrene-maleic anhydride copolymers are imidized and coated on paper to obtain hydrophobic and oleophobic surfaces with water contact angles of up to 148° [30]. Superhydrophobic and oleophobic paper can be prepared by spraying fatty acids, iron, and polystyrene compounds on the surface of filter paper [31].

The arrival of environmental pollution and the energy crisis has forced the hydrophobic modification of paper to favor the use of other green bio-based polymers (such as fatty acids, vegetable oils, chitosan, starch, and protein), which is also the trend of green and sustainable production. In the field of superhydrophobicity, the use of biopolymer-modified superhydrophobic paper has developed rapidly, and the nanoparticles encapsulated with vegetable oil, such as Samyn, have improved the hydrophobicity of the fibers, and the coated paper has a water contact angle of 160° [32].

6.2.1.3 Nanoparticle Deposition Method

Nanotechnology is also a common way to prepare hydrophobic and oleophobic paper-based materials by using nanotechnology to obtain uniform bi-scale micro/nanorough structures. Fragouli et al. impregnated the paper in a fresh mixture of supermagnetic nanoferrite ($MnFe_2O_4$) and cyanoacrylate to impart hydrophobic and oleophobic properties (contact angle between 120° ~ 140°), and if the paper is impregnated with a cured solution, a more layered rough structure can be formed, and a larger contact angle can be obtained, which is helpful for the application of cellulose-based paper in inductors and microfluidic devices [33]. TiO nanoparticles are used for superhydrophobic and oleophobic modification of paper, which can impart photocatalytic and pollutant degradation properties to paper. Ai et al. deposited flower-like TiO_2 nanoparticles on the surface of cotton fabrics by a simple spraying method and further modified them with 1H,1H,2H,2H perfluorooctyltrichlorosilane (FOTS) to construct a super-diphobic cotton fabric with excellent mechanical stability and antifouling properties [34]. Generally, when nanoparticles are deposited on the surface of paper, they need to have good compatibility and connectivity with fibers. Su et al. coated SiO_2 with epoxy groups (EP) or polyurethane (PU) and impregnated and deposited on the surface of cotton fibers to obtain superhydrophobic oleophobic fabrics with excellent friction resistance [35].

6.2.1.4 Chemical Modification Method

The methods for preparing hydrophobic and oleophobic paper-based materials by chemical modification mainly include the layer assembly method and the phase separation method.

The layer-by-layer assembly method is simple and easy to operate, and assembly is performed on the surface of the fibers by continuously using a polymeric electrolyte of opposite charges. Gustafsson et al. assembled polyallylamine hydrochloride and polyacrylic acid layer-by-layer on the fiber surface, adsorbed a layer of solid paraffin, and heat-treated at 160 °C to prepare superhydrophobic paper with a water contact angle of 150° [36]. Peng et al. assembled polyallyl ammonium chloride (PAH) and lignosulfonamide layer upon layer on the surface of the paper to prepare superhydrophobic oleophobic paper after heating [37]. This method opens up a new way for the high-value utilization of lignin and is environmentally friendly.

The phase separation method is a film-forming method in which the system produces two or more phases by controlling the conditions in the film-forming process to form a homogeneous or heterogeneous film. Li et al. used the phase separation method to prepare superhydrophobic and oleophobic paper by depositing low-cost, nontoxic, and hygroscopic carnauba wax on the surface of the paper, which is simple and economical and meets the requirements of industrial production [38]. In addition, the deposition of polyhydroxybutyrate on the surface of the paper by the phase separation method can also improve the micro/nanorough structure of the surface and reduce the surface energy at the same time, so as to prepare superhydrophobic and oleophobic paper. Joao et al. used a phase separation method to adsorb polyhydroxybutyrate on the fiber surface to form a micro/nanocoarse structure, and prepared a superhydrophobic paper with a water contact angle of 153° [39].

6.2.2 Application

Due to the excellent properties of superbiphobic surfaces and the increasing demand for superbiphobic materials, their multifunctional applications have received more and more attention, and great progress has been made. Superhydrophobic cellulose paper-based materials have excellent hydrophobic properties and are often used in packaging and other fields. For example, Li et al. assembled alginate and TiO_2 layers on the surface of the paper to produce fluorine-free, bacteriostatic superhydrophobic packaging paper [38]. Since the interfacial tension of oil is smaller than that of water (and in some cases much less), the fabrication of superbiphobic materials requires a more ingenious and delicate microstructure and a lower surface free energy. Superbiphobic materials can repel both water droplets and oil droplets, and have a wider range of applications than superhydrophobic materials, such as self-cleaning, anticorrosion, oil transportation, anti-bioadhesion devices, oil collection, antifouling, micro-droplet transfer, and oil–water separation. In recent years, the application of cellulose-based superbiphobic paper-based materials has shown diversification, such as the preparation of paper-based microfluidic devices and the acquisition of functional materials that intelligently respond to environmental conditions such as light, temperature, and pH.

6.2.2.1 Oil–Water Separation

Paper materials made from fibers have a high porosity, and when the paper is hydrophobic, it becomes a potential filter medium with oil–water separation function. The porosity and wetting properties of the material are used to separate the oil–water mixture more stably, and the separation effect is better than the traditional oil–water separation method. For example, Ge et al. coated paper towels with fluorinated SiO_2 nanoparticles as separators to create a super-diphobic paper-based material, which in turn serves as a versatile gravity-driven oil–water separator that can selectively remove water or oil from free oil–water mixtures and surfactant-stabilized emulsions. After pre-wetting the separator with ethanol, water can be removed from light oil–water mixtures and emulsions by gravity with high separation efficiency (99.9%) and separation flux. And vice versa, heavy oil can be removed by gravity on an ethanol-oil pre-wetted superbiphobic paper-based material [40].

6.2.2.2 Self-cleaning

Stanssens et al. applied imide-modified polymer nano-ions to the surface of the paper to form a hydrophobic coating, which was dominated by microstructure and had a water contact angle of 150°, while the surface of the paper with unmodified polymer nano-ions had a water contact angle of 135°. The rolling down experiment of water droplets and the hysteresis data of the contact angle indicate that the material has good self-cleaning ability [30]. Zhu et al. prepared fluorine-free hydrophobic paper by a simple impregnation method. First, they immersed the paper in polydimethylsiloxane to make the surface viscous, then soaked it in ZnO to obtain a rough surface, and then soaked it in polydimethylsiloxane again to obtain superhydrophobic paper. The paper prepared by this method has a contact angle of 160°, which has various functions such as self-cleaning, UV protection, and oil–water separation [41]. Geissler et al. prepared alkylated cellulose nanoparticles, immersed the base paper in a toluene solution of the nanoparticles three times with a contact angle of 120°, and then sprayed a dispersion of the nanoparticles with a contact angle of 150°, and the paper surface had the ability to self-clean [42].

6.2.2.3 Microfluidic Device

The precedent for paper-based microfluidic detection devices was pioneered by Martinez's team [43], which has developed rapidly in recent years. The paper-based microfluidic device controls the fluid flow through the microchannel structure and completes different chemical or biological reaction processes, which can achieve micro, convenient, and rapid detection, and provides a new research platform for creating a new situation in biochemical analysis. Songok teams used superhydrophobic and oleophobic paper to create a microfluidic detection device with closed channels, and found that the hydrophobic oleophobic transparent film obtained the maximum capillary dynamics when the gap between the hydrophobic and oleophobic transparent film and the paper base was 100 μm [44]. Compared with the open traditional paper-based channel, the flow system of the closed channel not only improves the flow speed of the liquid, but also effectively reduces the retention and evaporation of the liquid, which has certain significance for reducing the volume of the test sample and realizing the rapid detection of trace amounts.

6.2.2.4 Intelligent Response to Functional Materials

In order to expand the application field of superhydrophobic and oleophobic papers, there are more and more studies on intelligent superhydrophobic and lipophobic papers with responsive properties to temperature, pH, light, and heat. Zhao et al.

used the dip coating method to prepare a strong superbiphobic coating with thermal repair ability, which can firmly adhere to various substrates (fabric, sponge, wood, and filter paper) and also has the advantages of strong ultraviolet (UV) resistance, strong acid/alkali corrosion resistance, and good durability. In addition, the coating exhibits good thermal repair capabilities after thermal annealing [45]. Stepien et al. used TiO_2 nanoparticles to prepare a photosensitive superhydrophobic and oleophobic paper, which was transformed into a hydrophilic surface after UV irradiation and returned to a superhydrophobic surface after microwave heating (150 °C, 3 min). Analyses have shown that UV irradiation and microwave heating can induce changes in the number of CH+ and oxygen-containing molecules, resulting in a shift in wettability [46]. Yin et al. used tetrabutyl titanate and fluorosilane coupling agent to synthesize TiO sol. He studied the effects of ion species (Mg^{2+}, Ba^{2+}, N^+, F^-) and filling concentration on the hydrophobic/hydrophilic surface transition rate and found that the fastest transformation was achieved when the F^- addition was 2%. The ultraviolet irradiation reduces Ti^{4+} in TiO to Ti^{2+}, and the superhydrophobic surface changes to a hydrophilic state, which can be reverted to a superhydrophobic surface when placed in the dark [47]. Xu et al. prepared fluorine-free, pH-responsive superhydrophobic oleophobic fibers in an environmentally friendly manner. The coating layer was composed of SiO_2 and capric-modified TiO_2 nanoparticles, and the hydrophilic and lipophilic properties of the surface changed with the change of pH, thus achieving intelligent separation of oil–water mixtures [48].

6.3 Hydrophobic and Lipophilic Paper-based Functional Materials

6.3.1 Preparation

The preparation method of hydrophobic and lipophilic paper-based functional materials is similar to that of hydrophobic and oleophilic paper-based functional materials, but the difference lies in the affinity of paper-based functional materials for various oil products. In the previous section, the preparation methods of hydrophobic and oleophobic paper-based functional materials are mainly discussed: (i) physical and chemical modification method: impregnation method, template method, plasma method, papermaking filler method, etc.; (ii) surface coating method; (iii) nanoparticle deposition method; and (iv) chemical modification method: layer-by-layer self-assembly method and phase separation method. The results show that the surface tension of water at room temperature is about 0.072 N/m, while the surface tension of oil is about 0.020 ~ 0.035 N/m. Based on the preparation methods of hydrophobic and oleophobic paper-based materials in the previous section, this section mainly discusses how to achieve lipophilicity and its practical application while being hydrophobic.

Xu et al. immersed the filter paper in a methyl silicone resin solution and used the catalytic action of ammonia to further carry out the polycondensation reaction of the solution system, so that the filter paper became hydrophobic and lipophilic.

The filter paper itself has a large number of pores, and the scanning electron microscope (SEM) shows that the resin-impregnated filter paper still retains most of the interwoven fiber structure, but the pore size of the filter paper is reduced. This filter paper has the ability to reproducibly separate oil–water mixtures [49]. Wang et al. fully dispersed polystyrene in toluene reagent and then added a certain amount of hydrophobic SiO_2 nanoparticles to the system and soaked the filter paper with a mixed solution. The filter paper obtained after drying was hydrophobic (contact angle of 139° ~ 158°) lipophilic, and the filter paper could adsorb diesel fuel floating on the water surface or emulsified oil in water. When the volume ratio of diesel to water is 1:1, the oil–water separation rate of the filter paper can reach 98.2% [50]. Huang Xiangxuan used bisphenol A phenolic epoxy resin as the base resin, modified it with hydrophobic chemistry and then introduced fluorine-containing low-surface energy groups into the aqueous epoxy resin obtained by the anionic method. Finally, the superhydrophobic SiO_2 particles prepared by the sol–gel method were blended with fluorine-containing aqueous epoxy resin emulsion to impregnate the filter paper, which was not only hydrophobic and lipophilic (water contact angle of 152° and oil phase contact angle of 5°), but also had good physical and mechanical strength, achieving the effect of oil–water separation for removing contaminated water from the oil [51].

6.3.2 Application

Special wettability materials can separate oil in water or water in oil due to their different wettability of the oil phase and water phase, and hydrophobic lipophilic materials are a special type of wettability material. The paper obtained from plant fibers is a loose and porous structure, and hydrophobicizing endows it with a wide range of applications in different fields such as separation and filtration [52].

6.3.2.1 Oil–Water Separation

Superhydrophobic paper-based functional materials are rich in pores, with rough surfaces and pores at the micro-nano level, so hydrophobic and lipophilic paper-based functional materials can be used to separate oil–water mixtures. The porosity and wettability of the material are used to separate oil–water mixtures more robustly and with better separation results than traditional oil–water separation methods [53, 54].

Cao et al. used superhydrophobic superlipophilic filter paper (153° for water, 0° for oil, and about 12° for rolling) to separate Span 80's stable n-hexane water-coated emulsion and rolled the filter paper into a funnel shape for filtration. The superhydrophobicity, superlipophilicity, and internal three-dimensional structure of the filter paper have successfully broken the emulsion, turning the emulsion from its original milky white color to a clear and transparent state. The particle size of water droplets in the emulsion before and after filtration decreased from 1000 ~ 4000 nm to 0.5 ~ 4.0 nm, and the permeate flux of oil was above 5500 L/(m^2 h) [55]. Superhydrophobic and superlipophilic paper-based functional materials are more suitable for handling mixtures of heavy oil and water. Gravity causes the oil to sink below the water, come into contact with the solid surface, and continuously flow out of the membrane pores. If the mixture of light oil and water is poured into the filter column because the density of water is greater than that of light oil, a layer of water will be formed between the oil and the surface, and the oil cannot seep down from the surface, hindering the separation of oil and water. However, the adsorption capacity of superhydrophobic superlipophilic adsorption materials is generally limited, and the materials are sometimes susceptible to the invasion of corrosive water bodies and destroy the wettability of the surface.

6.3.2.2 Fuel Filter Paper

Vehicle filter paper is one of the main materials of the filter. It is a special filter paper, which is mainly divided into three categories: air filter paper, oil filter paper, and fuel filter paper. The purpose of the fuel filter paper is to purify the fuel that enters the vehicle system. There are a variety of materials used to produce fuel filters, including cotton yarn, nonwoven fabrics, metal wires, etc., but at present, fuel filters mostly use filter paper as filter elements. In recent years, the development of China's fuel filter paper industry has accelerated, and the water resistance and strength performance of fuel filter paper produced by many domestic manufacturers have been greatly improved compared with the past. At the same time, biomass diesel with high water volume is gradually moving toward daily production and life, although it has not yet been widely used, but based on its own series of advantages, biomass diesel will replace traditional diesel on a large scale in the future. The characteristics of biomass diesel put forward higher requirements for fuel filters, that is, more water resistance and higher oil–water separation performance. If the fuel filter paper is combined with the hydrophobic of the paper, the fuel filter paper has both porous structure and superhydrophobic lipophilic conditions, which can not only block the impurities in the oil, but also achieve the purpose of efficient oil–water separation.

Xin Jin et al. used cotton pulp and rag pulp as raw materials, copied them into loose and porous filter paper after certain pretreatment, and then impregnated the filter paper with water-based fluorocarbon emulsion, and the filter paper obtained after drying had good water resistance and fuel resistance [56]. Acrylic resin emulsion is a water-soluble system, which has many outstanding advantages, such as excellent weather resistance, ultraviolet light resistance, good heat resistance, corrosion resistance, excellent flexibility, strong adhesion, etc. Kong Shaoqi from Taiyuan University of Technology used synthetic acrylic resin and modified nano-SiO_2 particles to prepare a superhydrophobic coating and obtained the optimal coating process conditions [57]. This highly water-resistant composite coating is expected to be used for the impregnation of fuel filter paper. Shicheng Yan et al. modified acrylic monomers with vinyl triethoxysilane and octamethylcyclotetrasiloxane to obtain a resin with excellent water resistance and scrub resistance, which has low curing temperature requirements, room temperature is sufficient, and no additional heating and other curing operations are required [58]. This resin also has great potential for fuel filter paper applications.

6.4 Hydrophilic and Lipophilic Paper-based Functional Materials

6.4.1 Preparation

6.4.1.1 Physical Modification Method

Physical modification of fibers refers to the modification of the structure and morphology of fibers by external force without changing the chemical structure and composition of the fiber surface, thereby enhancing their compatibility with polymers [59–62]. Commonly used physical methods mainly include heat treatment, steam blasting treatment, and discharge treatment.

Because the cellulose of plant fiber itself contains a large number of hydroxyl groups, it has natural hygroscopic properties, and water exists in plant fibers in the form of free water and bound water, so it shows strong hydrophilicity, and its oil absorption performance is relatively weak. As a result, the fibers can be heat-treated to remove moisture from a part of their interior. Maosheng Jiang et al. used kenaf rod as raw material and prepared it into an oil-absorbing material by pyrolysis treatment, and analyzed the pyrolysate by elemental analysis, infrared spectroscopy, and scanning electron microscopy. The results showed that with the increase of temperature, the content of carbon in the material gradually increased, the content of oxygen gradually decreased, and the proportion of lipophilic aromatic ring structure increased significantly [63]. Through this experiment, the appropriate temperature can be adjusted to prepare paper-based materials with both water and oil absorption.

The steam blasting treatment method is that the water vapor enters the amorphous area of the fiber through the micropores on the surface of the plant fiber under high temperature and high pressure, and then the pressure decreases sharply, and the wood fiber expands, the fiber cell wall is broken, and the fiber morphology and structure change under the condition that the force changes violently instantaneously. The method does not require the addition of chemicals, is low-cost, and does not pollute the environment. Kotoro steam blasted bamboo fibers at 170 °C, 0.7 ~ 0.8 MPa, and formed composites with polylactic acid (PLA), and found that the bending strength, impact strength, and interfacial shear strength of the blasted composites were significantly improved except for partial tearing [64]. Inspired by this, hydrophilic and lipophilic paper-based functional materials can be prepared by adjusting the blasting parameters.

Discharge treatment is to activate the aldehyde group on the fiber surface under the action of an external electric field, so as to improve the oxidation activity of the fiber surface, and the surface energy of the fiber after discharge treatment can be changed, so as to improve the interface bonding ability between the fiber and the polymer matrix. Commonly used discharge methods include corona discharge, plasma treatment, and radiation discharge. Sang et al. used electron beams of different intensities to irradiate jute fibers and then form composites with PLA, and found that the interfacial shear strength of the radiated composites was higher than that of the unmodified composites [65]. This method is simple and easy to operate, but the equipment cost is high, and it is difficult to realize industrial applications. Park used plasma to treat polydioxanone fibers and prepare composites with PLA, and the results showed that the treatment time had a great impact on the properties of the composites, and the strength of the fibers could be improved, the surface polarity could be reduced, and the stress concentration could be reduced by low-temperature plasma modification, and the hydrophilic and lipophilic paper-based functional materials could be prepared [66].

6.4.1.2 Surface Coating Method

Surface coating is a simple method. Japanese Patent JP 2003-204824A discloses an oil-absorbing paper composed of plant fibers and a hydrocarbon resin derived from petroleum: the plant fiber slurry is first prepared into paper, and then a transparent agent (e.g., a hydrocarbon resin derived from petroleum) and an oil absorbent (e.g., silicate-based) are applied to the surface of the paper in the form of a printed pattern [67]. U.S. Patent US 4643939 discloses composite oil blotting paper made from hemp fibers in a certain proportion with polyolefin resin fibers. The composite oil blotting paper becomes transparent after oil absorption, and because it uses traditional papermaking technology, the prepared composite oil blotting paper has a rough feel, so that it can absorb water while absorbing oil [68].

Fu et al. reported a novel strategy to prepare a superamphiphilic fabric that is biphilic in both air and underwater environments. Using polyester fiber as the substrate, a specially designed cross-linkable hydrophilic and lipophilic functional polymer is used as the coating material, and the coating material is applied to the fabric matrix by a one-step wet chemical coating method. The contact angle of the treated fabric to water is 0°, and the water completely penetrates into the fabric within 1 second, and sixteen commonly used oils (surface tension of $18.4 \sim 50.8$ mN/m, immiscible with water) are selected for the superlipophilicity test. Regardless of whether the surface is dry (in air), premoistened with water, or immersed in water, the average contact angle of the oil is 0°. In water, the oil can be spread over fabrics that are completely wetted with water within 1 min [69].

Fu et al. prepared a novel underwater-oriented oil-transfer fabric using an impregnated coating method, in which a cross-linkable polymer composed of lipophilic and hydrophilic functional groups was applied to the fabric matrix, and then the fabric was coated with ultraviolet light on one side of the fabric. Polyester (PET) fabric is used as the model fabric backing. This treatment gives the fabric a super-friendly surface with a contact angle of 0° to both water and oil (surface tension of $18.4 \sim 72.8$ mN/m). The fabric exhibits almost identical surface wettability characteristics on both sides [70].

Makoto et al. prepared a thin layer of TiO_2 polycrystalline film on a glass substrate with anatase sol and annealed at 773 K to prepare a highly amphiphilic (hydrophilic and lipophilic) photogenerated TiO_2 surface [50]. The unique properties of this surface are attributed to the microstructural composition of the hydrophilic and lipophilic phases produced by UV radiation. UV irradiation may form surface oxygen vacancies at the bridging site, resulting in an associated conversion that favors the dissociation of water-adsorbed Ti^{4+} sites to Ti^{3+} sites. These defects may affect the affinity of the chemisorption of water at its surrounding sites, forming a hydrophilic region while the rest of the surface remains lipophilic, and the nanoscale separation between the hydrophilic and lipophilic phases is responsible for the high hydrophilicity of the TiO_2 surface. Because the droplet is much larger than the hydrophilic (or lipophilic) region, it immediately diffuses on such a surface, similar to the two-dimensional capillary phenomenon. The sun's ultraviolet exposure is sufficient to maintain the amphiphilic surface, so hydrophilic or lipophilic contaminants on the surface can be easily dispersed.

6.4.1.3 Chemical Modification Method

Compared with physical modification, chemical modification allows the functional groups to be chemically bonded to the surface of the paper base, so that the substance will not be diluted when it passes through the paper and will not cause the loss of functional groups, thus avoiding the defects caused by physical methods [71].

Chemical modification method to improve water absorption

Natural cellulose itself has a certain water absorption performance, but because of the highly dense crystalline structure, most of the hydroxyl groups are in a hydrogen bond association state, which limits its own water absorption ability. Therefore, people destroy this crystal structure by connecting a series of chemical reactions such as hydrophilic hydroxyl groups, carboxyl groups, and amide groups, and then ensure the gel strength after water absorption through moderate cross-linking.

1) Etherification and cross-linking

Although the etherification and cross-linking of cellulose can improve water absorption to a certain extent, the single effect is limited. Etherification alone is not enough; a certain amount of cross-linking is also required. The early cross-linking agents include formaldehyde, amino acids, citric acid, etc., among which the most widely used is epichlorohydrin cross-linking, first using sodium hydroxide/isopropanol to treat the sulfate slurry, and then etherifying the reaction with monochlorobutyrate, and then using epichlorohydrin cross-linking, the resulting product absorbs 0.9% sodium chloride solution with a rate of up to 121 g/g. The above operations belong to the method of first etherification and then cross-linking, and there are also reports of first cross-linking and then carboxymethylation (etherification reaction). It has been reported that carboxymethyl cellulose is cross-linked by epichlorohydrin and then freeze-dried, and the water absorption rate can reach 860 g/g.

2) Graft copolymerization

Graft copolymerization is another important method for the preparation of cellulose-based superabsorbent materials, which can be divided into direct grafting of cellulose and grafting after etherification of cellulose (hydroxyalkyl cellulose and carboxyalkyl cellulose). When grafting hydrophilic groups such as acrylonitrile, the commonly used cross-linker is *N*, *N*-methylenebis-acrylamide. A Japanese research team modified the multi-skin chain and prepared an intelligent multi-skin film. They first covalently bonded the pararosic and pararosic anthocyanin groups to the side chains of polyglutamic acid and then formed a film. In addition, they also formed the copolymer of leucine and glutamate esters and performed aminolysis to obtain an amphiphilic polydermal film. Grafting modification can be achieved by chemical grafting, photoinitiation grafting, plasma treatment, and other methods.

Cellulose can also be directly grafted with acrylic, using cotton fiber or microcrystalline cellulose as raw materials, which can obtain more than 100 times the water absorption rate. Compared with acrylonitrile, grafted acrylic acid has a strong water absorption capacity and a fast speed. Using cellulose as a slurry, adding appropriate cross-linker, emulsifier, and initiator, the superabsorbent material can be prepared by emulsion polymerization. There is a direct relationship between the water absorption rate and the grafting rate of cellulose grafted acrylic acid products, and the important factor affecting the grafting rate is the activation performance of cellulose, which includes oxidation, fibrillation, and radiation.

3) Cellulose is compounded with polymers

Studies have shown that cellulose and its derivatives can be combined with chitin, chitosan, starch, corn straw, and other cellulose-based absorbent materials for food, medical and health, daily necessities, and other fields. Cellulose and chitin were mixed in the total solution to prepare an effective phase-separated composite water-absorbing material, which could adsorb heavy metal ions in a low-concentration solution, and its adsorption capacity for heavy metals was greater than that of pure chitin tablets, which was mainly due to its good hydrophilicity and micro-network morphology. Semi-interpenetrating network technology is also an effective method for the preparation of cellulose-polymer composites. On the one hand, the applied cerium ammonium nitrate, wheat straw, potassium hydroxide, and acrylic acid endow the absorbent material with the ability to provide nitrogen and potassium for crops, and on the other hand, vinyl alcohol can improve the toughness of the absorbent material, so the product can be used in crops to improve the water retention performance of the soil. It has been reported that chitosan and carboxymethylcellulose solutions can be cross-linked with glutaraldehyde to generate an amphoteric water-absorbing material, and it can also be selected close to the anode or cathode by pH, a property that is affected by electric field forces and ionic strength [72].

4) Cellulose–inorganic complex

In recent years, cellulose–inorganic composite multifunctional materials have been widely used and have good application prospects in the fields of optics, electronics, and biology. Kaolin is a hydrophilic layered clay mineral material, which has a large surface area after deep processing and contains a large number of hydroxyl groups on the surface, so it can also interact with the hydrogen bonds or chemical bonds of polymers to form an organic–inorganic semi-interpenetrating three-dimensional network structure. Lin Songbai et al. used cerium amine nitrate as the initiator to composite carboxymethylcellulose grafted acrylamide copolymer (AM-g-CMC) with kaolin by the aqueous solution polymerization method to prepare superabsorbent materials, and the results showed that the ratio of acrylamide and carboxymethyl cellulose was 5 : 1, the initiator dosage was 1.6%, the cross-linker was 0.08%, the kaolin was 10%, and the superabsorbent material absorbed distilled water at a rate of 1182 g/g and a 0.9% sodium chloride solution at a rate of 92 g/g. The addition of kaolin helps to form cross-linking points, thus forming an effective network structure, which is conducive to the improvement of the strength of hydroabsorbent gels [73]. Montmorillonite is a layered aluminosilicate with active hydroxyl functional groups on the surface, good plane strength and stiffness, and high aspect ratio, so it is widely used in the study of superabsorbent materials [74].

1) Chemical modification method to improve oil absorption

Based on the large number of active hydroxyl groups in the molecular structure of cellulose and their reaction characteristics, alkali immersion, coupling agent treatment, and esterification grafting treatment were mainly used to change the chemical polarity and structure of the fibers, so as to improve the interfacial compatibility between the fibers and the polymer matrix, and ultimately improve the overall performance of the composites [75].

a) Lye treatment

Lye treatment is a relatively common and easy-to-operate method for fiber modification. By controlling the concentration of the lye, the components of the fiber soaked in the lye, including hemicellulose, lignin, pectin, and other impurities, were removed by chemical etching. After lye etching, the fiber surface will become rough, and many furrows will appear. The lye can weaken the hydrogen bonding between cellulose molecules, so that the distance between cellulose molecules will become larger, and various reagents can more easily immerse between the internal molecular structures of fibers.

b) Coupling agent modification

The surface energy of fibers is closely related to their natural hydrophilicity, and some studies have focused on ways to reduce their hydrophilicity. The silane coupling agent molecule contains two chemical groups with different polarities, which can act as a bridge between the hydrophilic fiber and the hydrophobic polymer, and effectively enhance the interfacial bonding performance between the fiber and the polymer.

c) Esterification modification

Esterification treatment includes acetylation and alkylation of fibers, which can enhance the interfacial compatibility of fibers and polymer matrices by changing the chemical composition and polarity of the fiber surface, thereby improving the overall properties of composite materials. A large number of hydroxyl groups contained on the surface of the fiber are easy to react with esterification reagents, which leads to the reduction of the number of hydroxyl groups on the fiber surface, reduces the probability of hydrogen bond formation between the hydroxyl groups, changes the polarity of the fiber surface, and weakens the hydrophilicity of the fiber. Because wood pulp fiber contains more lignin and hemicellulose, it has a high degree of esterification reaction. The modification of the fiber by esterification reaction, the esterification group exists on the fiber surface, the chemical composition of the fiber surface changes, and the crystallinity of the fiber decreases slightly due to the esterification reaction. It has been reported that cellulose higher fatty acid esters (above C_4) prepared by esterification have excellent nonpolar solvent solubility and great compatibility with hydrophobic polymers, which provides a basis for the development of preserved oil-absorbing materials.

d) Graft modification

Graft modification is an effective method to enhance the interfacial adhesion between fibers and polymer matrices. First, the fibers are pretreated to produce a living point on their surface, and then the polymer macromolecules are grafted onto the fibers under certain reaction conditions. Due to the large number of hydroxyl groups on the surface of the fiber, it is generally used as an activation point to improve the grafting rate. In recent years, graft copolymerization has been used to make fibers graft hydrophobic groups, making them hydrophobic and lipophilic.

1) Chemical modification method to improve both water and oil absorption

The chemical modification method to improve amphiphilicity is to use the interaction between ions and the fiber surface to introduce new groups on the fiber surface, which significantly changes the affinity/hydrophobicity, adhesion, printing, and dyeing properties of the fiber surface, but because it does not change the internal structure of the material, it can give the fiber surface excellent wetting properties without changing the properties of the fiber body. The surface of the modified fiber is grafted with hydroxyl (—OH) and aldehyde groups (—CHO) and carbonyl (C=O) and other hydrophilic functional groups, and the surface morphology of the fiber has changed to a certain extent, and the surface microroughness has increased. These changes ultimately result in an effective improvement in the wettability of the fiber surface.

6.4.2 Application

6.4.2.1 Industrial Wiping Paper

Industrial wipes are generally used in various industries such as industrial electronic instruments, printing, and precision machinery. In recent years, due to the development of China's industry and automobile manufacturing industry, industrial wipes have been rapidly promoted and applied in the machinery manufacturing industry. Some domestic enterprises have also started the production of industrial wiping paper, most of which use wood pulp and polyester fiber composite, then use the preparation process of paper—spunlace composite forming, and then, after fine processing, finally form industrial water and oil absorption wiping paper. Although the prepared products have many excellent properties, they cannot avoid the high cost of raw materials, the complex preparation process, and the high cost of preparation. Therefore, people began to use 100% virgin wood pulp as raw material and the wet paper machine manufacturing process, but the product has low wet strength, is easy to retain dust when used, has poor absorption performance and poor quality, and is easy to damage the wiped surface when used.

In view of the above problems, an invention patent application in China in 2007 disclosed a manufacturing method for optical lens wipes [76]. This method produces delicate, soft, high-quality, low-cost, and nonpolluting industrial wiping paper (made from wood pulp, made by a wet paper machine, and wrinkled by a dry process), which is particularly suitable for use in machinery manufacturing industries such as automobiles. And the application of the invention patent to prepare industrial wiping paper has the following advantages:

1) Using the traditional wet paper machine, it can be produced through process and equipment adjustment, with small equipment investment, easy control of the production process, and low production cost.
2) Made of 100% virgin wood pulp, the industrial wiping paper made of paper is smooth and delicate, will not scratch the surface of the swab, the fiber is densely combined, and the wiping does not leave dust, has good liquid absorption performance and decontamination ability, high wet strength, and is not easy to break after wiping.
3) The product is 100% recycled fiber, with no pollution.

6.4.2.2 Kitchen Absorbent Oil-absorbing Paper

With the rapid development of the economy and the improvement of people's living standards, people's demand and requirements for kitchen paper are getting higher and higher. Compared with traditional dishwashing towels and rags, kitchen washing oil blotting paper has a strong water absorption ability, no fragile, no slag, no residual water stains, etc., and also has a super oil absorption ability, which can clean kitchen utensils, remove oil stains, absorb excess oil from food, etc., and one-time use is to give it convenience, cleaning, and comfort. However, the disposable kitchen paper on the market is mostly made of polyester fiber as raw material and processed by spunlace composite technology, and the cost of raw materials is high and the preparation process is complex, which affects the popularization and application of disposable kitchen paper, so it is of great significance to develop a disposable kitchen paper with low cost and good water and oil absorption effect.

The research group of Wang Haisong of Dalian Polytechnic University used bleached softwood kraft pulp as the main raw material, and explored the production process of water-absorbing oil-absorbing paper for the kitchen on the ZQJ2 paper sheet former by compounding cotton pulp, synthetic fiber, absorbent resin, etc. Under the condition of bleached softwood kraft pulp (71%), cotton pulp (10%), synthetic fiber (15%), and absorbent resin (4%), the hand-copied sheets prepared by adding 1.5% wet-strength agent and 0.1% cationic polyacrylamide can replace disposable kitchen paper produced by spunlace composite technology using polyester fiber as raw material on the market [77]. In today's increasing demand for disposable kitchen paper, the use of wet forming technology to manufacture kitchen paper towels with high water and oil absorption capacity at low cost, instead of disposable kitchen paper produced by spunlace composite technology with polyester fiber as raw material on the market, is of great significance for improving people's quality of life and promoting the transformation and upgrading of the traditional paper industry.

6.4.2.3 Humidity Regulation Cling Paper

Superabsorbent paper has been widely used in the packaging, storage, and transportation of vegetables, fish, meat, and fruits. For small packaging, the plastic chassis is generally used, and several layers of superabsorbent paper are placed in the tray first, and then washed vegetables, fish, fruits, etc., and finally wrapped in transparent film. These superabsorbent papers can not only absorb excess water, oil, residual blood, etc., but also maintain proper humidity. This packaged item is able to remain moist and fresh even when stored in the refrigerator for long periods of time [78].

6.4.2.4 Medical Absorbent Paper

When autoclaving, packaging is an important part of ensuring successful sterilization. Improper packaging can produce wet packs, which can lead to sterilization failure. Wet packs appear in sterilization, most of which are related to condensation. That is, after a sterilization cycle and proper cooling, if there are visible or hidden moisture, water droplets, and water columns on the surface or inside the package, it is considered to be a wet pack. The wet bag can be effectively controlled by strictly abiding by the packaging specifications, controlling the operation of disinfection personnel, correctly loading and unloading the sterilization bag, checking the steam quality, prolonging the drying time, etc. At present, most hospitals use double-layer cotton cloth to absorb moisture in the sterilization instrument package, and there are many deficiencies such as incomplete moisture absorption, residual flocculents, and inconvenient management for repeated use. Therefore, Luo Yi et al. verified the efficacy of medical absorbent paper by comparing the two packaging methods of paper-based composite materials prepared from 100% virgin wood pulp and double-layer cotton cloth on the number and rate of wet bags in the bag, and found that the medical absorbent paper has the characteristics of high efficiency and water absorption, which can absorb more than 400% of its own mass. In addition, the medical absorbent paper is in close contact with the instrument, and its flocculent residue is almost zero compared with the cloth by the naked eye, although it does not directly come in contact with the patient, but the amount of fiber desquamation is small, which can avoid the instrument from being contaminated with fine fibers and foreign bodies, causing aseptic inflammation of the wound and prolonging the healing time, granulation hyperplasia, and other drawbacks. Medical absorbent paper has excellent properties such as physical and chemical properties, and color does not change after high temperature and high pressure, sufficient water absorption, antibacterial, and no lint.

6.4.2.5 Facial Oil Blotting Paper

In an environment with relatively high temperature and humidity, people's faces, especially the forehead and nose, are more likely to secrete oil, which brings inconvenience to people's lives and becomes a problem that plagues everyone. The emergence of oil blotting paper effectively solves people's troubles about this kind of problem, easy to carry, as long as you gently press and wipe to keep the facial skin fresh, become a popular product. Oil blotting paper first appeared in Japan. With the progress of society and the improvement of people's living standards, oil blotting paper has become a commonly used facial oil blotting wipe.

Li Jie [79] used hemp fiber and wood fiber and mixed the two in a certain proportion to prepare oil blotting paper. All materials are natural fibers, and the oil blotting paper made of oil blotting paper has good oil absorption, softness, and toughness, and can effectively absorb excess oil on the face. Japanese Patent JP 6-319664 discloses an oil blotting paper made from a mixture of plant fibers and inorganic fillers [79]. Another Japanese patent, JP 4-45 591, first uses plant fibers to make paper sheets, and then adheres porous spherical beads to the surface of the paper to solve the problems caused by calendering or coating the surface of the paper with powders such as calcium carbonate, which are thought to improve the absorption capacity of oil-absorbing paper [80].

6.5 Hydrophilic and Oleophobic Paper-based Functional Materials

6.5.1 Preparation

Materials with both superhydrophilic and superoleophobic properties in gas are ideal materials for oil–water separation, but the classical surface free energy theory considers superhydrophilic-superoleophobic to be an "abnormal" wettability, so the research on this type of surface has been progressing slowly, and it has not been successful until a small number of researchers have succeeded in recent years, among which there are few reports on hydrophilic and oleophobic paper-based functional materials.

Okada et al. used fluoroalkyl acrylic acid oligomers (FAAO) to prepare a surface with a contact angle greater than water for the first time and explained this "abnormal" wettability as a kind of material composed of molecules with two different wettability groups. When in contact with water, under the attraction of water, the hydrophilic group flips to the surface of the material and makes the material show hydrophilicity, but this flipping phenomenon does not occur when the material is in contact with the oil, so the contact angle of the oil is larger, that is, "Flip-Flop" theory, also known as surface remodeling or surface recombination. Some of the subsequent superhydrophilic/superoleophobic surfaces have been explained by this theory [81]. Yang et al. successively prepared superhydrophilic and superoleophobic surfaces by adsorbing anions of sodium perfluorooctanoate using PDDA and chitosan, and successfully applied them to oil–water separation [82].

6.5.2 Application

6.5.2.1 Grease-proof Paper

As one of the main pillars of modern packaging materials, paper has the advantages of a wide source of raw materials and a relatively cheap price like plastic, is light and easy to carry, and has excellent printing adaptability and easy transportation compared with glass and metal, so it has been widely used in daily life and industrial production. As we all know, packaging materials with different functions can be obtained due to the nature of the material, the manufacturing process, and the use of the product, such as moisture-proof packaging materials, mildew-proof packaging materials, oil-proof packaging materials, etc. In recent years, with the rapid development of the economy, the application of paper packaging materials in transportation packaging and food packaging has been increasing, and paper packaging materials have become the largest packaging material.

In order to achieve the effect of oil and oil resistance, people will use plastic packaging materials such as polyethylene, polypropylene, and polyvinyl chloride to make films, but these film materials are difficult to degrade or degrade very slowly in nature. Despite their ability to resist the penetration of grease, the waste from these plastic packaging materials is difficult to recycle and causes great pollution to the human environment. Easily degradable polymer materials and paper products can replace the packaging materials that are difficult to degrade in the natural environment, among which the easily degradable polymer materials can be divided into three types: one is the polymer prepared by polymerization by bioengineering methods, which can be gradually degraded into small molecule carbohydrates in the natural environment; the second is to add starch to the polymer resin, using the easy degradation of starch, this packaging material under the action of rain or microorganisms, due to the decay of starch and the packaging material is split into fine particles, but these fine particles remain in the soil, will not decompose for many years, and do not really solve the problem of environmental pollution; the third is a double-degradable plastic, which is a catalyst that can be photodegraded by adding to the polymer resin, through the decomposition of microorganisms and the photosynthesis of plants, the polymer chain of the polymer material is broken and gradually degraded into small molecule compounds, but it is still in the stage of scientific research and experiments, and further investigation is needed. The paper material is different; it is easy to degrade, environmentally protected, and recyclable. The paper material after use does not easily cause pollution to the environment, and the raw material source is wide, the price is cheap, which can be accepted by the majority of consumers, and can be applied to a variety of printing technologies. However, due to the hydroxyl group of the paper fiber itself and the porous structure of the paper, the ordinary paper packaging material does not have oil-proof performance, and must be treated with oil-proof treatment in order to meet the requirements of oil-proof packaging.

6.5.2.2 Oil Repellent

Oil-repellent agents that make paper oil-repellent are broadly divided into two categories: fluorine-containing and non-fluorinated. The surface of a solid is inevitably in contact with the outside world, and wetting is a common interfacial phenomenon when in contact with a liquid, so surface tension becomes an important parameter of the solid surface. The oil repellent is mainly used to improve the surface tension of the paper, so that the paper is not wetted by oil during use. Whether or not the paper can be wetted by oil depends on the interfacial tension between the surface of the paper and the oil droplets. When the tension on the surface of the paper is large enough, the interfacial tension between the paper surface and the oil droplets is greater than the interfacial tension between the paper surface and the air, and the oil droplets cannot wet the paper. Conversely, if the surface tension of the paper is not large enough and the interfacial tension between the paper and the oil droplets is less than the interfacial tension between the paper and the air, the paper will be wetted by the oil droplets.

6.5.2.3 Greaseproof Paper for Food Packaging

In order to obtain oil repellency, synthetic polymers such as polyethylene (PE) and polypropylene (PP) were used to coat the paper to give the paper certain oil repellency, but after a large amount of use, they will cause environmental pollution. At present, fluorine-containing compounds are commonly used in the industry to treat paper, and the oil-proof paper obtained has good oil-proofness, heat resistance, weather resistance, and air tightness, but recent studies have found that fluorine-containing compounds pose a hidden danger to human health. The commonly used oil repellent GPI produces PFOS when exposed to heat, which has been shown to be toxic and may cause liver cancer, disorder fat metabolism in the human body, delay the normal ossification of children, and bring harm to human health. In order to cope with severe environmental problems and food safety issues, food packaging greaseproof paper based on safe, degradable, and renewable materials has attracted the interest of many scholars.

Kjellgren and Giillstedt studied the oil repellent effect of chitosan on paper and compared the oil repellent properties of base paper with different air permeability after coating. The results show that when the air permeability is less than 0.001 nm/(Pa·s), the oil repellency of the grease repellent paper is good with the increase of the coating amount, but when the coating weight is less than 2.49 g/m^2, the oil repellent effect of the paper with higher air permeability is not good, and only the oil repellent paper with lower air permeability has better oil repellent performance [83].

U.S. Patent US2008/0171213A1 proposes a method to improve the oil repellent properties of grease repellent paper by changing the number of coating layers when the coating amount is fixed. The specific method is to pre-coat the surface of the paper with a primer coating, the main component of which is a starch derivative modified with $C_2 \sim C_6$ alkyl oxide (amylose content greater than 70%), and glycerin, borax, urea, etc. can be added to the coating to increase the elasticity of the coating and improve the oil repellency of the grease-proof paper [84].

Pichavant et al. used chitosan instead of fluorocarbons to study the oil repellency of coated paper in detail. When only chitosan (concentration 3%, coating amount 2.2%) is used to coat the base paper, the oil-proof effect of the paper is comparable to that of fluorocarbon oil repellent, but at this time, the concentration of chitosan oil repellent is higher, and the viscosity of the oil repellent is relatively large, which is not conducive to actual production, and the cost of fluorocarbon oil repellent is higher. In order to reduce the production cost, the mixed use of natural polymer materials and chitosan was studied, and the addition of carboxymethyl cellulose to the chitosan solution had no effect on the oil-proof performance of the paper, while the oil-proof effect of the oil-proof paper was comparable to that of fluorocarbons and fluorocarbons when mixed with sodium alginate, but there were also defects of excessive viscosity, which was not conducive to actual production [72].

Park et al. studied the oil repellency and mechanical properties of coated paper by using isolated soy protein for paper coating, and found that the mechanical properties of the paper changed when a mixture of glycerol and polyethylene glycol was added to the solution [85]. Trezza and Vergano studied zein-coated paper and found that the coated paper had excellent oil repellency and was comparable to that of some coated papers [86]. Butkinareel et al. mixed stearic acid with a hydrophobic starch solution for paper coating, and the contact angle of the grease on the paper surface was increased, and the oil repellency was better [87]. There are also researchers who use corn gluten, dispersed whey protein, concentrated whey protein, and carrageenan and other natural polymer materials for paper oil repellency research, and the research shows that many natural polymer materials used in paper coating can make the paper obtain oil repellency, but there are still differences in oil repellency between them.

Similarly, studies have shown that MFC, a nano-eco-friendly material, can also be used for paper coating to improve the oil-repellent properties of the paper. For example, Aulin et al. found that MFCs not only have good gas barrier properties, but also have good oil repellent properties in their barrier properties on carboxymethylated MFCs. They applied the paper with MFC, and then studied the penetration time of castor oil and turpentine on the coated paper according to the Tappi T-454 standard, and concluded that the oil resistance of the paper was enhanced when the air permeability of the paper decreased, and analyzed that the enhanced oil repellency was due to the formation of a continuous dense film on the surface of the paper by carboxymethylated MFC after coating, which sealed the paper voids [88]. Lavoine et al. used an enzymatically treated MFC suspension for paper coating to study the effects of the two coating methods on the mechanical and barrier properties of the paper, and experimentally verified that MFC has oil repellency, and the higher the coating amount, the better the oil repellency [89].

6.5.2.4 Oil–water Separation Membrane

A large amount of oily sewage is generated in industrial production and people's daily lives. At present, the treatment of oily wastewater has always been a worldwide problem, especially the treatment of emulsified oily wastewater in complex environments. The use of membrane separation technology to achieve oil–water separation is considered to be one of the most effective separation methods. Wu Weibing et al. developed a preparation method for paper-based functional materials for oil–water separation. Lignocellulose was used as raw material, citric acid as cross-linking and functionalization reagent, and sodium hypophosphite as catalyst, and paper-making, impregnation and drying, and maturation processes were used to prepare paper-based functional materials with good wet strength and oil–water separation properties. The paper-based functional material has a wet tensile index of up to 18 N·m/g, and the contact angle of oil droplets such as chloroform and castor oil can reach more than 150° under water, and it has underwater superoleophobic properties, which can efficiently filter mixtures of organic solvents such as chloroform, castor oil, toluene, and ethyl acetate with water, and the oil–water separation efficiency can reach more than 99% [90].

6.6 Prospect

With the development of modern science and technology, the demand for functional materials is getting higher and higher, but most of the functional materials have the shortcomings of high cost and complex production process, which can no longer meet the increasing application needs. Paper-based functional materials are of great significance to the national economy, civilization development, and national construction. It includes contemporary high-tech and related industries such as bioengineering technology, energy development methods, nanotechnology, environmental protection technology, space technology, computer technology, marine engineering science and technology, etc., which not only plays an important role in the promotion of high and new technology, but also plays a significant role in promoting the improvement of relevant traditional technologies in China and the realization of cross-era development.

At present, in the practical application process, the application of paper-based functional materials with special wettability is still limited by some factors, specifically in the following aspects: on the one hand, the existing methods mostly involve expensive instruments and equipment or complex technological processes, it is difficult to realize the preparation of large areas, and some materials are easily affected by external conditions, can not meet the requirements of long-term use, for example, the high surface energy of superhydrophilic materials is easy to be converted to the direction of low surface energy to achieve a stable state, and this conversion process will cause it to lose its superhydrophilic properties, and its potential application field is still in the exploration stage; on the other hand, the modification of some materials is inseparable from fluoride. This leads to a significant reduction in the surface properties of the material, so the wear resistance and permeability still need to be further improved. In addition, the preparation method is mostly carried out in air, which limits its further application, and the material exhibits different wetting properties due to the different media of water and air. The researchers found that bird feathers have good hydrophobic properties, but they are not immune to marine oil pollution, while fish scales are resistant to oil pollution and have superhydrophilic-superoleophobic properties in the air. Therefore, it is of great practical significance to study superhydrophilic-superoleophobic surfaces.

Overall, while some studies have made breakthroughs in the preparation of superhydrophilic-superoleophobic surfaces, there is no experimentally proven theoretical explanation for the formation of superhydrophilic-superoleophobic "anomalous" wettability. However, the lack of a unified theory leads to the lack of direction for the preparation of superhydrophilic-superoleophobic surfaces, and the corresponding research also stays aimlessly trying different preparation materials and preparation processes, and the development is very slow. At the same time, some problems, such as short service life and low mechanical strength of separation membranes, have not been effectively solved in the application of superhydrophilic-superoleophobic surfaces in oil–water separation. Therefore, it is of great significance to find a theoretical model that can be verified by experiments and guide surface preparation to promote the development of superhydrophilic-superoleophobic surfaces and promote their practical application in oil–water separation.

References

1 Liu, H., Wang, Y.D., Huang, J.Y. et al. (2018). Bioinspired surfaces with superamphiphobic properties: concepts, synthesis, and applications. *Advanced Functional Materials* 28 (19): 1707415.

2 Ganesh, V., Raut, H., Nair, A. et al. (2011). A review on self-cleaning coatings. *Journal of Materials Chemistry* 21 (41): 16304–16322.

3 Du, C., Wang, J., Chen, Z. et al. (2014). Durable superhydrophobic and superoleophilic filter paper for oil-water separation prepared by a colloidal deposition method. *Applied Surface Science* 313 (9): 304–310.

4 Qu, M.N., Ma, X.R., He, J.M. et al. (2017). Facile selective and diverse fabrication of superhydrophobic, superoleophobic- superhydrophilic and superamphiphobic materials from Kaolin. *ACS Applied Materials & Interfaces* 9 (1): 1011–1020.

5 Piltan, S., Seyfi, J., Hejazi, I. et al. (2016). Superhydrophobic filter paper via an improved phase separation process for oil/water separation: study on surface morphology, composition and wettability. *Cellulose* 23 (6): 3913–3924.

6 Wang, J., Geng, G., Liu, X. et al. (2016). Magnetically superhydrophobic kapok fiber for selective sorption and continuous separation of oil from water. *Chemical Engineering Research & Design* 115 (11): 122–130.

7 Ni, S.Z., Wang, C.J., and Dai, H.Q. (2017). Research progress on cellulose-based superhydrophobic paper. *Cellulose Science and Technology* 25 (2): 58–68.

8 Su, B., Tian, Y., Jiang, L. et al. (2016). Bioinspired interfaces with superwettability: from materials to chemistry. *Journal of the American Chemical Society* 138 (6): 1727–1748.

9 Artus, G., Zimmermann, J., Reifler, F. et al. (2012). A superoleophobic textile repellent towards impacting drops of alkanes. *Applied Surface Science* 258 (8): 3835–3840.

10 Brown, P. and Bhushan, B. (2016). Durable, superoleophobic polymer–nanoparticle composite surfaces with re-entrant geometry via solvent-induced phase transformation. *Scientific Reports* 6: 21048.

11 Tuteja, A., Choi, W., Mabry, M. et al. (2008). Robust omniphobic surfaces. *Proceedings of the National Academy of Sciences* 105 (47): 18200–18205.

12 Im, M., Im, H., Lee, J.H. et al. (2010). A robust superhydrophobic and superoleophobic surface with inverse-trapezoidal microstructures on a large transparent flexible substrate. *Soft Matter* 6 (7): 1401–1404.

13 Mirvakili, M., Hatzikiriakos, G., and Englezos, P. (2013). Superhydrophobic lignocellulosic wood fiber/mineral networks. *ACS Applied Materials & Interfaces* 5 (18): 9057–9066.

14 Jiang, L., Tang, Z.G., Clinton, M.R. et al. (2016). Fabrication of highly amphiphobic paper using pulpdebonder. *Cellulose* 23: 3885–3899.
15 Zhang, J., Wang, H., Liu, M. et al. (2016). Controlling the evaporation lifetimes of sessile droplets on superhydrophobic paper by simple stretching. *RSC Advances* 6 (16): 12862–12867.
16 Li, L., Breedveld, V., and Hess, D.W. (2013). Design and fabrication of superamphiphobic paper surfaces. *ACS Applied Materials & Interfaces* 5 (11): 5381–5386.
17 Jiang, L., Tang, Z.G., Clinton, M.R. et al. (2017). Two-step process to create "Roll-Off" superamphiphobic paper surfaces. *ACS Applied Materials & Interfaces* 9 (10): 9195–9203.
18 Arbatan, T., Zhang, L., Fang, X.Y. et al. (2012). Cellulose nanofibers as binder for fabrication of superhydrophobic paper. *Chemical Engineering Journal* 210 (6): 74–79.
19 Mertaniemi, H., Laukkanen, A., Teirfolk, J.E. et al. (2012). Functionalized porous microparticles of nanofibrillated cellulose for biomimetic hierarchically structured superhydrophobic surfaces. *RSC Advances* 2 (7): 2882–2886.
20 Yun, T.T., Wang, Y.L., Lu, J. et al. (2020). Facile fabrication of cellulosic paper-based composites with temperature-controlled hydrophobicity and excellent mechanical strength. *Paper and Biomaterials* 5 (2): 20–27.
21 Qu, M.N., Liu, L.L., Liu, Q. et al. (2021). Highly stable superamphiphobic material with ethanoltriggered switchable wettability for high-efficiency on-demand oil–water separation. *Journal of Materials Science* 56: 2961–2978.
22 Li, J., Yan, L., Ouyang, Q.L. et al. (2014). Facile fabrication of translucent superamphiphobic coating on paper to prevent liquid pollution. *Chemical Engineering Journal* 246: 238–243.
23 Tang, Z., Li, H., Hess, D.W. et al. (2016). Effect of chain length on the wetting properties of alkyltrichlorosilane coated cellulose-based paper. *Cellulose* 23 (2): 1401–1413.
24 Huang, W., Xing, Y., Yu, Y. et al. (2011). Enhanced washing durability of hydrophobic coating on cellulose fabric using polycarboxylic acids. *Applied Surface Science* 257 (9): 4443–4448.
25 Liu, J., Huang, W., Xing, Y. et al. (2011). Preparation of durable superhydrophobic surface by sol-gel method with water glass and citric acid. *Journal of Sol-Gel Science and Technology* 58 (1): 18–23.
26 Cunha, A.G., Freire, C.S.R., Silvestre, A.J.D. et al. (2010). Preparation and characterization of novel highly omniphobic cellulose fibers organic-inorganic hybrid materials. *Carbohydrate Polymers* 80 (4): 1048–1056.
27 Liu, K.F., Li, P.P., Zhang, Y.P. et al. (2019). Laboratory filter paper from superhydrophobic to quasisuperamphiphobicity: facile fabrication, simplified patterning and smart application. *Cellulose* 26: 3859–3872.
28 Zhang, J., Feng, H., Zao, W. et al. (2014). Flexible superhydrophobic paper with a large and stable floating capacity. *RSC Advances* 4 (89): 48443–48448.
29 Zhang, W., Peng, L., Qian, L. et al. (2014). Fabrication of superhydrophobic paper surface via wax mixture coating. *Chemical Engineering Journal* 250 (6): 431–436.

30 Stanssens, D., Abbeele, H.V.D., Vonck, L. et al. (2011). Creating water-repellent and super-hydrophobic cellulose substrates by deposition of organic nanoparticles. *Materials Letters* 65 (12): 1781–1784.

31 Lee, W. and Ahn, Y. (2016). Spray coating of hydrophobic iron fatty acids/PS composite solutions for the preparation of superhydrophobicpaper. *Bulletin of the Korean Chemical Society* 37 (11): 1862–1865.

32 Samyn, P., Schoukens, G., Stanssens, D. et al. (2012). Incorporating different vegetable oils into an aqueous dispersion of hybrid organic nanoparticles. *Journal of Nanoparticle Research* 14 (8): 1–24.

33 Fragouli, D., Bayer, I.S., Corato, R.D. et al. (2012). Superparamagnetic cellulose fiber networks via nanocomposite functionalization. *Jmaterchem* 22 (4): 1662–1666.

34 Ai, J.X. and Guo, Z.G. (2020). Facile preparation of superamphiphobic fabric coating withhierarchical TiO_2 particles. *New Journal of Chemistry* 44: 19192–19200.

35 Su, C. and Li, J. (2010). The friction property of super-hydrophobic cotton textiles. *Applied Surface Science* 256 (13): 4220–4225.

36 Gustafsson, E., Larsson, P.A., and Warburg, L. (2012). Treatment of cellulose fibres with polyelectrolytes and wax colloids to create tailored highly hydrophobic fibrous networks. *Colloids & Surfaces A: Physicochemical & Engineering Aspects* 414 (46): 415–421.

37 Peng, L., Meng, Y., and Li, H. (2016). Facile fabrication of superhydrophobic paper with improved physical strength by a novel layer-by-layer assembly of polyelectrolytes and lignosulfonates-amine. *Cellulose* 23 (3): 1–13.

38 Li, H., Yang, J., Li, P. et al. (2016). A facile method for preparation superhydrophobic paper with enhanced physical strength andmoisture-proofing property. *Carbohydrate Polymers* 160 (6): 9–17.

39 Joao, F., Obeso, C.G. et al. (2013). Modification of paper using polyhydroxybutyrate to obtain biomimetic superhydrophobic substrates. *Colloids & Surfaces A: Physicochemical & Engineering Aspects* 416 (1): 51–55.

40 Ge, D.T., Yang, L.L., Wang, C.B. et al. (2015). A multi-functional oil–water separator from aselectively pre-wettedsuperamphiphobic paper. *Chemical Communication* 51 (28): 6149–6152.

41 Zhu, T.X., Li, S.H., Huang, J.Y. et al. (2017). Rational design of multi-layered superhydrophobic coating on cotton fabrics for UV shielding, self-cleaning and oil-water separation. *Materials Design* 134: 342–351.

42 Geissler, A., Loyal, F., Biesalski, M. et al. (2014). Thermo-responsive superhydrophobic paper using nanostructured cellulose stearoyl ester. *Cellulose* 21 (1): 357–366.

43 Martinez, A.W., Phillips, S.T., and Whitesides, G.M. (2008). From the cover: three-dimensional microfluidic devices fabricated in layeredpaper and tape. *Proceedings of the National Academy of Sciences of the United States of America* 105 (50): 19606–19611.

44 Songok, J. and Toivakka, M. (2016). Enhancing capillary driven flow for paper-based microfluidic channels. *ACS Applied Materials & Interfaces* 8 (44): 30523–30530.

45 Zhao, D.D., Pan, M.W., Yuan, J.F. et al. (2020). A waterborne coating for robust superamphiphobic surfaces. *Progress in Organic Coatings* 138: 105368.

46 Stepien, M., Saarinen, J.J., Teisala, H. et al. (2013). ToF-SIMS analysis of UV-switchable TiO_2-nanoparticle-coated paper surface. *Langmuir the ACS Journal of Surfaces & Colloids* 29 (11): 3780–3790.

47 Yin, Y., Ning, G., Wang, C. et al. (2014). Alterable superhydrophobic-superhydrophilic wettability of fabric substrates decoratedwith ion-TiO_2 coating via ultraviolet radiation. *Industrial & Engineering Chemistry Research* 53 (37): 14322–14328.

48 Xu, Z., Zhao, Y., Wang, H. et al. (2016). Fluorine-free superhydrophobic coatings with pH-induced wettability transition for controllable oil-water separation. *ACS Applied Materials & Interfaces* 8 (8): 5661–5667.

49 Xu, Z., Jiang, D., Wei, Z. et al. (2018). Fabrication of superhydrophobic nano -aluminum films on stainless steel meshes by electrophoretic deposition for oil-water separation. *Applied Surface Science* 427: 253–261.

50 Wang, S.H., Li, M., Lu, Q.H. et al. (2010). Filter paper with selective absorption and separation of liquids that differ in surface tension. *ACS Applied Materials & Interfaces* 2 (3): 677–683.

51 Huang, X.X. (2012). *Preparation of Superhydrophobic/Superlipophilic Aqueous Epoxy Resin Emulsion Coating and Its Application in Oil-Water Separation Filter Paper*. Guangzhou: South China University of Technology.

52 Wang, Y.T. (2017). *Preparation of Hydrophobic and Oil-Hydrophilic Paper-Based Composites and Study on Oil-Water Separation Characteristics*. Guangzhou: South China University of Technology.

53 Sun, T.L., Feng, L., and Gao, X.F. (2005). Bioinspired surfaces with special wettability. *Account Chemical Research.* 38 (8): 644–652.

54 Adebajo, M.O., Frost, R.L., and Kloprogge, J.T. (2003). Porous materials for oil spill cleanup: a review of synthesis and absorbing properties. *Journal of Porous Materials* 10 (3): 159–170.

55 Cao, C. and Cheng, J. (2018). A novel Cu (OH) 2 coated filter paper with superhydrophobicity for the efficient separation of water-in-oil emulsions. *Materials Letters* 217: 5–8.

56 Jin X. Fuel filter filter paper with good safety, environmental protection and stability and preparation method thereof. CN 201410170104. 7. 2014-08-27.

57 Kong, S.Q., Yan, G.C., Yue, B. et al. (2014). Research on new acrylic resin superhydrophobic materials. *Journal of Taiyuan University of Technology* 1: 71–75.

58 Yan, S.C., Shi, B.Z., Zheng, Y. et al. (2013). Synthesis process of silicone-modified acrylic resin. *Elastomer* 23 (4): 43–46.

59 Aguilera, F., Méndez, J., Pásaro, E. et al. (2010). Review on the effects of exposure to spilled oils onhuman health. *Journal of Applied Toxicology: An International Journal* 30 (4): 291–301.

60 Zhu, H. and Guo, Z. (2016). Understanding the separations of oil/water mixtures from immiscible toemulsions on super- wettable surfaces. *Journal of Bionic Engineering* 13 (1): 1–29.

61 George, S., Ott, A., and Klaus, J. (1996). Surface chemistry for atomic layer growth. *The Journal ofPhysical Chemistry* 100 (31): 13121–13131.

62 Suntola, T. (1989). Atomic layer epitaxy. *Materials Science Reports* 4 (5): 261–312.

63 Riatla, M. and Leskelä, M. (1999). Atomic layer epitaxy-a valuable tool for nanotechnology. *Nanotechnology* 10 (1): 19–24.

64 Kim, S.W., Han, T.H., Kim, J. et al. (2009). Fabrication and electrochemical characterization of TiO_2 three-dimensional nanonetwork based on peptide assembly. *ACS Nano* 3 (5): 1085–1090.

65 Xia, L., Li, C., Zhou, S. et al. (2019). Utilization of waste leather powders for highly effective removal of dyes from water. *Polymers* 11 (11): 1786.

66 Jiang, M.S., Huang, B., and Zhou, H.H. (2009). The formation mechanism of high oil absorption characteristics of red hemp stem pyrolyze. *Chinese Hemp Science* 31 (2): 143–147.

67 Tokoro, R., Vu, D.M., Okubo, K. et al. (2008). How to improve mechanical properties of polylacticacid with bamboo fibers. *Journal of Materials Science* 43 (2): 775–787.

68 Ji, S.G., Cho, D., Park, W.H. et al. (2010). Electron beam effect on the tensile properties and topology of jute fibers and the interfacial strength of jute-PLA green composites. *Macromolecular Research* 18 (9): 919–922.

69 Park, J., Kim, D., and Kim, S. (2004). Nondestructive evaluation of interfacial damage properties for plasma-treated biodegradable poly (*p*-Dioxanone) fiber/poly (L-Lactide) composites by micromechanical test and surface wettability. *Composites Science and Technology* 64 (6): 847–860.

70 Luo, Y. and Shu, C. (2016). Practical application of medical absorbent paper to control the wet bag in the bag. *Medical Information* 29 (18): 391–393.

71 Fu, S.D., Zhou, H., Wang, H.X. et al. (2013). Amphibious superamphiphilic fabrics with self-healing underwater superoleophilicity. *Materials Horizons* 6: 122–129.

72 Fu S D, Zhou H, Wang H X, et. al. Superhydrophilic, underwater directional oil-transport fabrics with a novel oil trapping function. *ACS Applied Materials & Interfaces*, 2019, 11: 27402–27409.

73 Prathap, A. and Sureshan, K.M. (2017). Organogelator-cellulose composite for practical and eco-friendly marine oil-spill recovery. *Angewandte Chemie International Edition. in English* 56 (32): 9405–9409.

74 Feng, X.J. and Jiang, L. (2006). Design and creation of superwetting/antiwetting surfaces. *Advanced Materials* 18 (23): 3063–3078.

75 Zhang, W., Shi, Z., Zhang, F. et al. (2013). Superhydrophobic and superoleophilic PVDF membranes for effective separation of water-in-oil emulsions with high flux. *Advanced Materials* 25 (14): 2071–2076.

76 Tursi, A., Beneduci, A., Chidichimo, F. et al. (2018). Remediation of hydrocarbons polluted water by hydrophobic functionalized cellulose. *Chemosphere* 201: 530–539.

77 Guo, J.T., He, S.H., and Wang, H.S. (2017). Research on the production process of absorbent and oil-absorbing paper for kitchen. *Research and development* 22: 35–38.

78 Braga, E.R., Huziwara, W.K., Martignoni, W.P. et al. (2015). Improving hydrocyclone geometry for oil/water separation. *Brazilian Journal of Petroleum and Gas* 9 (3): 115–123.

79 Li J. Cleansing blotting paper [P] CN00120630. 3. 2001-07-24.

80 Gu, J., Xiao, P., Chen, J. et al. (2014). Janus polymer/carbon nanotube hybrid membranes for oil/water separation. *ACS Applied Materials & Interfaces* 6 (18): 16204–16209.

81 Kota, A.K., Li, Y., Mabry, J.M. et al. (2012). Hierarchically structured superoleophobic surfaces with ultralow contact angle hysteresis. *Advanced Materials* 24 (43): 5838–5843.

82 Tian, Y. and Jiang, L. (2013). Wetting: intrinsically robust hydrophobicity. *Nature Materials* 12 (4): 291–292.

83 Kjellgren, H., Gallstedt, M., Engstrom, G. et al. (2006). Barrier and surface properties of chitosan-coated greaseproof paper. *Carbohydrate Polymers* 65 (4): 453–460.

84 Ji, S.G., Cho, D., and Lee, B.C. (2010). Chemical and thermal characterizations of electron beam irradiated jute fibers. *Journal of Adhesion and Interface* 11 (4): 162–167.

85 Park, H.J., Kim, S.H., Lim, S.T. et al. (2000). Grease resistance and mechanical properties of isolated soy protein- coated paper. *Journal of the American Oil Chemists'Society* 77 (3): 269–273.

86 Guo, K., Jiang, B., Zhao, P. et al. (2021). Review on the superhydrophilic coating of electric insulator. *IOP Conference Series: Earth and Environmental Science* 651 (2).

87 Chen, B.B., Dong, Z., Zhang, M.J. et al. (2020). A novel sepiolite-based superhydrophilic/superoleophobic coating and its application in oil-water separation. *Chemistry Letters* 49 (12).

88 Bordenave, N., Grelier, S., Pichavant, F. et al. (2007). Water and moisture susceptibility of chitosan and paper-based materials: structure-property relationships. *Journal of agricultural and food chemistry* 55 (23): 9479–9488.

89 Lavoine, N., Desloges, I., Khelifi, B. et al. (2014). Impact of different coating processes of microfibrillated cellulose on the mechanical and barrier properties of paper. *Journal of Materials Science* 49 (7): 2879–2893.

90 Wu W B, Tian H, Jiang S, et al. Paper-based functional material for oil-water separation, preparation method and application thereof. CN111218853A. 2020-06-02.

7

Flame-retardant and Heat Insulation Paper-based Functional Materials

7.1 Flame-retardant Paper-based Functional Materials

Flame-retardant papers are a kind of specialized paper-based functional materials characterized by their flame resistance and high-temperature durability. They are widely used in fields, such as construction, packaging, interior decoration, automotive filtration, and cable insulation. Due to the fact that the main component of paper products is natural cellulose fiber, a considerable portion of real-life fires are caused by paper products. With the increasing awareness of fire prevention, the demand for flame-retardant properties of paper is growing, and research on flame-retardant paper has attracted significant attention [1].

7.1.1 Classification of Flame Retardants for Paper-based Functional Materials

Materials capable of imparting flame-retardant properties to a paper are referred to as flame retardants. Flame retardants are divided into two main categories based on their chemical composition: organic flame retardants and inorganic flame retardants. Organic flame retardants mainly include halogen-based flame retardants, phosphorus-based flame retardants, and nitrogen-based flame retardants, while inorganic flame retardants mainly include inorganic phosphorus- and boron-based flame retardants, metal hydroxides, metal oxides, alkali metal salts, and ammonia salts [2]. In recent years, a new type of expanding flame retardants have emerged, which are mostly phosphorus–nitrogen compounds or complexes. Currently, the most widely used flame retardants in industry are halides, phosphates, antimony trioxide, aluminum hydroxide, zinc borate, and others. In the research of environmentally friendly flame retardants for paper-based functional materials, there is a focus on ammonium polyphosphate and hydrotalcite. Additionally, there is considerable research on improving flame-retardant efficiency through surface modification and synergistic blending.

Paper-Based Functional Materials: Preparation and Applications, First Edition.
Xianping Yao, Genrong Chen, Zhen Yao, and Liping Zheng.

7.1.1.1 Organic Flame Retardants

1) Halogen-based Flame Retardants

Halogen-based flame retardants are one of the largest categories of organic flame retardants in terms of global production. However, during combustion, halogen-based flame retardants release irritating and corrosive hydrogen halide gases [3]. When used in conjunction with synergists, such as antimony trioxide, a large amount of smoke containing irritating and corrosive gases is released during combustion, posing a threat to life safety. Currently, halogen-based flame retardants are gradually being replaced with environmentally friendly flame retardants. Globally, there is a concerted effort to transition flame-retardant product structures from halogen-based to environmentally friendly types [4]. There are over 70 varieties of chlorine-based and bromine-based flame retardants. Among them, chlorinated paraffins and brominated flame retardants, such as decabromodiphenyl ethane, decabromodiphenyl ether, and tetrabromobisphenol A, are most widely used.

2) Organic phosphorus-based flame retardants

Organic phosphorus-based flame retardants include phosphates, hypophosphites, phosphine oxides, and phosphorus–nitrogen compounds, among others [5]. Phosphates are typically derived from the reaction of the corresponding alcohol or phenol with phosphorus oxychloride or from the reaction of the corresponding alcohol or phenol with phosphorus trichloride, followed by chlorination and hydrolysis.

Phosphates are abundant in resources, low in cost, compatible with other materials, and widely used [6]. The presence of carbon–phosphorus bonds enhances the chemical stability of phosphates, making them resistant to water and solvents and ensuring long-lasting flame retardant performance. Phosphate-based flame retardants produce fewer toxic and corrosive gases during combustion compared to halogen-based flame retardants. Major products include dimethyl methylphosphonate, alkyl phosphates, amide phosphates, cyclic phosphates, etc.

7.1.1.2 Inorganic Flame Retardants

1) Metal hydroxide flame retardants

Metal hydroxide flame retardants mainly include magnesium hydroxide, aluminum hydroxide, layered double hydroxides, etc. Due to their environmental friendliness, non-toxicity, non-production of corrosive gases, good stability, low smoke generation, and effective flame retardancy, they have been continuously gaining attention. Their flame-retardant mechanisms primarily involve dilution, isolation, and heat absorption effects.

Layered hydroxide-based intercalation flame retardants are characterized by high efficiency, non-toxicity, halogen-free, and low-smoke generation. The application of layered hydroxide flame retardants in paper-based functional materials has become a hot topic. A study by Li Chao focused on a flame-retardant system mainly composed of magnesium–aluminum layered hydroxide [7]. A flame-retardant paper was prepared using a coating process, and it was found that using magnesium–aluminum layered hydroxide alone resulted in poor flame-retardant performance, but when combined with red phosphorus, they exhibited good synergistic effects [8].

Li Xianhui and others used magnesium–aluminum layered hydroxide as flame-retardant filler in papermaking. Flame-retardant paper was prepared by adding flame retardants to the pulp [9]. When cationic polyacrylamide (CPAM) was not added and only 40% magnesium–aluminum layered hydroxide was added, the limiting oxygen index of the paper reached 25%. When 0.025% CPAM and 30% magnesium–aluminum layered hydroxide were added, the limiting oxygen index reached 26.2%, achieving flame-retardant level [10].

Zhao Huifang and others developed a modified ammonium polyphosphate/aluminum hydroxide/zinc borate composite flame retardant prepared from polyphosphoric acid, phosphoric acid, urea, aluminum hydroxide, and zinc borate. After modifying the flame retardant with a silane coupling agent, it was added to the pulp to prepare flame-retardant paperboard. The composite flame retardant has low water solubility, and a concentration of less than 10% achieves flame-retardant effects, with good smoke suppression and low production costs [11].

2) Inorganic phosphorus-based flame retardants

Inorganic phosphorus-based flame retardants mainly include ammonium polyphosphate, red phosphorus, and phosphates. They are characterized by good thermal stability, non-volatility, non-production of corrosive gases, and long-lasting effects, thus receiving wide applications.

Ammonium polyphosphate is an important high-efficiency inorganic flame retardant that has rapidly developed over the past two decades. It can be used alone or in combination with other flame retardants and is widely used as an additive flame retardant in the paper-making industry [12].

At high temperatures, ammonium polyphosphate rapidly decomposes into ammonia and polyphosphoric acid. Ammonia can dilute the oxygen concentration in the gas phase, thereby inhibiting combustion, while polyphosphoric acid acts as a strong dehydrating agent, promoting polymer dehydration and char formation, isolating the polymer from oxygen, and inhibiting combustion in the solid phase.

Hong Li et al. used ammonium polyphosphate to prepare air filter paper with excellent flame-retardant performance.

Mo Ziyue et al. used melamine–formaldehyde–resin–microencapsulated ammonium polyphosphate. Flame-retardant paper was prepared by adding fillers to the pulp, and when the filling amount reached 10%, the flame-retardant paper met the requirements of flame retardancy [13].

Sha Lizheng prepared a three-component composite filler of ammonium polyphosphate–silica–al synergist through *in situ* polymerization. Flame-retardant paper prepared by adding fillers to the pulp achieved non-flammability [14].

Phosphorus-based flame retardants often employ nitrogen–phosphorus combination synergistic enhancement, resulting in excellent flame-retardant effects, high thermal stability, and self-extinguishing properties. Expandable flame retardants with phosphorus, nitrogen, and carbon as their core have become a hot topic in flame-retardant research in recent years. Expandable flame retardants are a type of additive flame retardants, typically consisting of an acid source (dehydrating agent), a carbon source (char-forming agent), and a gas source (foaming agent), without halogens.

In the combustion process, the acid source, dehydrating agent, and carbon source in the expandable flame retardants undergo esterification dehydration reactions in the air, generating a large amount of char, non-combustible gases, and producing bubbles. These substances form an expanded porous char layer covering the material surface, forming a foam–char layer. During this process, the flame-retardant substrate melts and is protected by the expanded porous char layer formed by the flame-retardant material, providing insulation, oxygen isolation, smoke suppression, and drip prevention, thereby inhibiting heat and mass transfer between the condensed and gas phases, and achieving flame retardancy. Therefore, expandable flame retardants exhibit excellent low toxicity, low smoke, and high-efficiency flame-retardant properties.

Huang Gaoneng et al. used a combination of internal addition and surface sizing to add single-molecule expandable flame retardants, such as pentaerythritol–polyphosphate–melamine salt, to paper. When the content of melamine salt reached 30%, the limiting oxygen index of the paper increased to 33.3%, achieving good flame-retardant levels.

3) Nitrogen-based flame retardants

Nitrogen-based flame retardants are highly desirable due to their minimal volatility, non-toxicity, excellent compatibility with polymers, high decomposition temperature, and suitability for processing. They are typically found in the form of ammonium salts (such as monoammonium phosphate, diammonium phosphate, and ammonium sulfate). In recent years, melamine-based polymer flame retardants have been successfully developed, offering excellent economic benefits.

4) Expanded graphite-based flame retardants

Expanded graphite is produced by treating natural graphite with concentrated sulfuric acid, followed by washing, filtering, and drying, and then expanding it at 900 to 1000 °C. When exposed to temperatures above 200 °C, the compounds trapped in the layered lattice of graphite decompose instantly, causing the graphite to expand along its structural axis by several hundred times. It reaches its maximum volume at around 1100 °C, with the final expanded volume reaching up to 280 times its initial volume. This characteristic allows expanded graphite to rapidly increase in volume during a fire, extinguishing the flame.

7.1.1.3 Nano Flame Retardants

1) Carbon Nanotubes

Carbon nanotubes can be conceptualized as hollow tubular structures formed by the curling of graphene layers. Based on the number of layers in the tube wall, carbon nanotubes can be categorized into single-walled carbon nanotubes (SWCNTs) and multiwalled carbon nanotubes (MWCNTs). Carbon nanotube flame retardants require small addition amounts, typically less than 5%, to achieve excellent flame-retardant effects [15]. Research indicates that the flame-retardant effectiveness of carbon nanotubes is enhanced after ball milling, with the ignition time of the material prolonged by approximately twice. This is because the ball milling process causes carbon nanotubes to fracture, increase their surface area, and may generate more free radicals, thereby extending the ignition time.

2) Graphene

Graphene is a two-dimensional carbon nanomaterial with an extremely large surface area, ultra-strong elastic modulus, high thermal conductivity, and other properties. Flame retardancy of graphene has been a research hotspot in the past decade. Kim et al. directly ignited high-purity graphene in air and observed that although the ignited portion of graphene turned red, the flame did not spread; upon moving the flame away, the graphene quickly extinguished [16]. Han et al. investigated the difference in flame retardancy between graphene oxide (GO) with varying degrees of oxidation and found that the thermal stability and power reduction of composite materials decreased with increasing oxidation degree, with unoxidized graphene exhibiting the best flame-retardant effect [17].

The flame-retardant performance of graphene mainly depends on the structure and yield of the residual carbon layer. First, the two-dimensional structure of graphene itself can act as a protective layer. Second, the formation of "twisted channels" between the graphene layers also slows down the transfer of flammable materials to a certain extent. Finally, the addition of graphene significantly increases the graphitization degree of the residual carbon layer, enhancing its thermal stability and structural integrity.

Overall, carbon-based nanomaterial flame retardants have high efficiency and are non-toxic, but they face challenges such as difficult preparation techniques and high costs, which limit their broader applications. The development prospects of such flame retardants mainly depend on improvements in relevant preparation technologies, such as the simplified and low-cost preparation of carbon nanotubes and graphene.

3) Nano silicon dioxide

Nano silica dioxide is a powdery substance with a three-dimensional chain-like structure, possessing various advantages, such as optical properties against ultraviolet radiation, super plasticity, high strength, large magnetostriction, large specific surface area, and low thermal conductivity. After being added to polymer materials, the mechanical properties and thermal stability of composite materials are both improved to some extent, making it suitable for flame retardant modification of materials.

7.1.2 Flame-retardant Mechanism of Paper-based Functional Materials

7.1.2.1 The Combustion Process of Paper-based Functional Materials

Plant fiber paper is primarily composed of cellulose $(C_6H_{10}O_5)_n$, along with hemicellulose and lignin. The combustion process of cellulose mainly involves pyrolysis and oxidation reactions, which can be classified into flaming and non-flaming combustion. Non-flaming combustion, also known as glowing combustion, occurs based on flaming combustion, where the products of flaming combustion continue to oxidize, requiring much higher temperatures than flaming combustion [18].

Generally, when the temperature is below 300 °C, cellulose undergoes dehydration and carbonization, producing H_2O, CO, CO_2, and solid residual carbon. This process is endothermic and cannot sustain itself, requiring external heat input. Cellulose exhibits some flame retardancy in this temperature range. When the temperature exceeds 300 °C, cellulose's pyrolysis reactions begin, producing nonvolatile liquid levoglucosan. This nonvolatile liquid levoglucosan then decomposes to generate flammable gases and volatile liquids, accelerating the fire spread. Flammable gases and volatile liquids undergo combustion in the presence of oxygen, generating a large amount of heat and transferring the flame to adjacent materials. The adjacent paper surface is heated to the pyrolysis temperature, initiating further pyrolysis reactions, sustaining the combustion process.

During the combustion of cellulose, a series of intense radical reactions also occur. During combustion, cellulose materials undergo pyrolysis, breaking chemical bonds to produce hydroxyl radicals. These hydroxyl radicals then catalyze the decomposition of cellulose, generating hydrocarbon radicals, and so forth. Under suitable conditions, hydrocarbon radicals can generate new hydroxyl radicals, continuing the radical reaction cycle until complete combustion of the paper occurs.

7.1.2.2 Flame-retardant Mechanism of Paper-based Functional Materials

Flame-retardant treatment of paper-based functional materials involves impeding the thermal decomposition of fibers, suppressing the generation of flammable gases, or isolating heat and air while diluting flammable gases. From the combustion mechanism of paper, it can be seen that imparting flame retardancy to paper generally involves disrupting the combustion process. The flame retardation process works by slowing down or preventing six key factors: enhancing material thermal stability, capturing free radicals, forming non-flammable protective films, absorbing heat, creating a layer of heavy gases for isolation, and diluting oxygen and flammable gases. Therefore, flame-retardant mechanisms are classified into gas-phase flame retardation mechanism, condensed-phase flame retardation mechanism, and interrupting heat exchange flame retardation mechanism based on the approaches by which flame-retardant materials terminate combustion.

1) Gas-Phase Flame-Retardant Mechanism

The gas-phase flame retardation mechanism refers to the flame-retardant effect that interrupts or delays the chain reaction of combustion in the gas phase. For example, nitrogen-based flame retardants release gases such as CO_2, NH_3, N_2, and H_2O when heated, reducing the concentration of oxygen and flammable gases produced during the thermal decomposition of polymers in the air. The non-flammable gases generated carry away some of the heat, reducing the temperature of the polymer surface. The generated N_2 captures free radicals, suppressing the chain reaction of combustion, thus preventing combustion.

2) Condensed-Phase Flame-Retardant Mechanism

The condensed-phase flame retardation mechanism refers to the retardant effect of flame-retardant materials that delay or prevent the thermal decomposition of substances that can produce flammable gases and free radicals in the solid phase. It involves the formation of a flame-retardant, heat-insulating, and oxygen-blocking porous carbon layer on the surface, preventing flammable gases from entering the combustion gas phase and interrupting combustion.

For example, inorganic flame retardants release crystalline water when heated, which evaporates, decomposes, and releases water vapor. This reaction absorbs a significant amount of heat from combustion, reducing the surface temperature of the material and greatly decreasing the thermal decomposition and combustion rate of the polymer material. The large amount of water vapor produced during decomposition dilutes the concentration of flammable gases, also acting as a flame retardant and providing some cooling effect. The products of decomposition, such as magnesium oxide and aluminum oxide, combine with the charred products on the surface of the plastic during combustion to form a protective film. This film effectively blocks the infiltration of heat and oxygen while preventing the escape of small flammable gases, thereby achieving flame retardation.

3) Interruption of heat exchange flame-retardant mechanism

The mechanism of interrupting heat exchange flame retardation refers to the removal of part of the heat generated by the combustion of flame-retardant materials, preventing the material from maintaining the thermal decomposition temperature, thereby preventing the continuous generation of flammable gases, and causing self-extinguishing of the combustion.

For example, the flame retardation mechanism of halogen-based flame retardants includes three aspects: thermal insulation and cooling, termination of chain reactions, and cutting off the heat source. When halogen-based flame retardants decompose under heat, they absorb some of the heat, lowering the temperature. Simultaneously, the density of hydrogen halide gas generated by decomposition is greater than that of air, which can expel the surrounding air, forming an oxygen-blocking barrier. Hydrogen halide can also react with the free radicals generated by polymer decomposition to produce halogen radicals. These halogen radicals can further react with polymer chains to generate hydrogen halide, thereby cutting off the reaction between the free radicals generated by polymer decomposition and oxygen. Additionally, the presence of flame retardants weakens the van der Waals forces between polymer chains, causing the material to be in a viscous flow state when heated. At this stage, the material becomes fluid, and when heated, it can carry away some of the flame and heat, thereby achieving flame retardation [19].

7.1.3 Preparation of Flame-retardant Paper-based Functional Materials

There are two main methods for preparing flame-retardant paper-based functional materials. One method involves using natural fibers as raw materials and utilizing traditional papermaking techniques to produce flame-retardant paper. Flame retardants can be added during the papermaking process or applied to the surface of the paper. The other method involves using flame-resistant or non-flammable fibers as raw materials to produce flame-retardant paper. Common flame-retardant fibers include carbon, glass, ceramic, aramid, and asbestos fibers, among others.

7.1.3.1 Internal addition

The method of adding flame retardants to the pulp is divided into two types. One method involves adding flame retardants during pulping or in the pulp supply system. This method typically uses water-insoluble, powdery flame retardants. Commonly used water-insoluble fine powdery flame retardants include antimony trioxide, insoluble oxides, aluminum hydroxide, magnesium hydroxide, etc. This method is commonly used in the production of pulp-molded products, insulation boards, and hardboards.

The advantage of this method is its simple operation, uniform distribution of flame retardants in the paper, and relatively uniform flame-retardant function of the paper. However, the disadvantage is that the loss of flame retardants is severe, and achieving a retention rate of 80% for flame retardants is very difficult. The best way to address this loss is to add suitable retention aids.

The addition method is more effective when used to produce fireproof cardboard or high-basis-weight paper. The dosage of flame retardants should be determined based on factors such as the purpose of the paper and paper products, and the effectiveness of the flame retardants used. Generally, for low-basis-weight paper, about 20% (based on dry fiber mass) of flame retardants are used, while for papers with larger basis-weights, 5–10% of flame retardants are used. In practical applications, to endow paper with good flame retardancy, the flame retardants added to the paper are often a combination of several types, such as phosphorus–halogen systems, antimony–halogen systems, and phosphorus–nitrogen systems. These composite or compound flame retardants can synergistically enhance the flame-retardant effect, which is much better than that of single-element flame retardants.

Zhao Huifang et al. used plant fibers, persulfates, and melamine polyphosphate as raw materials to prepare flame retardant plant fibers through grafting reactions. After adding other papermaking aids, they produced flame retardant paper through beating, pressing, surface sizing, and drying. The basis weight of the paper was 60–80 g/m^2, the longitudinal dry tensile strength was 25–30 N/15 mm, the burst index was 2.6–3.5 $kPa{\cdot}m^2/g$, the limiting oxygen index was ≥27%, and it exhibited good flame retardancy [20].

Sha Lizheng et al. used persulfate as an initiator to graft-modify melamine polyphosphate onto plant fibers, obtaining flame-retardant plant fibers. The flame-retardant paper made from these fibers exhibited excellent flame-retardant effects [21].

Sha Lizheng et al. also prepared ammonium polyphosphate–silica aerogel composite fillers by an *in situ* polymerization method and added them to the interior of paper sheets. The results showed that phosphoric acid and urea underwent *in situ* polymerization on the surface and within the micropores of silica aerogel to form ammonium polyphosphate. The ammonium polyphosphate–silica aerogel composite filler exhibited better thermal stability than ammonium polyphosphate and significantly promoted the charring of the filled paper sheets.

7.1.3.2 Surface Coating

The surface coating method involves applying flame retardants to the surface of paper to produce flame-retardant paper. This method is suitable for insoluble or difficult-to-solubilize flame retardants. In this method, the flame-retardant powder is uniformly dispersed in a binder to form an emulsion coating. Then, the coating is applied to the surface of the paper using a coating method, and the coated paper is dried by heating to obtain flame-retardant paper [22].

The advantage of the surface coating method is that most of the flame retardants are concentrated on the surface of the paper, minimizing the impact on the paper's physical properties. Particularly, for processing papers that require coating, adding flame retardants to the coating is sufficient, and this method can be completed on a coating machine. To achieve good flame-retardant effects, it is preferable to use flame retardants with small and uniform particle sizes.

The disadvantage of this method is that the flame-retardant effect may not be ideal, as it does not provide flame retardancy to the interior of the paper. However, it is quite effective in providing flame resistance to the surface of the paper. It can serve as an oxygen barrier and is suitable for surface treatment of hot-pressed hard fiberboards, wallpaper boards, and corrugated box-lining boards.

7.1.3.3 Impregnation

The impregnation method is suitable for water-soluble flame retardants. Paper is immersed in a solution containing water-soluble flame retardants, and after heating and drying, flame-retardant paper is obtained. The impregnation method is an external treatment process, characterized by short processing time and easy operation. It requires paper with high absorbency and wet strength and is suitable for the production of cotton short fiber paper, decorative crepe paper, special wallpaper, non-woven fabrics, and impregnated paper for construction use. However, flame-retardant paper produced by this method tends to have poor water resistance, strong moisture absorption, significant reduction in paper strength, and is prone to yellowing and stiffening [23].

Through vacuum suction, flame retardants on the surface of the original paper can be drawn into the fibers between the paper, allowing the flame retardants to penetrate not only the surface fibers but also the layers within the paper. At the same time, vacuum suction brings plant fibers closer together, enhancing the hydrogen bond strength between them and improving the tensile strength of the original paper itself. A comparison of scanning electron microscopy images of paper produced by impregnation and vacuum suction methods is shown in Figure 7.1. The impregnation method allows flame retardants to be wrapped around the surface fibers of the paper, but fewer flame retardants can penetrate into the interior of the paper. The vacuum suction method allows flame retardants to adhere to the fiber surface and fill the interior of the paper, effectively wrapping the fibers with flame retardants.

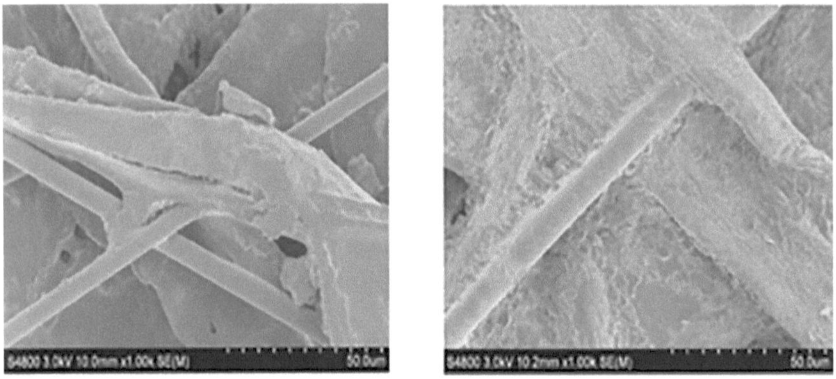

(a) (b)

Figure 7.1 Scanning Electron Microscope image of impregnated paper (a) and vacuum suction paper (b).

As a result, more flame retardants are attached to the paper surface, leading to better flame-retardant effects with the vacuum suction method.

7.1.4 Application of Flame-retardant Paper-based Functional Materials

In the late 1960s, researchers worldwide made fruitful achievements in the study of flame-retardant paper. In China, research on flame-retardant paper began in the late 1980s. Although it started relatively late, China has also developed paper varieties with excellent flame-retardant properties, which are widely used in many industrial fields, such as packaging, decorative construction, automotive filtration, aerospace, and aviation.

7.1.4.1 Wallpaper

Wallpaper, as an interior decoration material, holds a significant share in the decoration market. However, paper-based wallpapers are prone to flame propagation in case of fire, leading to rapid fire spread. Flame-retardant wallpaper is produced by treating the original paper with flame-retardant agents, significantly reducing the incidence of indoor fires [24].

Currently, the most commonly used wallpaper products are made of polyvinyl chloride (PVC). The surface layer of PVC wallpaper uses polymer organic materials, while the base layer is made of combustible paper, resulting in fast combustion and the release of toxic gases, posing significant risks to human health and the environment. Therefore, it is necessary to add flame retardants to the base paper or surface layer of wallpaper to enhance its flame-retardant properties. The comparison of wallpaper combustion performance is shown in Figure 7.2.

7.1.4.2 Packaging Materials

Paper packaging materials play a crucial role in the entire packaging industry. Untreated paper packaging materials pose a fire risk during storage, transportation,

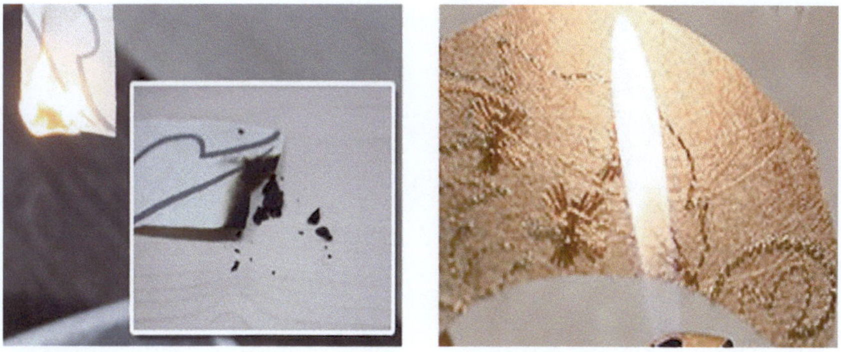

Figure 7.2 Experimental phenomenon of flame-retardant wallpaper.

and use. Therefore, flame-retardant treatment of paper packaging materials is essential.

There are two main methods for producing flame-retardant packaging paper: impregnation and coating [25]. Additionally, by refining wood pulp through phosphorylation, modified pulp with a positive charge in water can be obtained. Under low pH conditions, it can attract glass or ceramic fibers, making it possible to produce flame-retardant packaging paper with higher strength. Sun Junjun and others used a composite flame retardant consisting of expandable graphite, magnesium hydroxide, and sodium silicate to prepare flame-retardant paperboard, achieving a flame-retardant rating.

Zhao Zhifang et al. utilized melamine polyphosphate-treated secondary fibers to prepare flame retardant leather suitcase paperboard with good physical strength, a tear index of 2.1–2.3 $kPa \cdot m^2/g$, a breaking length of 3.5–3.8 km, and a limiting oxygen index of ≥27% [26]. Sha Lizheng et al. prepared a flame-retardant and smoke-suppressant packaging paperboard using ammonium polyphosphate–silica/nano-titanium dioxide composite flame retardant. This flame-retardant and smoke-suppressant plant fiber-based material exhibited good flame-retardant and smoke-suppressant properties, low production cost, and environmental friendliness.

Zhao Huifang et al. used softwood pulp board, hardwood pulp board, and aramid pulp to produce paper through compounding and then applied ammonium polyphosphate–silica slurry mixed ink to the surface of the original paper using a gravure printing machine. The flame-retardant paper produced by this method exhibited good flame-retardant effects and maintained the main physical strength indicators of the original paper, making post-processing easy [27]. Wu Min et al. invented a flame retardant made by reacting chitosan with ammonium dihydrogen phosphate, which, when coated on paper, not only had good flame-retardant effects but also exhibited high strength and good antibacterial properties. Toray Industries, Inc. developed a flame-retardant paper containing 10–35% pulp, 40–70% alumina, and 0.1–10% guanyl phosphoric acid and amino sulfonic acid.

7.1.4.3 Honeycomb Paper

Flame-retardant honeycomb paper is made from corrugated paper cores using impregnation and coating methods to add highly efficient and environmentally

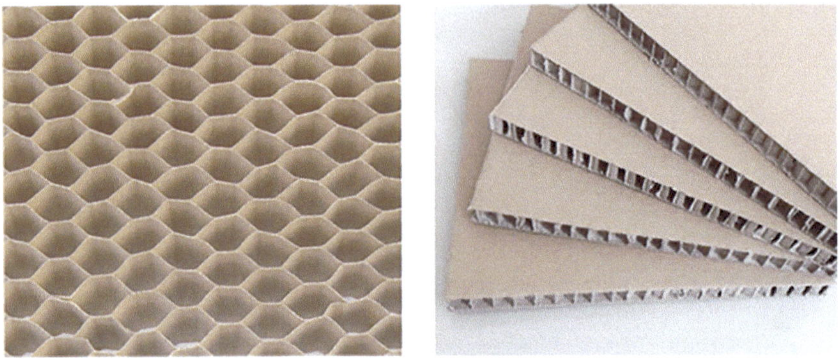

Figure 7.3 High-strength corrugated base paper for flame-retardant honeycomb core paper.

friendly expanding flame retardants to produce flame-retardant paper. It is then processed into honeycomb paper cores. It has the characteristics of high load-bearing capacity, good elasticity, light weight, and low cost. It is widely used in industries such as tobacco, ceramics, and glass warehousing and can also be used as filling materials for partitions, doors, windows, cabinets, furniture, and partitions (Figure 7.3). Lu Xianggen et al. used domestically produced waste corrugated cardboard (OCC) waste paper and melamine polyphosphate as reactive flame retardants, with persulfate as the initiator, to graft 40–70% flame-retardant OCC pulp at 80–90 °C. Then, the flame-retardant OCC pulp was formed into ultra-low basis weight corrugated isolation paper, and the graft-modified waste liquid was used to prepare flame-retardant surface adhesives for the surface of the ultra-low basis weight corrugated isolation paper [28].

7.1.4.4 Filter Paper

With the development of the automotive industry, engine manufacturers have increasingly stringent requirements for air filter paper used in internal combustion engines. High-performance automotive air filter paper not only demands good flame retardancy of the paper, but also requires excellent physical properties.

Currently, companies such as Korean Oslon, Japanese Apo, and German Gessner have introduced environmentally friendly air filter papers with outstanding flame-retardant properties. As the demand for flame retardancy continues to rise, researchers have begun studying flame retardants used in automotive air filter papers. Sha Lizheng and others proposed that a single flame retardant is no longer sufficient to meet the demand for flame retardancy. They advocate for the combination of multiple flame retardants to enhance the flame-retardant effect through the synergistic interaction of different flame-retardant elements, which has become a hot topic in the flame-retardant industry.

Hong Li, Hu Jian, and others used ammonium polyphosphate to prepare flame-retardant air filter paper and found that the flame retardant performance of air filter paper containing ammonium polyphosphate was good. Moreover, the greater the amount added, the greater the enhancement of flame retardancy.

In recent years, researchers have begun modifying flame retardants before compounding them with emulsions. For example, Liu Wenbo et al. synthesized styrene emulsion by copolymerizing vinyltrimethoxysilane with styrene and acrylic acid ester. They found that as the content of vinyltrimethoxysilane in the styrene emulsion increased, the mechanical strength and water resistance of the filter paper improved.

Wei Jianxiong and others prepared a nitrogen–phosphorus epoxy resin emulsion and studied its application in air filter paper, finding excellent flame-retardant performance.

Hong Li used red phosphorus and melamine cyanurate separately as flame retardants and found that flame-retardant paper treated with compound processing had higher flame retardancy [29].

Li Jing and others used itaconic acid as a functional monomer to modify styrene latex and compounded the modified styrene latex with flame retardant to improve the performance of flame-retardant air filter paper. After compounding the flame retardant with modified styrene latex in a mass ratio of 1:1.4 and impregnating the original air filter paper, it was found that the stiffness, burst resistance, and limiting oxygen index of the paper reached 4.07 mN/m, 301 kPa, and 30.9%, respectively, meeting the performance requirements of flame-retardant air filter paper industry standards.

7.1.4.5 Electrical Insulation Paper

Flame-retardant paper products have broad prospects for application in electrical insulation. Because their resistance tends toward infinity in a "dry" state, flame-retardant paper can be used as molding material for high-voltage transformers and for cable wrapping. Flame-retardant paperboard can serve as insulating paperboard, widely used in electrical components such as appliances, instruments, switches, and transformers. Sun Yikun successfully developed a new type of supporting paper for composite lightweight panels—flame-retardant honeycomb board paper. The development of this new product has played a positive role in the successful development of defense and military products—shipboard composite lightweight panels, contributing to the production and development of shipboard composite lightweight panel compartment systems to replace fire-resistant compartment systems made of composite rock wool panels. Chen Jiwei used 5000–6000 mesh aluminum hydroxide and adopted production technology of pulp addition and surface impregnation of flame retardants to prepare environmentally friendly flame-retardant insulation paperboard.

7.1.5 Prospect

Currently, flame-retardant technology has entered a new development stage, and the flame-retardant technology of paper-based functional materials is moving towards the direction of high efficiency, economy, and environmental friendliness. Among them, flame-retardant paper prepared with halogen-free and inorganic flame retardants has the characteristics of low flame spread rate, low smoke, low toxicity, and high flame retardancy efficiency, making it the direction of future development. To maximize the role of inorganic flame retardants in paper flame-retardant treatment, various new flame-retardant technologies, such as ultra-fine, surface modification, synergistic compounding, and multifunctionalities of inorganic flame retardants, will be the focus of development in recent years. The production of flame-retardant paper generally requires a considerable amount of flame retardants. Composite flame retardants possess the characteristics of various flame retardants, and how to maximize the advantages of each flame retardant in composite flame retardants is also an important direction.

7.2 Thermal Insulation Paper-based Functional Materials

Thermal insulation paper-based functional materials, primarily composed of fibers, are novel materials with thermal insulation properties prepared through paper-making forming technology. These materials exhibit the following characteristics during their preparation process: water serves as the primary dispersing medium; the raw materials primarily consist of short fibers, typically with a length less than 30 mm. Thermal insulation paper-based functional materials are produced using paper-making forming technology; the bonding between fibers in this material is primarily achieved not through hydrogen bonding, but rather through mechanical force, adhesives, hot pressing, solvent swelling, or other reinforcement techniques.

7.2.1 Classification of Thermal Insulation Paper-based Functional Materials

Thermal insulation materials refer to materials or material composites with insulation properties that can shield thermal flux. They typically possess characteristics such as lightweight, porousness, and low thermal conductivity. Based on their composition, they can be categorized into three main types: organic thermal insulation materials, inorganic thermal insulation materials, and metallic thermal insulation materials.

7.2.1.1 Organic Thermal Insulation Materials

Natural organic thermal insulation materials include cork, plant fibers, animal hair, etc. Artificially synthesized materials are produced using organic polymers as raw materials, with foaming agents and stabilizers added during the synthesis process. This causes the organic polymers to rapidly generate a large number of bubbles within a fixed mold, increasing their volume several times, and then drying to form insulation. Commonly used materials include polyurethane, polystyrene foam plastics, etc. Organic thermal insulation materials have the advantages of good thermal insulation performance, low-temperature resistance, lightweight, sound absorption, etc. However, their disadvantages, such as flammability, poor aging resistance, difficult degradation, and potential toxicity, limit their application.

7.2.1.2 Inorganic Thermal Insulation Materials

Inorganic thermal insulation materials typically have a porous structure, with solid phases surrounded by gas phases, making their heat transfer mechanism complex. Common natural inorganic thermal insulation materials include asbestos, perlite, diatomaceous earth, etc. Common artificial inorganic thermal insulation materials include foam glass concrete, ceramic fibers, glass fibers, and porous thermal insulation bricks. Compared to organic thermal insulation materials, inorganic thermal insulation materials have lower thermal conductivity, high temperature resistance, non-combustibility, wide availability of raw materials, and lower cost.

7.2.1.3 Metal Thermal Insulation Materials

Metal thermal insulation materials have high infrared thermal radiation reflectivity. These materials include metal foils, such as copper, aluminum, and nickel, as well as composite materials formed by sandwiching metal foils with organic or inorganic materials (such as honeycomb structures). The surface of the material is smooth and flat, with high reflectivity, and it has high tensile strength in both longitudinal and transverse directions. It can be used to reduce heat transfer through convection, conduction, and radiation, but such materials tend to be relatively expensive.

7.2.1.4 Aerogel Thermal Insulation Materials

With the development of nanotechnology, research on nano-porous aerogel super-insulating materials is continuously progressing towards practical and industrial applications. Aerogels have highly developed porosity (80–99.8%) and can have a density as low as 0.03 g/cm^3. Their nanoscale pore size significantly reduces the thermal conductivity of gas molecules and heat convection. Additionally, aerogel materials possess characteristics such as softness, ease of cutting, hydrophobicity, and environmental friendliness, gradually replacing traditional flexible thermal insulation materials like fiberglass. Fu Ping, using a mechanical foaming method, employed aerogel with low density, low thermal conductivity, and high porosity as a filler material and short-cut glass fibers as reinforcement to produce a new type of ultra-light fiber aerogel foam concrete. Test results show that fiber aerogel foam concrete has better thermal insulation properties than traditional materials, presenting great prospects in the field of building wall thermal insulation. Chen Fa utilized hydroxyapatite with ultra-long nanowire structure as a support skeleton and silica aerogel particles as fillers to prepare a new type of nano-composite insulation paper with low thermal conductivity, excellent fire resistance, and flexibility through secondary forming method.

7.2.2 Thermal Transfer Principle of Paper-based Functional Materials

The second law of thermodynamics states that wherever there is a temperature difference, heat will spontaneously flow from a hotter object to a colder object, or from a hotter part of an object to a cooler part. According to the thermal insulation mechanisms of materials, heat transfer primarily occurs through conduction, convection, and radiation. In nature, these three modes of heat transfer are usually carried out through carriers, including molecules, atoms, electrons, and phonons. From a macroscopic perspective, solids primarily transfer heat through conduction, liquids through convection and conduction, and gases primarily through radiation. Differentiating by environmental conditions, the main methods of heat transfer vary; at higher temperatures, radiation is predominant, while at lower temperatures, conduction is the main method. In gases and liquids, convection is consistently the primary mode of heat transfer.

At room temperature, the thermal conductivity of gases is much lower than that of solid-phase materials. Therefore, current thermal insulation materials are mainly composed of a combination of gas- and solid-phase structures, with the presence of gas-phase structures greatly weakening the contribution of solid-state heat transfer in thermal insulation materials. The main forms of solid- and gas-phase structures in thermal insulation materials include continuous gas-phase structure, such as powdery thermal insulation material filling layer; continuous solid-phase structure, such as organic foam plastics; and homogeneous continuous solid- and gas-phase structure, such as fibrous thermal insulation materials, etc.

Macroscopically, the following heat transfer modes exist within thermal insulation materials: solid-phase heat conduction, gas-phase heat conduction, heat conduction between solid and gas phases, heat convection between gas phases, and heat radiation transmitted to the outside world in the form of electromagnetic waves, etc. The total thermal conductivity of thermal insulation materials should be the superposition of the effects of thermal conductivity of each part of the heat transfer. At higher temperatures, the influence of heat radiation on thermal conductivity predominates, while at room temperature, heat radiation can be negligible.

7.2.3 Preparation of Thermal Insulation Paper-based Functional Materials

There are many methods for preparing thermal insulation paper-based functional materials, including direct casting, wet end addition, fiber casting, and surface processing methods.

7.2.3.1 Direct Casting Method for Thermal Insulation Fibers

The direct casting method involves casting fibers with thermal insulation properties directly into paper. While the process is relatively simple, it requires fiber modification before casting. Fiber modification alters the inherent properties of the fibers, such as their charge characteristics, which significantly impacts the bonding between fibers. Moreover, fiber modification processes are complex, time-consuming, and relatively costly, limiting their industrial application to some extent.

Shen Xiaoping et al. produced silica–alumina fiber paper with a thickness ranging from 0.2–5 mm using a paper machine. This paper is utilized as lining for high-temperature furnaces and gas pipelines, as well as insulation layers for electric heating appliances and filters for high-temperature gases.

Lu Zhaoqing et al. used polyimide (PI) and its precursor polyamide acid as the main fiber raw materials. They added a certain amount of polyethylene oxide (PEO) and aramid 1414 pulp to cast polyimide fiber raw paper. Subsequently, they impregnated and thermally pressed the raw paper, studying the effects of different casting processes on the strength, electrical properties, and thermal resistance of PI fiber paper substrates.

7.2.3.2 Wet End Addition

The wet end addition method involves mixing pulp with thermal insulation materials to produce paper. This method affects some properties of the pulp itself while adding insulation materials. During the papermaking process, insulation materials may be lost along with fine fibers and fillers, resulting in waste and pollution of the water, thereby increasing the cost of subsequent water treatment. Hence, it is common to add retention aids during the papermaking process to retain more insulation materials in the paper. Despite various challenges, this method is relatively simple to operate in actual production, and the distribution of insulation materials in the paper is relatively uniform.

7.2.3.3 Papermaking Method with Mixed Fibers

The papermaking method with mixed fibers involves first preparing fibers with thermal insulation properties, and then mixing these fibers with natural plant fibers to produce paper. The pulp, processed by a pulper, is stored in a stock chest. Thermal insulation materials are added to the pulp for blending, and then the paper is made using the conventional papermaking method. After passing through the headbox and forming on the wire, the paper is processed through pressing, drying, calendaring, and winding to obtain the final product.

This method is simple to operate and highly practical. However, the thermal insulation paper produced by this method may experience a decrease in mechanical properties due to the disruption of hydrogen bonds within the paper. Mo Jicheng et al. synthesized magnesium aluminum silicate *in situ* in the cell cavity and cell wall of hardwood pulp fibers using sodium silicate, aluminum sulfate, and magnesium sulfate as raw materials. They then used this composite fiber to cast low-temperature thermal insulation paper, determining the appropriate synthesis conditions.

7.2.3.4 Surface Coating Method

The surface coating method involves applying a coating containing thermal insulation materials or a small amount of added thermal insulation materials onto the surface of a material through spraying, gluing, coating, immersion, etc., to produce materials with thermal insulation functionality. Among these methods, the coating method is commonly used in practical production to prepare thermal insulation paper-based functional materials.

The surface coating method involves coating the surface of the original paper with a layer of thermal insulation coating.

The coating process is simple, economically efficient, and conducive to resource utilization and environmental protection. Thermal insulation paper produced by the surface coating method does not undergo internal structural damage, and its mechanical properties and barrier properties can be improved to some extent.

Nowadays, many researchers are committed to developing coatings with highly efficient thermal insulation effects to achieve energy conservation and emission reduction.

Shen Binhua prepared epoxy polysiloxane resin composite thermal insulation coatings containing alumina and silica hollow spheres by surface coating, reducing the thermal conductivity from 0.92 W/(m·K) to 0.30 W/(m·K). Ye Xiufang et al. studied the effects of three different thermal insulation mechanisms of functional fillers, namely diatomaceous earth, hollow glass microspheres, and antimony-doped tin oxide nanoparticles (ATO), on the thermal insulation effect of polyvinylidene fluoride coating (PVDF light oil), and prepared a composite thermal insulation and insulation fluorocarbon coating with synergistic thermal insulation effects. According to tests, the thermal insulation temperature difference of this composite thermal insulation and insulation fluorocarbon coating can reach 16.9 °C.

7.2.4 Applications of Thermal Insulation Paper-based Functional Materials

Thermal insulation paper-based functional materials have gained a prominent position in the field of thermal insulation materials due to their lightweight, easy production, processing, portability, environmental friendliness, and safety. They can be utilized in various areas, including oven baking trays, fast food packaging, insulation layers for electric hot pots, insulated cups, and linings for high-temperature furnaces. Figure 7.4 illustrates some common application areas of thermal insulation paper-based functional materials.

7.2.4.1 Insulation Thermal Paper Used in the Power Sector

During the operation of electrical transformers, insulation paper is subject to thermal aging, and the moisture generated during aging further accelerates the thermal aging of the insulation paper. Therefore, the aging condition of insulation paper largely determines the service life of transformers. Thus, improving the thermal

Figure 7.4 Common application fields of heat-insulating paper-based functional materials *Source:* Forgem / Adobe Stock Photos.

aging resistance of insulation paper is of paramount importance for the safe operation of transformers.

Qi Yixin et al. used silane coupling agent KH-550 and phenolic resin as modifiers to impregnate insulation paper that had been treated with drying, and then obtained modified insulation paper after heat curing. The results showed that KH-550 reacted with phenolic resin, introducing a small amount of Si–O bonds, which improved the mechanical properties and thermal resistance of the modified insulation paper.

Ma Jun et al. added modified nano SiO_2 particles to polyamide acid precursor solution and prepared nano SiO_2/PI composite thermal insulation paper by impregnation method. The composite insulation paper was aged at 150 °C, 160 °C, 170 °C, 180 °C, 190 °C, and 200 °C for 32 hours. The insulation paper was taken out every 4 hours for mechanical property testing and microscopic morphology analysis, and the thermal aging resistance of the composite insulation paper was studied.

7.2.4.2 Thermal Insulation Paper-based Functional Materials in Food Packaging

With the acceleration of people's lifestyles, an increasing number of disposable products inundate daily life, such as disposable take-out containers and paper cups. To prevent burns during use, these disposable paper products need to be endowed with certain thermal insulation functionality.

Benders Paper Cup Company has improved the heat resistance and strength of the thermal insulation paper cups produced by the company. Relevant tests have shown that even when the paper cups are filled with hot water, consumers can comfortably hold them for more than 3 minutes, thereby avoiding the risk of burns.

In order to enhance the thermal insulation and cooling capabilities of existing chocolate packaging materials and alleviate the problem of chocolate melting at high temperatures, Liu Jinyang et al. utilized thermal insulation and cooling coatings. They combined these coatings with chocolate packaging materials in the form of coating, demonstrating the feasibility of this method.

7.2.4.3 Thermal Insulation Paper-based Functional Materials in Daily Life

The global warming phenomenon has led people to rely more on air conditioning, refrigerators, and other cooling devices in their daily lives. However, the use of these appliances inevitably releases greenhouse gases, exacerbating environmental degradation.

Mao Tengli utilized titanium dioxide (TiO_2) and hollow glass microspheres to prepare thermal insulation coatings. Through surface coating processes, these coatings were compounded with cellulose base paper to produce thermal insulation paper. The results indicate that this thermal insulation paper effectively reflects ultraviolet, visible, and some near-infrared light within the wavelength range of 200 to 920 nm. This property aids in reducing indoor temperatures and decreasing energy consumption.

In Australia, a company used thermal insulation paper in its 110-story building and found that it not only saved $150,000 in expenses but also blocked over 96% of ultraviolet rays.

References

1 Wei, W. (2017). *Study on the preparation process and flame retardant properties of efficient halogen-free flame retardant paper*. Fuzhou: Fujian Agriculture and Forestry University.

2 Shi, Y. and Wang, G. (2016). Research progress on synthesis of organic phosphorus flame retardants and their application in flame retardant polymer materials. *Polymer Materials Science and Engineering* 32 (5): 167–175.

3 Shaw, S.D., Blum, A., Weber, R. et al. (2010). Halogenated flame retardants: Do the fire safety benefits justify the risks? *Reviews on Environmental Health* 25 (4): 261–305.

4 Inagaki, N., Onishi, H., Kunisada, H. et al. (2010). Flame retardancy effects of halogenated phosphates on poly (ethyleneterephthalate) fabric. *Journal of Applied Polymer Science* 21 (1): 217–224.

5 Tian, N., Wen, X., Gong, J., and ea al. (2013). Synthesis and characterization of a novel organophosphorus flame retardant and its application in polypropylene. *Polymers for Advanced Technologies* 24 (7): 653–659.

6 Nguyen, T.M.D., Chang, S.C., Condon, B. et al. (2012). Development of an environmentally friendly halogen-free phosphorus-nitrogen bond flame retardant for cotton fabrics. *Polymers for Advanced Technologies* 23 (12): 1555–1563.

7 Chen, M., Liu, Z., and Li, L. (2010). Development of magnesium hydroxide flame retardant and crystallization mechanism. *Guangzhou Chemical Industry* 38 (7): 17–19.

8 Zhou, H., Liu, Z., and Wei, Y. (2009). Research on preparation of flame retardant paper using magnesium hydroxide as flame retardant. *China Pulp & Paper* 28 (1): 13–16.

9 Yuan, B., Zhang, J., Yu, J. et al. (2016). Transparent and flame retardant cellulose/aluminum hydroxide nanocomposite aerogels. *Science China Chemistry* 59 (10): 1335–1341.

10 Li, X. and Qian, X. (2008). Study on the application of magnesium aluminum hydroxide as flame retardant filler in papermaking. *China Pulp & Paper* 27 (12): 16–19.

11 Zhao, H., Sha, L. Environmentally friendly flame-retardant paperboard and its preparation method: CN106868916A. 2017-06-20.

12 Veen, I.V.D. and Boer, J.D. (2012). Phosphorus flame retardants: Properties, production, environmental occurrence, toxicity and analysis. *Chemosphere* 88 (10): 1119–1153.

13 Mo, Z., Zhao, H., Liang, W.C. et al. (2016). Preparation of flame-retardant paper by microencapsulation of melamine-formaldehyde resin with ammonium polyphosphate. *Paper and Paper Making* 35 (7): 35–38.

14 Sha, L., Zhao, H. Flame-retardant and smoke-suppressing plant fiber paper-based material and its preparation method: CN106120431B. 2019-01-15.

15 Dong, Z., Zhu, Z., Rui, W. et al. (2009). Research progress on flame retardancy of carbon nanotube/polymer composite systems. *Journal of Textile Research* 30 (3): 136–142.

16 Kim, S. and Wilkie, C.A. (2008). Transparent and flame retardant PMMA nanocomposites. *Polymers for Advanced Technologies* 19 (6): 496–506.

17 Chen, S., Zhu, J., Wu, X. et al. (2010). Graphene oxide MnO_2 nanocomposites for supercapacitors. *Acs Nano* 4 (5): 2822–2830.

18 Tian, H., Wu, Y., Yin, Y. et al. (2016). Study on combustion kinetics of cellulose-like herbaceous energy plants. *Journal of Combustion Science and Technology* 22 (4): 335–341.

19 Liu, J. (2013). *Study on the flame retardant performance and mechanism of wood fiber-based ultra-lightweight materials*. Fuzhou: Fujian Agriculture and Forestry University.

20 Zhao, H., Sha, L. Flame-retardant paper-based material prepared using reactive flame-retardant fibers and its preparation method: CN109853289A. 2019-06-07.

21 Sha, L., Ma, C., Zhao, H. A reactive flame-retardant plant fiber and its preparation method and application: CN109811571A. 2019-05-28.

22 Zhang, M. (2010). *Processed Paper and Specialty Paper*, 82. Beijing: China Light Industry Press.

23 Huang, G. (2017). *Synthesis and Application Research of Novel Nitrogen-Expanding Flame Retardants*. Guangzhou: South China University of Technology.

24 Li, C. (2011). *Research on Preparation of Flame Retardant Paper by Magnesium Aluminum Hydrotalcite*. Tianjin: Tianjin University of Science and Technology.

25 Sha, L. (2016). *Preparation of Ammonium Polyphosphate-Silica Aerogel Composite Flame Retardant Filler and Study on Flame Retardant Mechanism of Paper*. Guangzhou: South China University of Technology.

26 Zhao, Z., Sha, L., Gu, J. Flame Retardant Cowhide Box Paperboard Prepared from Flame Retardant Secondary Fiber and Its Preparation Method: CN 109853282B. 2020-05-29.

27 Zhao, H., Sha, L. Composite Flame Retardant Paper-Based Material and Its Preparation Method: CN103510429A. 2014-01-15.

28 Lu, X., Sha, L., Gu, J.Ultra-low Quantitative Flame Retardant Corrugated Isolation Paper and Its Preparation Method: CN109811577A. 2019-05-28.

29 Li, Y., Sha, L., Zhao, H. et al. (2019). Preparation and flame-retardant mechanism of flame retardant air filter paper. *BioResources* 14 (4): 8499–8510.

8

Paper-based Functional Materials for Filtration and Separation

8.1 Overview

Filtration materials are one of the most important foundational materials for industrial purification and separation. With the continuous advancement of modern science and technology, various industries have undergone significant improvements, and public awareness of environmental protection has increased. The global demand for filtration materials has been increasing year by year. Currently, the development trend is the composite application of multiple materials, which can offer better filtration performance.

Paper-based functional materials are new materials primarily made from fibers, using papermaking forming technology to create a three-dimensional network structure. These materials differ entirely from traditional paper in both structure and performance, offering flexible, designable structures and morphological properties, as well as characteristics such as breathability, temperature resistance, corrosion resistance, water and oil repellency, and filtration performance. As a result, paper-based functional materials have broad applications in oil–water separation, filtration, adsorption, and other areas.

Within paper-based functional materials, there is a category known as filtration paper. Filtration paper or paperboard can separate unwanted substances or impurities from gases or liquids, recovering useful materials, and is commonly used to separate mixtures composed of different phases. Special filtration paper (and paperboard) has the following characteristics: good physical strength, high porosity, excellent chemical stability, and high filtration efficiency.

The performance and applications of filter paper vary depending on the types of raw materials used, additives, and forming conditions. The choice of raw materials for paper-based filtration materials plays a decisive role in the mechanical, chemical, and thermal properties of the product during the filtration process. Commonly used fibers in filter paper include the following types:

- Polytetrafluoroethylene fiber

Known for its excellent resistance to high temperatures and chemical corrosion, polytetrafluoroethylene (PTFE) fiber can withstand continuous temperatures

Paper-Based Functional Materials: Preparation and Applications, First Edition.
Xianping Yao, Genrong Chen, Zhen Yao, and Liping Zheng.

Figure 8.1 Polytetrafluoroethylene sieves.

of 260 °C and instantaneous temperatures up to 280 °C. It is widely used in high-temperature gas filtration. PTFE fiber is highly resistant to acids and alkalis and is not easily dissolved. Its density is twice that of polyester. Through heat treatment and shrinkage at temperatures below 300 °C, PTFE fiber achieves high dimensional stability. However, compared to other fibers, PTFE has relatively poor tear strength, corrosion resistance, and tensile performance. To overcome these drawbacks, PTFE fiber can be subjected to special needle felt reinforcement treatments, particularly enhanced shrinkage treatments, which provide sufficient strength and extend its service life to three to five years. As shown in Figure 8.1, PTFE mesh demonstrates these properties.

- Polyacrylonitrile fiber

Exhibits good thermal and chemical resistance under both dry and wet conditions. It remains stable in hot acidic environments and is suitable for filtration applications in such conditions. Under dry heat conditions, polyacrylonitrile (PAN) outperforms polyamide (nylon) and natural fibers. The maximum recommended operating temperature for PAN homopolymers is 130 °C, while for copolymers, it is 120 °C. PAN exhibits strong resistance to most inorganic and organic acids, performing better than aromatic polyamide fibers and polyamide fibers. Additionally, PAN has excellent alkali resistance, comparable to that of polyamide fibers.

- Aromatic polyamide fiber

A non-thermoplastic fiber that does not burn or support combustion. It can endure high temperatures and is used in high-temperature filtration applications. At high temperatures, aromatic polyamide (aramid) fibers, such as Nomex (a type of commercial aromatic polyamide polymer), will only carbonize or break down into small molecules rather than soften suddenly like typical thermoplastic fibers. Nomex fibers can withstand operating temperatures of up to 200 °C in dry conditions.

Nomex fibers are hydrolysable and can degrade quickly in high-temperature environments with the presence of chemicals and water vapor. In weakly acidic and neutral environments with water vapor concentrations below 10%, Nomex fibers are suitable for operating temperatures up to 190 °C with a lifespan of up to two years.

If the water vapor concentration increases to 20%, the temperature must be reduced to below 165 °C to maintain a life of two years. If the temperature remains around 190 °C, the lifespan decreases to just over six months.

At room temperature, Nomex fibers exhibit excellent bending resistance, comparable to polyester fibers. Aging at temperatures above 135 °C enhances Nomex fibers' abrasion resistance and bending strength. Nomex fibers can withstand most inorganic acids, organic acids, weak alkalis, and most hydrocarbons' chemical effects. Their performance is not affected by dust and gases produced by the metallurgical and mining industries.

- Polyphenylene sulfide fiber

A high-performance thermoplastic engineering plastic with excellent heat resistance and chemical stability. Polyphenylene sulfide (PPS) fiber can continuously operate at temperatures up to 190 °C and is used in environments with hot gases and chemical exposure. In an autoclave at 160 °C, PPS filter material can retain more than 90% of its strength. PPS filter materials exhibit excellent chemical resistance, particularly against sulfur and acids. This makes PPS an ideal choice for filtration applications in environments with harsh chemical conditions.

- Polyimide fiber

Known for its high-temperature stability and excellent filtration efficiency. Polyimide (PI) fiber is used in high-efficiency filter media, suitable for continuous use at temperatures up to 260 °C. It also has flame-retardant properties, strong resistance to all common organic solvents, good resistance to inorganic and organic acids, and average resistance to alkali at pH 5 to 9.

- Polypropylene fiber

Polypropylene (PP) fibers for gas or dust filtration are available in both staple and filament forms. PP fiber maintains good strength at temperatures between 100 and 110 °C and offers excellent chemical resistance and durability. Moreover, as shown in Figure 8.2, it is more resistant to corrosion and mold than most man-made fibers.

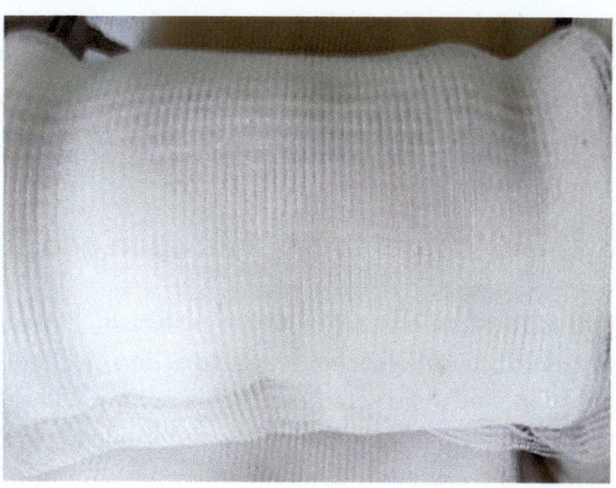

Figure 8.2 Polypropylene gas–liquid filters.

- Polyester fiber

Commonly used in filtration applications for its strength and thermal stability at temperatures up to 135 °C. However, its performance decreases at higher temperatures and in high-humidity environments. Polyester (PET) fiber has poor hydrolysis resistance and is not suitable for use in high-alkali and high-moisture working conditions (more than 5% water vapor, high temperature above 100 °C). For these shortcomings, the solution is as follows: if the temperature exceeds 135 °C, use Nomex and other high-temperature-resistant fiber filter media; if water vapor exceeds 5%, the temperature should be lowered to less than 100 °C; high alkali, high humidity, can be used to resist hydrolysis of polyacrylonitrile fibers and other fiber filter media.

- Glass fiber

Distinguished by its high-temperature resistance and excellent filtration performance, glass fiber is ideal for high-temperature filtration applications and offers significant advantages in terms of strength, chemical resistance, and durability. Moreover, glass fibers are chemically inactive and do not absorb moisture, so they do not swell. The high temperature resistance and thermal stability of glass fibers enable them to achieve good filtration results at temperatures higher than room temperature without the risk of fire. They exist in the form of small columns, which is the most desirable shape for a filter material. Glass fibers have the following advantages: first, glass fibers have excellent heat resistance. The surface chemical treatment of glass fiber filter media can be used at temperatures up to a maximum of 280 °C, which is very suitable for dust removal projects. So, at present and for some time to come, glass fiber remains an important high-temperature filter material. Second, glass fibers have high strength and low expansion rate. Glass fiber tensile strength than a variety of other natural, synthetic fibers are higher, the elongation is only 2% to 3%, and this feature is sufficient to ensure that the use of its design and production of large diameter ratio of the filter bag has sufficient tensile strength and dimensional stability. Third, glass fiber has excellent corrosion resistance. At present, commonly used glass fiber is divided into two types: alkali glass fiber and medium-alkali glass fiber. Alkali glass fiber has a high degree of stability at room temperature when exposed to water, wet air, and weak alkali solutions, but it is not resistant to higher concentrations of acid or alkali erosion. On the other hand, medium-alkali glass fiber has better water and acid resistance. Therefore, the appropriate type must be selected according to the composition of the glass fiber in order to achieve optimal performance as a filtration material.

8.2 Fundamentals of Filtration and Separation

Filtration paper-based functional materials, like other filtration materials, can be classified according to different phases into mechanisms such as gas filtration, liquid filtration, and solid separation. This section will introduce several common filtration principles, most of which are applicable to paper-based functional materials in filtration processes.

8.2.1 Gas Filtration and Separation

8.2.1.1 Mechanism

During filtration, several mechanisms govern the deposition of particles in a fluid. The most important mechanisms are discussed below:

1) Diffusion deposition

Due to Brownian motion, fine particles deviate from the flow lines of the fluid. As particle size decreases, Brownian motion increases, raising the probability of particle collision with fibers. This mechanism is particularly significant for particles smaller than 1 μm, especially those smaller than 0.2 μm, which move like gas molecules and are separated when they contact the filter fibers. Diffusion deposition increases with lower flow speeds and smaller fiber and particle sizes.

2) Direct interception

Larger particles collide and adhere to the surface of the filter material, forming a dust layer. The efficiency of this mechanism depends on particle size, velocity, and factors like electrostatic, gravitational, or thermal forces.

3) Inertial deposition

Larger particles (greater than 1 μm) do not follow curved flow lines in compressed air but continue along their original path, eventually impacting the filter fibers. Inertial deposition becomes stronger with larger particles and higher airflow speeds, improving filtration efficiency.

4) Gravitational deposition

Particles fall due to gravity, deviating from the fluid's flow lines and depositing on fibers. This mechanism mainly affects larger particles with higher settling speeds.

5) Electrostatic deposition

Friction between airflow and filter fibers generates static electricity, while dust particles also become charged. The electrostatic force (Coulomb force) attracts dust to the fibers, enhancing adsorption and filtration efficiency.

6) Van der Waals deposition

When particles are very close to the fibers, intermolecular Van der Waals forces cause deposition.

Various filtration mechanisms often work together, and their effectiveness depends on factors like fiber porosity, airflow speed, and particle size. These combined mechanisms enable fiber filters to achieve filtration efficiencies greater than 99%, making them widely used in gas filtration.

8.2.1.2 Filtration Process

The gas filtration process can be described by several parameters, such as pressure drop (Δp), particle amount (G), and filtration efficiency (E). The pressure drop (Δp)

of the filter is determined by the following formula:

$$\Delta p = p_1 - p_2 \tag{8.1}$$

where p_1 is the gas pressure at the filter inlet (Pa) and p_2 is the gas pressure at the filter outlet (Pa).

For a "clean" filter, the pressure drop (Δp) depends only on the fluid's characteristics and the filter's porous material. As filtration continues, the pressure drop also depends on the characteristics of particles deposited in or on the filter.

Regarding particle amount (G), if G_1 represents the amount of particles entering the filter per unit time, G_2 represents the amount of particles leaving the filter per unit time, and G_3 represents the amount of particles remaining in the filter per unit time, the equation can be obtained from the conservation law:

$$G_1 = G_2 + G_3 \tag{8.2}$$

The filtration process is theoretically complex and can be categorized into steady-state and unsteady-state filtration based on fluid characteristics.

1) Steady-state filtration

In steady-state filtration, particles accumulate on the "clean" filter, and the structural changes caused by particle deposition are minimal, meaning they do not significantly affect the primary parameters, Δp and E. During this stage, Δp and E do not change over time, so the process is considered steady-state. Practically, for gases with low concentrations of suspended particles, the filtration process can be approximated as steady state in the initial stages. This process assumes a collision efficiency of 1 between particles and the filter's structural components, so particles that contact the surface remain captured and are not re-released.

2) Unsteady-state filtration

The filtration process is much more complex, especially in later stages. Particle deposition causes structural changes in the filter, and the key parameters Δp and E change over time (either increasing or decreasing). Eventually, the filter becomes clogged, and this stage is referred to as unsteady-state filtration. Research in unsteady-state filtration focuses on the collision efficiency of particles with the capture surface, as well as particle adhesion. These processes, including secondary effects, cause changes in Δp and E over time.

8.2.2 Liquid Filtration and Separation

8.2.2.1 Filtration Types

Solid–liquid filtration refers to processes in which a suspension passes through a porous medium, removing some or all of the solid particles from the liquid. The structure of a typical solid–liquid filter is shown in Figure 8.3. Many paper-based filter cartridges and filter papers themselves serve as the support structure. The choice of filtration process depends on several factors, such as the characteristics of the

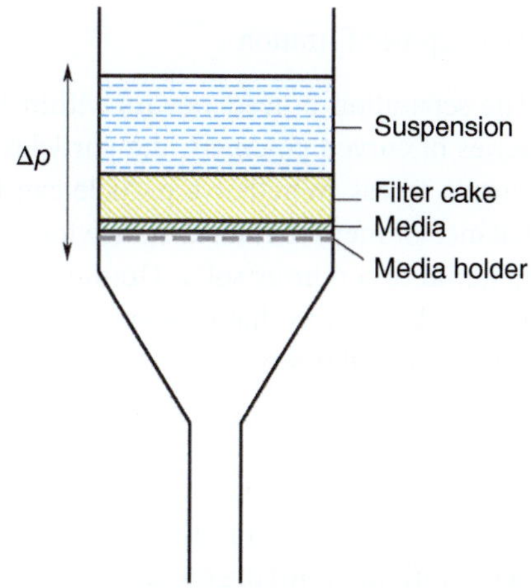

Figure 8.3 Structure of an ordinary solid–liquid filter.

solid particles, the fluid properties, the amount of material to be processed, whether the filtration is continuous, the dryness of the filter cake, and whether the retained material is solid or liquid.

An important design consideration for filters is the driving force, which can be gravity, suction, positive pressure, or centrifugal force. The selection of the appropriate force source depends on the factors mentioned above.

To facilitate the analysis of relevant factors, solid–liquid filtration is typically categorized based on the type of filtering medium used. The common types include screen filtration, microporous filtration, membrane filtration, deep-bed filtration, and cake filtration.

1) Screen filtration

This filtration uses a grating or mesh to remove coarse suspended solids such as weeds, rags, fibers, and pulp. Typical equipment includes grates, screens, and microfilters. This method is generally suitable for large flow rates where high filtration precision is not required.

2) Microporous filtration

Involves shaped filter media like filter cloth or sheets, and sometimes a layer of filter aid (such as diatomaceous earth) is applied to the medium to form a fine-pored filter cake that removes fine particles. The choice of filtering material depends on the required filtration precision.

3) Membrane filtration

Uses semi-permeable membranes as the filtering medium, driven by a certain force (such as pressure or electric field). Due to the extremely small pores and selective permeability, membrane filtration can remove bacteria, viruses, organic matter, and dissolved solutes from water. Main equipment includes reverse osmosis, ultrafiltration, and electrodialysis. While membrane filtration offers excellent resistance to high temperatures and corrosion, its relatively high cost and technical limitations prevent widespread adoption.

4) Deep-bed filtration

The separation process occurs within the medium. If the medium is viewed as a series of curved channels, the particles collide with the walls and are retained by certain forces. Whether a particle can be removed from the flow depends on the balance of inertial and resistance forces acting on it. The medium can be granular material or a porous solid. Granular filters include thick sand filters or precoated filter aids, such as diatomaceous earth, while porous solids include felt filters and sintered metal filters.

5) Cake filtration

In cake filtration, solid particles accumulate on the surface of the medium. After a short initial period, filtration continues through the deposited solid layer. This process continues until the pressure drop across the cake exceeds a maximum allowable value, or the available pores are filled. Cake filtration is widely used in processing industries, especially for filtering concentrated suspensions and recovering large amounts of solids. The most important factor in cake filtration is the permeability or resistance of the cake. This can be controlled by adjusting the particle size distribution of the material, adding other solids to the cake, or altering the agglomeration state of the particles.

As mentioned, the actual filtration process is typically a combination of several mechanisms, with some or all occurring simultaneously or sequentially.

8.2.2.2 Fundamentals of Filtration

This section further explores the theory behind cake filtration and deep-bed filtration, which are commonly encountered in daily life.

1) Cake filtration

The cake filtration process involves separating fine solid particles. Typically, it deals with particles that, when captured on the surface of a porous medium that only allows liquid to pass through, do not easily settle out of the liquid in which they are dispersed. As the process continues, additional particles accumulate on the surface of the initial solid particle layer, forming a "cake."

Cake filtration can be categorized into two types: incompressible cake filtration and compressible cake filtration.

Incompressible cake filtration: when the particles forming the cake are rigid and do not deform, increasing the pressure difference across the cake does not significantly change the particle shape or the voids between them. Such cakes are termed incompressible.

Compressible cake filtration: if the pressure difference increases and the permeability and porosity of the cake change, the cake is classified as compressible.

For truly incompressible cakes, the permeability and porosity remain constant regardless of pressure difference or flow rate. However, in practice, most cakes will experience changes in permeability or porosity when filtration conditions change. It is evident that the moisture content of the cake influences its porosity. For an

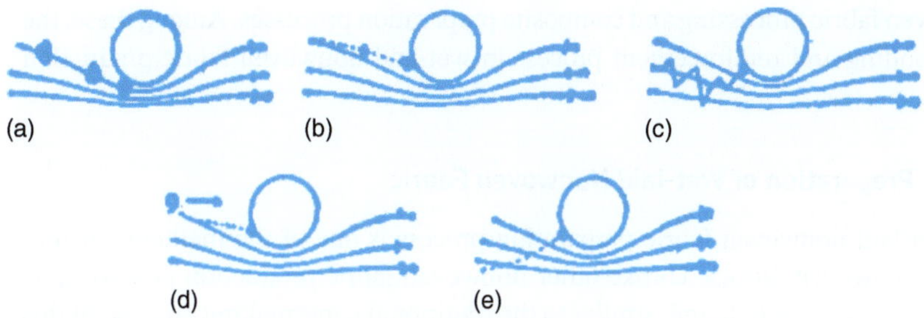

Figure 8.4 Forms of particle capture in depth filtration.

entirely incompressible cake, it is impossible to reduce the moisture content by simply increasing filtration pressure.

Moderate compressibility is beneficial as it allows for relatively low moisture content without significantly reducing filtration rate. Conversely, highly compressible slurries exhibit high resistance even under minimal pressure, leading to low filtration rates and high moisture content, making them inefficient for filtration. In such cases, pretreatment techniques like filter aids must be used to reduce compressibility.

2) Deep-bed filtration

Deep-bed filtration involves passing liquid through a layer of particles or fibers to purify it. It is widely used in water purification and the treatment of wastewater containing hydrocarbons, alcohols, and esters. The filter medium in deep-bed filtration has a three-dimensional porous structure that captures solid particles smaller than the pore size. Even particles far smaller than the pores can be captured deep within the medium.

Deep-bed filtration operates through a complex mix of mechanisms, where particles, under the influence of inertial force, liquid pressure, or Brownian motion, first encounter the walls of the pores. The particles then either adhere to the pore walls or aggregate due to Van der Waals forces or other surface forces.

The mechanisms for particle capture in deep-bed filtration include interception, inertia, diffusion, gravitational settling, and various fluid effects (as shown in Figure 8.4). Notably, Weigelsvagen demonstrated that particles with diameters between 2 and 10 μm have a significant gravitational settling rate that influences rapid filtration [1]. Due to collisions between suspended fluid molecules, fine particles undergo random diffusion, with smaller particles exhibiting more erratic motion, increasing the chance of collision with obstacles, and thus improving the filtration efficiency.

8.3 Preparation of Paper-based Functional Materials for Filtration and Separation

The preparation processes for paper-based functional materials used in filtration and separation are diverse. Currently, the main preparation methods are wet-laid nonwoven fabric processing and composite preparation processes. Among these, the fiber bonding and reinforcement process in wet-laid nonwoven fabric production is key.

8.3.1 Preparation of Wet-laid Nonwoven Fabric

The wet-laid nonwoven fabric preparation process is one of the methods for producing nonwoven fabrics. Unlike other nonwoven fabric production processes, its forming process is wet-based, similar to the traditional papermaking process. In this method, water is used as a carrier to uniformly disperse fiber materials into a slurry. The slurry is further dewatered on a forming mesh and then dried and bonded to form the fabric.

The wet-laid nonwoven fabric preparation process is well-established, offering controllable pore size and porosity. The fabric has good fiber uniformity and density, as well as strong tensile strength, making it widely used in filtration and separation materials.

8.3.1.1 Wet-Laid Nonwoven Fabric Production Process

The production process of wet-laid nonwoven fabric is similar to traditional papermaking, but there are significant differences due to the nature of the fiber materials, particularly in the raw material preparation and forming stages.

In the fiber preparation stage, wet-laid nonwoven fabrics primarily use longer fibers, with fibers having a length-to-diameter ratio greater than 300 making up more than 50% of the total fabric mass. These longer fibers tend to drain water quickly and are mostly hydrophobic, making them difficult to disperse and prone to clumping. To maintain good dispersion of the fibers in the suspension, the slurry concentration is usually much lower than in traditional papermaking, often up to 90% lower [2].

In addition to the differences in fiber preparation, there are also variations in the forming stage. The method of web formation and the bonding technique selected during this stage significantly influence the quality of the nonwoven fabric.

8.3.1.2 Types of Fibers

The wet-laid nonwoven fabric production process uses various fiber raw materials, including plant fibers and chemical fibers. Plant fibers such as wood pulp, cotton pulp, and hemp pulp are commonly used, with fiber lengths typically ranging from 4 to 10 mm. However, due to the limitations of natural fibers in providing certain properties (e.g., strength), a combination of plant fibers and chemical fibers is often used in production. This blend helps to enhance the strength and water absorption of the fabric while also providing chemical fibers' benefits, such as increased tensile strength, tear resistance, and specific properties like high air permeability and ion selectivity. Common chemical fibers used include polypropylene, polyethylene, polyester, polyvinyl alcohol, and viscose fibers.

8.3.1.3 Fundamentals of Wet-Laid Nonwoven Fabric Formation

The formation principle of wet-laid nonwoven fabrics is similar to that of traditional papermaking, both involving the uniform dispersion of fiber raw materials in a slurry suspension. Wet-laid nonwoven fabrics typically use longer synthetic fibers, which are hydrophobic, smooth-surfaced, and not fibrillated, making them prone to clumping and difficult to interlace. To address this issue during the dispersion of the fiber slurry, high-speed, low-concentration stirring and the addition of dispersing agents are employed to improve fiber dispersion. The dispersing agents work by increasing the slurry viscosity, restricting fiber movement, reducing aggregation, and slowing fiber sedimentation, thus ensuring uniform dispersion.

Unlike traditional papermaking, in which fibers settle on a forming wire, wet-laid nonwoven fabrics rapidly dewater on the forming wire. During the forming process, the fiber attachment state is controlled by filtering resistance. If fiber dispersion is uneven, areas with lower filtration resistance will have fewer fibers, allowing for faster water drainage. During the subsequent fiber sedimentation, fibers will first settle in these areas, resulting in a more uniform sheet formation [3].

8.3.1.4 Formation Equipment of Wet-Laid Nonwoven Fabric

The choice of forming equipment significantly impacts the performance of nonwoven fabrics. Currently, the primary forming equipment for wet-laid nonwoven fabrics includes the rotary screen former and inclined screen former.

1) Rotary screen former

The main equipment for rotary screen forming is the vacuum rotary screen former, which is equipped with a vacuum box in the forming area. A major issue with rotary screen formers in production is that the centrifugal force limits the speed, making it difficult to increase the machine speed. However, the vacuum rotary screen former overcomes this limitation by counteracting centrifugal force, enabling higher machine speeds and improving production efficiency [4].

When forming, the rotary screen former dewaters quickly in a short time and distance, resulting in higher longitudinal tensile strength in the formed paper. However, it is not ideal for making nonwoven fabrics from chemical fibers or long fibers, which is why inclined screen formers are more commonly used for wet-laid nonwoven fabric production.

2) Inclined screen former

The vacuum rotary screen former's design is limited in terms of the vacuum suction box and forming area compared to inclined screen formers. The key difference between an inclined screen former and a standard long wire former is that the inclined screen former is tilted at an angle (typically 10°–15°) relative to the bottom rail.

In the inclined screen former, the slurry is fed from the slurry pipe, flows through a channel into the forming area, and undergoes dewatering through the vacuum box below. The vacuum box is divided into several separate dewatering chambers, which can be individually adjusted. The slurry enters the forming area, where the flow direction is perpendicular to the dewatering box, allowing the fibers to orient freely. Fiber orientation is controlled by adjusting the mesh speed and slurry flow rate.

At the end of the forming area, there is a lip plate that covers part of the forming area to prevent the slurry from forming at the interface between air and water, which could cause fiber aggregation. Adjusting the length of the lip plate can control the flow speed of the slurry [5].

The inclined screen former has several advantages over traditional long-wire and rotary screen formers, particularly in terms of paper uniformity, air permeability, and water removal, making it more suitable for the formation of long-fiber slurries.

Advantages of the Inclined Screen Former:

1) Low pulp concentration for forming

The inclined screen former operates at a much lower pulp concentration compared to traditional systems, generally ranging from 0.01% to 0.05%. This low concentration ensures that long fibers have sufficient space to remain in suspension during the forming process, preventing fiber aggregation. Fiber lengths typically range from 3 mm to 25 mm in this process.

2) Superior paper uniformity and air permeability

The slurry is highly diluted during formation, which ensures that long fibers remain well-distributed and suspended. The inclined screen former enables rapid and efficient dewatering, synchronizing the dewatering and forming processes. This results in excellent paper uniformity and superior air permeability, which is crucial for producing materials such as automotive filters and coffee filters, where these properties are essential.

3) Low longitudinal/transverse tensile strength ratio

The inclined screen former's unique forming and dewatering process allows the fibers to form on the screen under vacuum suction without significant external forces altering their natural arrangement. The result is a fiber distribution with little directional orientation across the screen. This uniform distribution leads to a lower longitudinal/transverse tensile strength ratio compared to paper made on long-wire or rotary screen formers. While the typical ratio for inclined screen-formed paper is between 1.1:1 and 2.5:1, for conventional long-wire and rotary screen-formed paper, the ratio is typically between 2.5:1 and 5:1. This lower ratio improves the paper's overall strength and evenness.

With the development of filter materials and specialty papers, the inclined screen former has evolved from a single-layer to a multi-layer design, and now, some models are equipped with hydraulic screen boxes in front of the mesh. These improvements help in the production of high-performance filter materials and specialty papers. This evolution is shown in diagrams (Figures 8.5 to 8.8). The enhancements in equipment design further support the production of superior filtering materials, catering to specific performance requirements [6].

8.3.1.5 Fiber Bonding and Reinforcement Technology

In the preparation process of wet-laid nonwoven fabrics, plant fibers are often beaten to fibrillate the fiber surface, increasing the number of hydrogen bonds exposed

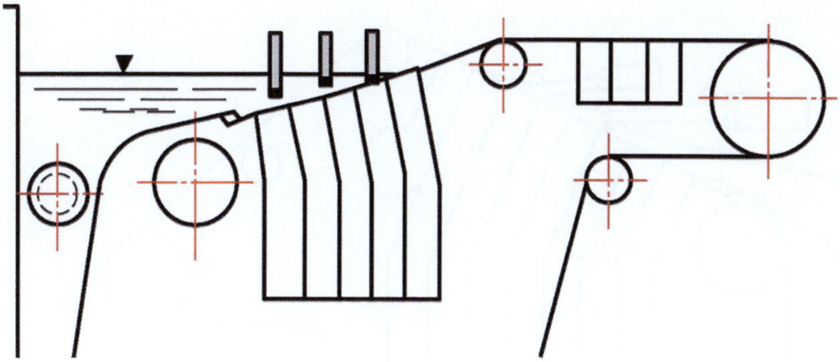

Figure 8.5 Initial slanting mesh former.

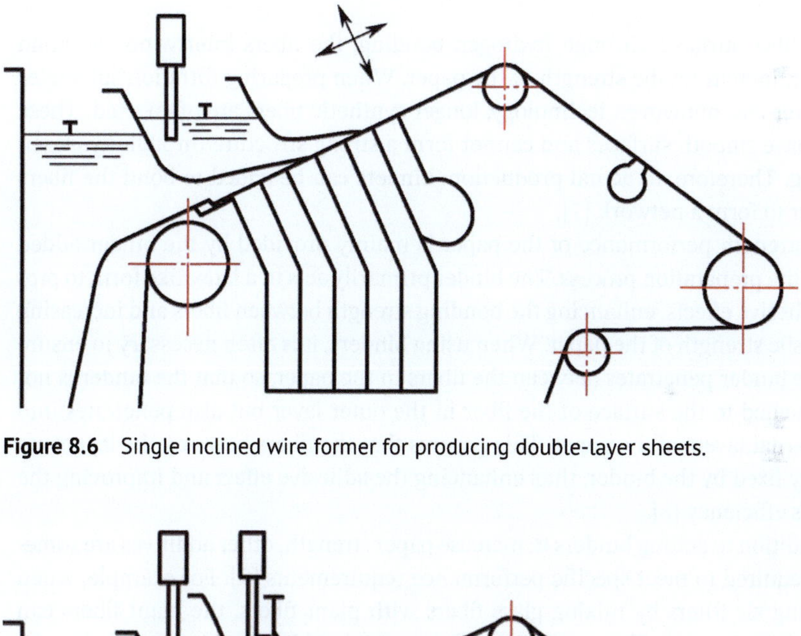

Figure 8.6 Single inclined wire former for producing double-layer sheets.

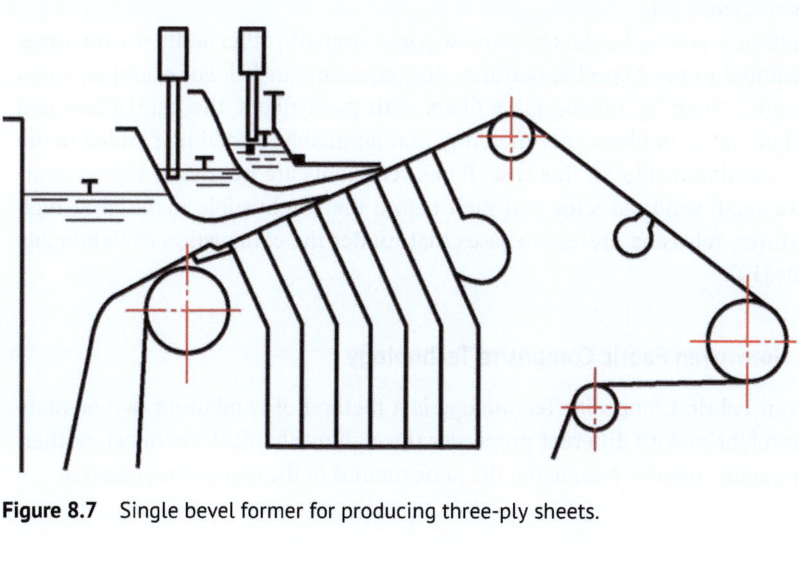

Figure 8.7 Single bevel former for producing three-ply sheets.

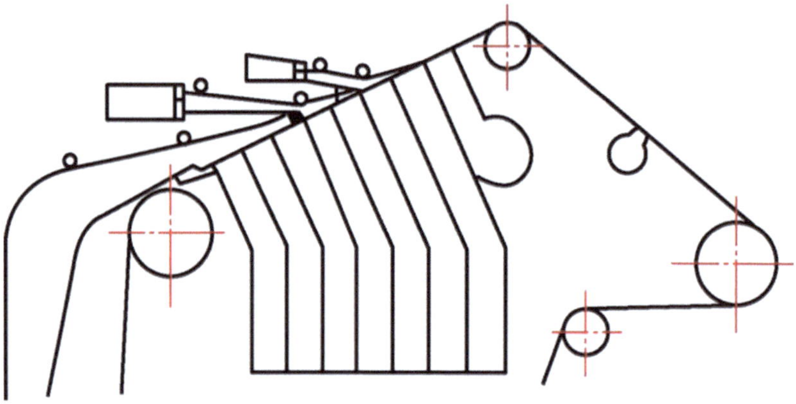

Figure 8.8 Three-ply inclined mesh former with hydraulic mesh front boxes.

on the fiber surface. Through hydrogen bonding, the fibers intertwine and bond together, increasing the strength of the paper. When preparing filtration substrates using wet-laid nonwoven technology, longer synthetic fibers are often used. These fibers have smooth surfaces and cannot form a strong structure through hydrogen bonding. Therefore, in actual production, binders can be added to bond the fibers together to form a network [7].

The strength performance of the paper is mainly provided by the binder added during the preparation process. The binder primarily acts in a latex-like form to provide adhesive effects, enhancing the bonding strength between fibers and increasing the tensile strength of the paper. When using binders, it is often necessary to ensure that the binder penetrates between the fibers in the paper, so that the binder is not just attached to the surface of the fiber in the outer layer but also penetrates into the internal layers of the paper. This ensures that the fibers between the layers are securely fixed by the binder, thus enhancing the adhesive effect and improving the binder's efficiency [8].

In addition to adding binders to increase paper strength, other additives are sometimes required to meet specific performance requirements [9]. For example, when preparing air filters by mixing glass fibers with plant fibers, the plant fibers can burn, while the glass fibers are inherently nonflammable. The binder added to the paper is also flammable. In this case, flame retardants are necessary. Flame retardants are small solid particles that melt before the combustible material at high temperatures, releasing protective gases that hinder the combustion of flammable materials [10].

8.3.2 Nonwoven Fabric Composite Technology

Nonwoven Fabric Composite Technology is a method of combining two or more nonwoven fabrics with different properties through mechanical, chemical, or thermal treatments, thereby enhancing the performance of the composite material.

8.3.2.1 Adhesive Composite

Adhesive composite is a method that uses chemical bonding, where adhesives are applied to fuse two or more layers of materials together, forming a composite filtration material. The advantage of an adhesive composite is that the bond strength between the materials is high, which is beneficial for post-processing of the product. However, the use of adhesives also reduces the pore size of the paper substrate, which may damage the material structure and shorten the product's final service life.

8.3.2.2 Heat Bonding Composite

Heat bonding composite is a process that uses heat or a combination of heat and pressure, with low-melting-point polymers as the outer layer and high-melting-point polymers as the core layer, to bond two or more layers of material. This method is advantageous because it is simple, fast, cost-effective, and does not involve chemical adhesives.

A common method used in heat bonding is ultrasonic bonding. Figure 8.9 illustrates the mechanism of ultrasonic bonding. When a 220 V voltage and 50 Hz frequency current enters the ultrasonic generator, the generator starts working, increasing the frequency from 50 Hz to 20,000 Hz, and maintaining this frequency. Under the combined effect of pressure and vibration frequency, the molecular motion inside the material intensifies, converting the high-frequency vibration energy into heat, causing the material to soften and melt, thus bonding two layers of polymer materials together.

Single filtration materials have many drawbacks, such as low filtration efficiency and short service life. By combining nonwoven fabric and polypropylene nonwoven fabric, a composite filtration material can be prepared, which offers multi-stage filtration effects, high purification capability, and improved filtration efficiency.

Figure 8.10 shows the structure of a nonwoven fabric/polypropylene nonwoven fabric composite filtration material prepared using the heat bonding composite method. The outer layer consists of a low-melting-point nonwoven fabric as the skin layer, with a high-melting-point polypropylene nonwoven fabric layer as the core, followed by a fiber filtration layer. The center of the structure features a breathable layer made from polymer resin fibers, which has a certain level of air permeability and does not cause resistance to airflow. A fiber filtration layer is placed between the polypropylene nonwoven fabric layer and the breathable layer. This multi-layer

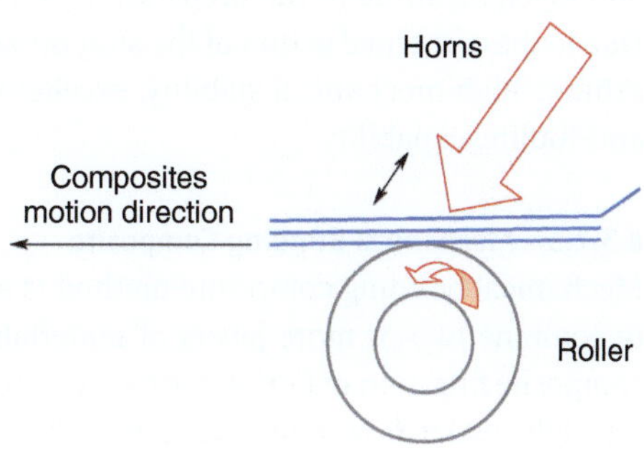

Figure 8.9 Ultrasonic compounding mechanism.

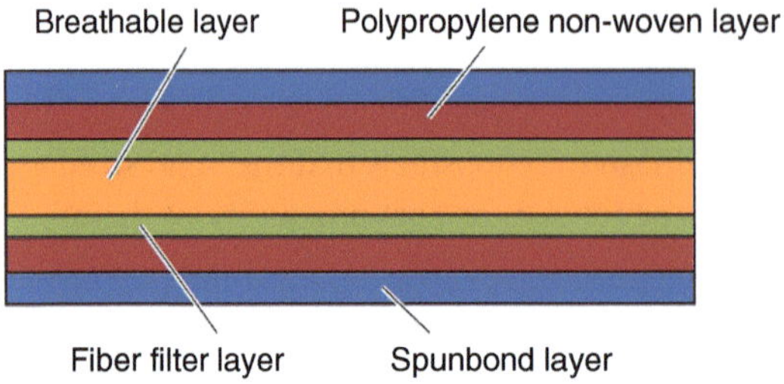

Figure 8.10 Nonwoven/polypropylene nonwoven composite filter material structure.

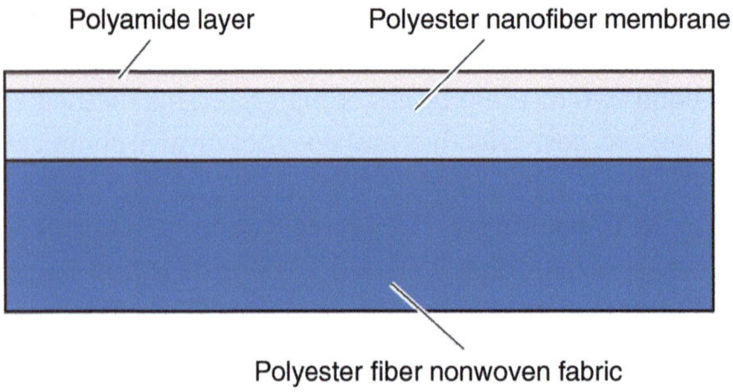

Figure 8.11 Polyamide/polyester fiber nonwoven composite structure.

composite filtration material can filter particles of different sizes and has strong purification capabilities.

Forward osmosis membrane technology is widely used in the field of seawater desalination due to its low energy consumption and high water production rate. The use of nonwoven fabric as a support layer, with an active layer composite on top, has become a research hotspot in recent years. Figure 8.11 shows the structure of a polyamide/polyester fiber nonwoven fabric composite material. In this structure, polyester fiber nonwoven fabric is used as the base layer. First, electrospinning technology is employed to form a super-thin polyester nanofiber membrane on the nonwoven fabric. Then, hot-pressing treatment is applied to fuse the fibers at the interface, enhancing the interfacial bonding strength between the nanofiber membrane and the nonwoven fabric. Following this, an interfacial polymerization process is used to form a polyamide active layer on the surface of the composite support layer, resulting in the preparation of a composite forward osmosis membrane. Due to the combined action of the support layer and the active layer, this membrane exhibits high mechanical stability, excellent water treatment efficiency, and strong anti-fouling capability.

8.3.2.3 Mechanical Bonding Composite

Mechanical bonding composite method is a technique that uses mechanical forces to combine two or more layers of materials. This method can be used to prepare composite filtration materials with excellent permeability, meeting the requirements for high carrier flow rates. Additionally, as no chemical adhesives are required, it expands the range of applications for the material.

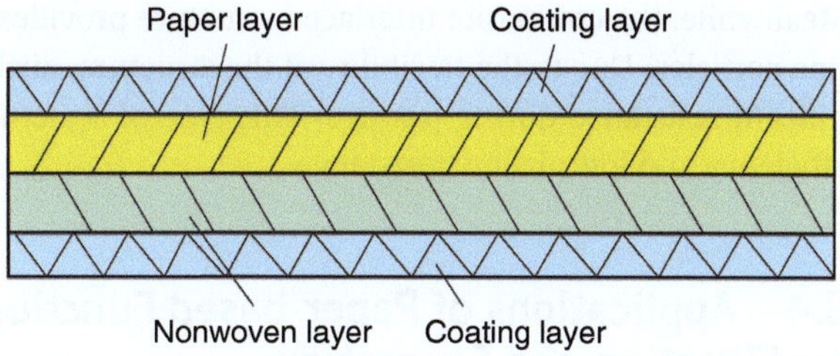

Figure 8.12 Paper/nonwoven composite filter material structure.

Many existing filtration materials are made from plant fibers to create paper-based filters, such as those used in automotive air filters. However, these materials have certain shortcomings, such as pore size permeability and paper burst resistance. On the other hand, nonwoven fabric filters made from chemical fibers, such as those used in boiler flue gas filters, offer higher strength and burst resistance, but their pore size and permeability variability are lower than those of paper-based filters. By combining these two types of materials, the disadvantages of single materials can be compensated for, improving the strength and filtration precision of the material.

Figure 8.12 shows the structure of a paper/nonwoven fabric composite filter material. The preparation of the paper layer involves dispersing plant fibers, pulping, and then forming the material through a wet papermaking process on the papermaking wire, followed by drying to form the fibrous paper sheet. The nonwoven fabric layer is made by melt-blowing polypropylene, and the fabric is then heat-pressed to strengthen it. By layering the two materials and applying high-pressure water flow, the fibers of the paper layer are implanted into the nonwoven fabric layer through micropores. Finally, the composite paper/nonwoven fabric filter material is produced by coating. This composite material compensates for the limitations of both paper-based and nonwoven fabric filters, further enhancing the overall performance.

Electrospun nanofibers and nonwoven fabrics are interlaced and stacked to prepare composite filtration materials, creating a sandwich structure between the fibers and nonwoven fabric. This forms a graded filtration mechanism, which can improve filtration efficiency.

Figure 8.13 shows the structure of the electrospun nanofiber/nonwoven fabric composite filter material. The multi-layer interlaced composite structure features coarse fibers in the main body of the composite material, which have a good filtration effect for larger particles. The larger pore sizes of these fibers facilitate airflow.

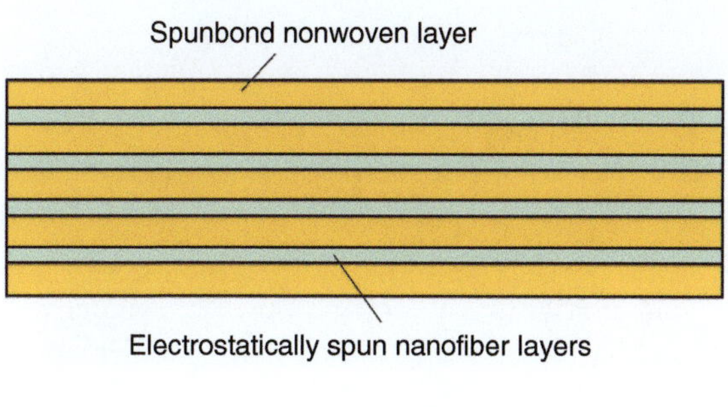

Figure 8.13 Structure of electrostatically spun nanofiber/spunbond nonwoven composite filter material.

Meanwhile, the composite interlaced structure provides a better filtration effect for fine particles. Due to the flexibility of the structure, airflow does not encounter significant resistance during passage. This type of structure results in high filtration efficiency and low airflow resistance.

8.4 Applications of Paper-based Functional Materials for Filtration and Separation

8.4.1 Automotive Filtration

Automotive filtration materials are key components in manufacturing automotive filters, primarily serving to filter impurities and moisture from air and fuel to prevent engine wear and extend its lifespan. These materials are essential for maintaining engine performance and operation. Automotive filters are categorized based on their use, including air filters (Figure 8.14), fuel filters, and oil filters. These filters can be either pleated or flat. Automotive filtration materials are permeable, allowing fluid to pass through while separating impurities.

The manufacturing process and performance of automotive filtration materials have improved over time. These materials are classified into vehicle filter paper, nonwoven fabric filters, and composite filters. Vehicle filter paper is made by impregnating and drying base paper with resin. It is primarily made from plant fibers, with synthetic or inorganic fibers added for performance enhancement. Nonwoven fabric filters have higher air permeability and a greater number of pores per unit area, improving filtration efficiency. This has led to a rapid increase in their usage.

Composite filters are made by combining two or more layers of different filter materials, with varying structures depending on the bonding process. Composite filters offer superior filtration performance compared to single-layer materials. Researchers have also proposed the concept of deep gradient filtration, where fibers are arranged in a gradient structure to continuously vary the pore size, balancing

Figure 8.14 Automotive air conditioning air filtration materials.

filtration efficiency and resistance. These filters are highly effective, with fine pores and excellent durability, making them ideal filtration materials.

In China, the development of automotive filtration materials began later, in the 1950s–60s, trailing over 100 years behind Western countries. However, with economic reforms, foreign investments, and rapid advancements in the automotive industry, China now has a number of leading automotive filter manufacturers. However, domestic filter papers still face challenges like lower filtration efficiency, shorter lifespan, and less stability compared to advanced foreign brands. With continuous technological advancements and significant investment, domestic companies are now targeting international benchmarks to innovate and improve product quality, efficiency, lifespan, and cost-effectiveness.

8.4.2 High-efficiency Air Filtration

High-efficiency air filters (HEPA) are widely used in various industries, including factory exhaust systems, hospitals, pharmaceuticals, food, cosmetics, and cleanrooms. When selecting air filtration materials, it is essential to meet national purification standards for filter efficiency and consider factors such as resistance (initial resistance, final resistance, and dust-holding capacity). A higher final resistance increases lifespan but reduces airflow. For air conditioning systems, filters with higher efficiency than the design level should be selected. Although the initial cost is higher, it prevents issues like fan performance degradation due to duct blockages, extends the filter's lifespan, reduces cleaning frequency, and saves costs.

Key considerations when selecting filter paper include:

Effective area: larger effective areas increase dust holding capacity, reduce resistance, and extend the filter's lifespan, though at a higher cost.

Fiber diameter: finer fibers provide better interception and higher filtration efficiency.

Binder content: higher binder content increases tensile strength and filtration efficiency, reduces shedding, and improves dust resistance, but increases resistance.

China's development of high-efficiency air filters began in the 1960s, with further advancements after the introduction of foreign filters in the 1980s. However, domestic filters still lag behind advanced countries in Europe and America. The performance improvement of high-efficiency filters significantly impacts filtration efficiency. With technological advancements, higher demands are placed on these filters: precision processing, miniaturization, and high purity. The higher these requirements, the better the filtration performance.

High-efficiency air filters can be categorized into with and without partition types (Figures 8.15 and 8.16). Filters with partitions use aluminum foil separators and pleated filter media, which accumulate larger dust particles at the bottom and effectively filter microdust on the sides. The deeper the pleats, the longer the lifespan. Filters without partitions use hot-melt glue instead of aluminum foil to separate the filter media. These filters offer similar performance to partitioned filters but have a smaller thickness (e.g., 50 mm vs. 150 mm) and are better suited for applications where space and energy efficiency are required. Both types are widely

Figure 8.15 Partitioned HEPA filter.

Figure 8.16 Compartmentless HEPA filter.

used in aerospace, electronics, pharmaceuticals, biotechnology, and cleanroom air purification.

The filtration mechanism of high-efficiency air filters (HEPA) is complex, with various mechanisms working together to effectively improve filtration efficiency. As air filters have evolved, there has been a shift from traditional HEPA filters with partitions to HEPA filters without partitions. Some industries are now using ultra-low penetration air filters (ULPA), which have higher filtration grades but more challenging technical requirements. ULPA filters represent the future trend in air filtration development.

8.4.3 Water Filtration

Membrane filtration, a key technology in water treatment, began development in the 1950s for seawater desalination. It separates solids from liquids by passing water through membrane pores, retaining impurities without chemical changes. However, the small size of membrane pores presents technical challenges.

The main research objectives in water filtration include improving filtration speed, increasing pollution retention capacity, enhancing filtration efficiency, and improving backwashing efficiency. Granular filters, due to their limited particle size, struggle to address these issues effectively. Fibrous filters, with their smaller diameters, larger surface areas, lower water resistance, and higher filtration precision, are emerging as an important alternative material. For instance, fibrous filtration technology has been applied in reverse osmosis seawater desalination pretreatment, showing promising results. In tests with coagulated seawater, fiber filters reduced turbidity to below 0.2 Nephelometric Turbidity Units (NTU), maintained removal efficiency over 93%, and kept Silting Density Index (SDI15) below 3.2. The removal efficiency for total iron exceeded 94%, with particle removal rates above 85%. Water head loss increased slowly, remaining under 0.02 Megapascal (MPa). The treated water met the requirements for reverse osmosis membranes.

In recent years, high-efficiency fiber filtration, with its large surface area, high porosity, low resistance, and easy cleaning, has gradually replaced traditional sand filtration and has been widely adopted in seawater desalination pretreatment.

8.4.4 Battery Separator Paper

Battery separator paper, a special functional material, is placed between the positive and negative electrodes of a battery. Its main function is to isolate the electrodes, allowing ions to pass through while preventing electron flow, thus avoiding internal short circuits. The quality of the separator paper directly impacts the battery's lifespan and safety. Depending on the type of battery, separator papers are categorized into zinc-manganese dry cell separator paper, alkaline-manganese cell separator paper, cadmium-nickel cell separator paper, nickel-metal hydride separator paper, and lithium-ion battery separator paper. Though performance requirements vary, the general specifications for separator paper are as follows:

- Good insulation to prevent direct contact between the electrodes and avoid internal short circuits.
- Adequate ion permeability with pores allowing ion transport.
- Low resistance and efficient ion conductivity.
- High chemical stability.
- Sufficient mechanical strength and toughness.

Paper-based separators are made from cellulose and its derivatives through papermaking processes. There are two main preparation methods: the pure cellulose method, which offers a high dielectric constant, thermal stability, and chemical resistance; and the blending method with other fibers, which provides customized performance.

Due to their excellent properties, cellulose-based separators have become a focus of research in the battery industry, potentially surpassing traditional polyolefin separators. Some studies have applied dopamine self-assembly on the separator surface, enhancing adhesion, mechanical strength, stability, and cycling performance for high-density batteries.

Although research on paper-based separators has shown promising results, commercial production is still some time away. Challenges remain in uniformity, thickness, and preventing modified cellulose from dissolving in organic electrolytes. Further experimental studies are needed, and high-pressure-resistant cellulose separators for advanced batteries are an emerging research trend.

8.4.5 Face Masks

In the past, China primarily used ordinary cotton yarn masks, which had a loose fabric structure, poor facial fit, and low filtration efficiency, offering minimal protection against viruses. With the improvement of healthcare standards, disposable masks made from nonwoven materials have now become the standard.

Most medical masks on the market consist of at least three layers. The middle layer is the core filtration layer, while the outer and inner layers serve as protective layers. These protective layers are typically made using spunbond nonwoven fabric technology, which provides the mask's framework and effectively intercepts larger particles (greater than 10 μm) to prevent early clogging of the core filtration layer. The core spunbond nonwoven fabric manufacturing process focuses on two key aspects: fiber network consolidation and raw material modification technology, as shown in Figure 8.17. During the thermal consolidation of the spunbond fiber web, techniques such as heat pressing or hot air are used. The fiber raw materials are subjected to high temperature and pressure from a pair of rollers, causing the filaments to deform, melt, and bond together to form a network.

The core filtration layer of the mask is typically formed using melt-blown nonwoven fabric technology, as illustrated in Figure 8.18. Melt-blown nonwoven fabric is characterized by fine fibers, small pore size, large surface area, and high porosity. Research in melt-blown nonwoven fabrics mainly focuses on reducing fiber diameter to the nanoscale and polymer modification. By improving the spinneret structure

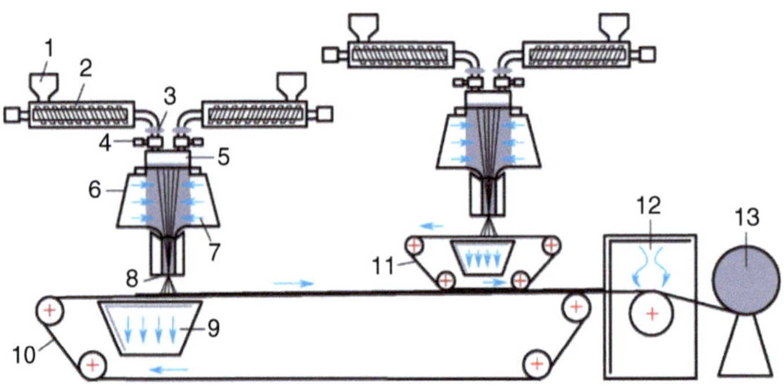

Figure 8.17 Spunbond nonwoven fabric molding technology process flow. 1-hopper; 2-screw extruder; 3-filter; 4-metering pump; 5-spinning box; 6-drawing unit; 7-cooling air; 8–9-suction device; 10, 11-web forming device; 12-reinforcement device; 13-rolling device.

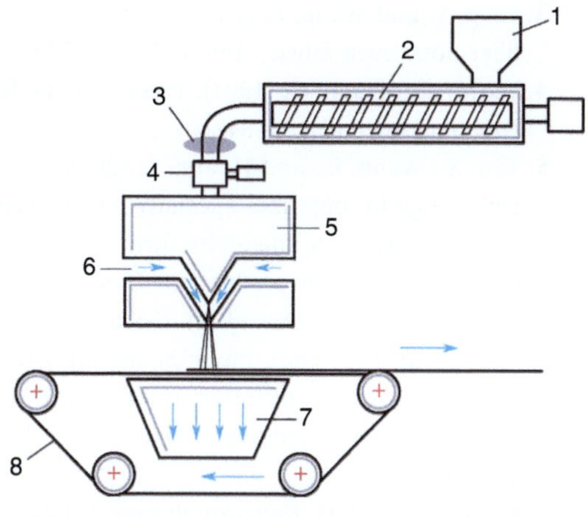

Figure 8.18 Meltblown nonwoven molding technology process. 1-hopper; 2-screw extruder; 3-filter; 4-metering pump; 5-spinning die head; 6-hot air; 7-suction device; 8-Web-forming device.

design and using high-melt-index polymer pellets, the fiber diameter can be reduced to enhance filtration efficiency. Polymer modification techniques can also address the common issue of short shelf life for melt-blown nonwoven fibers.

To meet growing health and safety demands, nonwoven materials for protective masks not only need to meet filtration standards but also exhibit enhanced antibacterial and antiviral properties. At Donghua University, Zhu Xiaoming modified titanium dioxide (TiO_2), a photocatalytic antibacterial material, to improve its stability. The modified TiO_2 was then incorporated into polypropylene melt-blown nonwoven fabric through online composite methods and into polyethylene/PET spunbond nonwoven fabric through offline methods. These materials were then combined to create TiO_2-modified antibacterial composite filters, showing excellent antibacterial performance and potential for medical mask production.

In response to the COVID-19 pandemic, companies in Japan, the United States, and China developed paper-based masks using wet-laid nonwoven fabric technology, which met N95 filtration standards and contributed significantly to global pandemic control.

As public health management improves and awareness of health and safety grows, the demand for protective masks with higher filtration efficiency, lower breathing resistance, and improved comfort will increase. Future developments will focus on enhancing the overall performance of mask materials to address issues such as breathing resistance and fit, ensuring better safety and reducing exposure risks.

References

1 Chang, J. and Vigneswaran, S. (1990). Ionic strength in deep bed filtration. *Water Research* 24 (11): 1425–1430.

2 Wang, Z., Dang, Y., and Li, H. (2006). Technology of wet-laid nonwoven fabric. *Household Paper* 28 (3): 35–37.

3 Yang, Y. and Wang, Z. (2017). Experimental study on wet-laying composite fiber nonwoven fabric. *Anhui Chemical Industry* 43 (6): 39–42.
4 Ye, M. and Duan, C. (1991). Pressure cylinder mould former. *Paper and Papermaking* 5 (2): 33–34.
5 Yao, X., Wang, L., and Huang, L. (2015). Application of inclined-wire forming technology in long-fiber specialty paper. *East China Paper* 46 (5): 30–36.
6 Guo, Z. (2002). Inclined-wire forming machine for wet-laid nonwoven fabric and specialty paper production. *Household Paper* 24 (12): 37–38.
7 Peng, Q. (2012). *Effects of Multi-Scale Structures of Fiber/Resin Composites on Mechanical Properties*. Jinan: Shandong University.
8 Zhou, T. (2013). *Application Research of Glass Fiber in Thin Page Substrate Produced by Papermaking Method*. Xi'an: Shaanxi University of Science and Technology.
9 Jinlong, X. (2011). *Preparation and Application Research of External Cross-Linked Acrylic Copolymer Emulsion Adhesive for Nonwoven Fabrics*. Nanchang: Jiangxi Normal University.
10 Wenjun, Y. (2017). *Research on Air Filter Paper and its Flame Retardant Modification*. Suzhou: Soochow University.

9

Paper-based Functional Materials for Analysis and Detection

9.1 Development of Paper-based Functional Materials for Analysis and Detection

The use of paper-based materials in analytical detection has a long history. As early as 1664, the renowned British chemist Boyle discovered that the purple liquid extracted from lichens turned red in acidic conditions and blue in alkaline conditions. He utilized this characteristic to invent litmus paper, which was used to determine the acidity or alkalinity of solutions. This innovation laid the foundation for modern pH paper used in chemistry laboratories [1]. In the following approximately 300 years, researchers began to explore the application of paper-based materials in various analytical detection fields. For example, in 1949, Müller and colleagues invented a paper-based detection device for eluting mixed pigments by impregnating filter paper with paraffin to form specific channels [2]. In 1956, Comer invented the first test paper for semiquantitative detection of glucose content in urine [3]. In 1952, British chemists Martin and Synge were awarded the Nobel Prize in Chemistry for their invention of paper chromatography [4]. Since then, paper-based analytical detection techniques have garnered widespread attention and recognition in the scientific research community worldwide.

With the continuous advancement of research in paper-based analytical detection technology, commercialization of its research outcomes has gradually developed. Since the mid-1960s, there has been a steady introduction of commercial products for paper-based analytical detection, stemming from earlier research efforts. These products include the first Dextrostix blood glucose testing strips introduced in 1964 [5], and the first "one-step" pregnancy test strips launched by Unilever in 1988 [6]. From the mid-1990s to the early 2000s, the commercialization of paper-based analytical detection technology experienced a rapid development phase. Successful products introduced during this period covered various aspects including diabetes testing, cholesterol testing, pathogen detection, infectious disease testing, and more [4]. The introduction of these products significantly improved the convenience of human health management and disease detection. It is worth noting that during this period, most of the paper-based analytical detection products primarily utilized lateral flow assays (LFAs) [7]. These types of detection products mainly include sample pads, conjugate pads, detection pads,

Paper-Based Functional Materials: Preparation and Applications, First Edition.
Xianping Yao, Genrong Chen, Zhen Yao, and Liping Zheng.

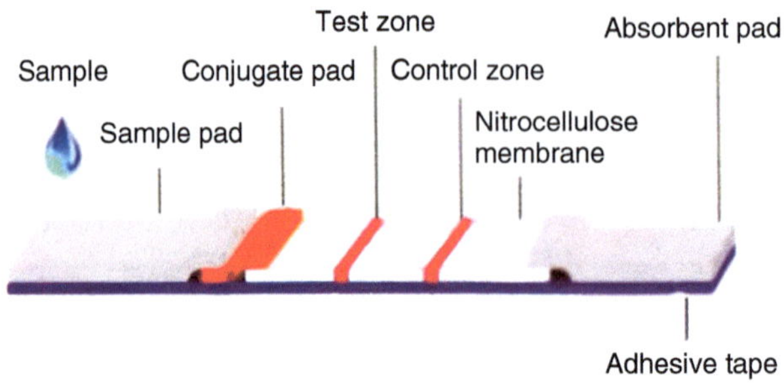

Figure 9.1 Schematic diagram of lateral flow assay paper-based analytical detection product structure.

and absorbent pads, with their structural diagram, as shown in Figure 9.1 [8]. The specific principle involves adding the sample onto the sample pad. The sample solution, under the capillary action of the paper-based material, flows through the conjugate pad where pre-added and dried antibodies with signal indicators bind to the target antigen, forming a complex of "signal indicator-antigen-antibody." This complex then flows together toward the detection pad, where it binds to the capture antibody fixed on the surface, resulting in signal output. The absorbent pad utilizes the capillary force of the paper-based material to drive the sample solution laterally toward the end of the test strip, thus increasing the sample volume and enhancing the detection's selectivity. In this process, colored latex microspheres or gold nanoparticles are often used as signal molecules. Such paper-based strip tests have achieved great success due to their simplicity, low cost, and portability. However, this detection method also has certain limitations, such as a requirement for relatively large sample volumes and unsuitability for multiplex analysis and quantitative analysis.

By the end of the 20th century, with the rapid development of science and technology across various fields worldwide, traditional analytical detection techniques also began to undergo a new round of transformation. Analytical detection devices gradually shifted toward miniaturization, integration, and portability [9]. In 1990, Manz teams introduced the concept of miniaturized total analysis systems (μ-TAS), also known as "Lab on a Chip" (LOC), which can be described as "a laboratory built on a chip" [10]. The ultimate goal is to transfer the functionalities of analytical laboratories into portable analytical devices or chips through miniaturization and integration of analytical detection equipment, achieving personalization and domestication of analytical systems. Microfluidic chip technology, by utilizing microscale or nanoscale channels to control minute volumes ($10^{-9} \sim 10^{-18}$ L) of liquid samples or reagents, allows for the construction of biochemical analysis processes on a chip measuring only a few square centimeters or even smaller. It enables sample preparation, reactions, detection, and other operations, making it the fastest-developing and most promising area within micro total analysis systems [11]. In 2007, Professor Whitesides' research group at Harvard University in the United States combined the characteristics of microfluidic technology and paper-based materials. They first proposed the concept of microfluidic paper-based analytical devices (μPADs), which utilize paper-based materials instead of traditional materials like quartz, glass, silicon, or polymer for microfluidic chip fabrication. These microfluidic devices have structured microfluidic channels processed on the surface of the paper. This innovation enables integrated detection functionalities for glucose and protein measurements [12]. Since then, the field of functional materials for paper-based analytical detection has initiated a new wave of research focused on designing microfluidic analytical detection chips using paper-based materials. Professor Whitesides' research group has also developed a paper-based microfluidic chip-mobile device integrated system, which has facilitated the application of paper-based microfluidic chips in telemedicine [13]. Additionally, they stacked multiple layers of planar paper-based chips and used double-sided tape to fix them, creating the first 3D paper-based microfluidic chip. This innovation enabled high-throughput, multitarget detection [14]. Following this, Professor Crooks' research group utilized a folding method to fabricate 3D paper-based microfluidic chips. Compared to the previous stacking method, the folding approach offers more convenience in operation and allows for greater versatility in design [15]. In 2009, Professor Pelton's research group at McMaster University in Canada introduced the concept of "bioactive paper." They aimed to establish a "lab on paper" platform to enable rapid biological detection under low-cost conditions [16].

The rapid development of paper-based microfluidic detection chip technology is closely intertwined with the inherent suitability of paper-based materials for use as carriers in analytical detection [9, 17]. As is well known, paper-based materials have wide availability, are renewable, cost-effective, easy to process, compatible with biological samples, and biodegradable. Additionally, paper-based materials possess porous and hydrophilic characteristics, allowing for fluid transport of samples or reagents under capillary action alone [4, 18]. The comparison details between paper-based materials and traditional materials such as quartz, glass, silicon, and polymer materials as substrates for microfluidic detection chips are provided in Table 9.1. Compared to microfluidic chips made from traditional materials, paper-based microfluidic detection chips have several very evident advantages:

1) Abundant raw material sources, low cost, and easy scalability for mass production.
2) No need for external pumps or other fluid delivery energy devices; fluids can flow and transport within the paper-based material via capillary action.
3) Lower consumption of required reagents and smaller sample volumes needed for testing.
4) Detection background does not produce interference, facilitating photometric detection.
5) Good biocompatibility and easy to modify chemically to alter the surface properties of paper-based materials.

Lightweight, foldable, easy to carry, and transport.

Based on the above advantages, paper-based microfluidic detection chips have become a powerful tool for point-of-care testing (POCT), also known as "instant testing"or "bedside testing" [19, 20]. In recent years, primary healthcare institutions in

Table 9.1 Comparison of paper-based materials as substrates for microfluidic detection chips and traditional materials.

	Material			
Property	**Glass**	**Silicon**	**PDMS**	**Paper**
Surface profile	Very low	Very low	Very low	Moderate
Flexibility	No	No	Yes	Yes
Structure	Solid	Solid	Solid, gas-permeable	Fibrous
Surface-to-volume ratio	Low	Low	Low	High
Fluid flow	Forced	Forced	Forced	Capillary action
Sensitive to moisture	No	No	No	Yes
Biocompatibility	Yes	Yes	Yes	Yes
Disposability	No	No	No	Yes
Biodegradability	No	No	To some extent	Yes
High-throughput fabrication	Yes	Yes	No	Yes
Fictionalization	Difficult	Moderate	Difficult	Easy
Spatial resolution	High	Very high	High	Low to moderate
Homogeneity of the material	Yes	Yes	Yes	No
Price	Moderate	High	Moderate	Low
Initial investment	Moderate	High	Moderate	Low

low-resource areas, such as remote rural regions, still face issues such as shortages of medical resources, limited equipment capabilities, and outdated diagnostic technologies. The detection of health and physiological indicators in regional populations has garnered widespread attention from governments and health organizations worldwide. The use of expensive and complex large-scale equipment in many diagnostic systems increases the complexity of the testing process, making it difficult for mobile medical units to conduct rapid screening and timely diagnosis, as well as for ordinary users to perform self-testing. To address these issues, the development of new POCT products is urgently needed [21, 22]. The World Health Organization (WHO) has proposed the ASSURED criteria for the design of POCT products, which stand for Affordable, Sensitive, Specific, User-friendly, Rapid and robust, Equipment-free, and Deliverable to end-users [22, 23], as shown in Figure 9.2. Technological innovations in paper-based microfluidic detection chips have received strong support from many government agencies and enterprises. Notably, the nonprofit organization "Diagnostics for All," founded by Professor Whitesides and the "SENTINEL" Bioactive Paper Network led by Professor Pelton, has received grants of USD 5.94 million and CAD 19 million, respectively,

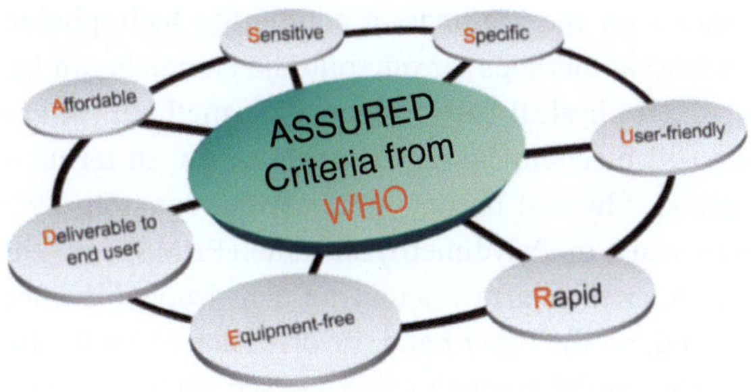

Figure 9.2 The ASSURED criteria for (POCT) products published by the WHO.

to conduct related fundamental research and accelerate the development and application of these technologies. The renowned Gates Foundation has also funded multiple studies on the use of paper-based microfluidic chips for low-cost diagnostics. Additionally, paper-based microfluidic detection chips provide robust technical support for portable testing and on-site real-time monitoring required in environmental monitoring, biochemical analysis, and food safety testing.

This chapter will detail the main fabrication methods, detection methods, and applications of paper-based microfluidic detection chips in the following sections. Additionally, it will provide an outlook on the future development prospects of paper-based microfluidic detection chips.

9.2 Fabrication Methods for Paper-based Analytical Detection Chips

Based on current research, paper-based detection chips can be classified into two main categories: paper-based microfluidic chips and bioactive paper. The working principle of paper-based microfluidic chips involves creating hydrophobic boundaries on the naturally hydrophilic paper substrate. This is achieved by methods such as closing the gaps in the paper fiber array or surface modification of the paper fibers. These processes result in patterned hydrophilic and hydrophobic contrasts, enabling the construction of sample transport channels, switches, branching channels, reaction wells, and other detection units in the hydrophilic regions. Using capillary action, the sample solution can autonomously flow unidirectionally or multidirectionally within the channels to the detection area. There, it reacts with prefixed indicators, allowing fluid handling and analysis directly on the paper. As for bioactive paper technology, it leverages the inherent hydrophilic, porous, and biocompatible properties of paper. The paper fiber array serves as a support substrate, and biological molecules or indicators are immobilized on the paper through physical modification, chemical covalent bonding, or loading of nano/microparticles. This modification endows the paper substrate with biological detection capabilities. Through these approaches, paper-based detection chips can perform various analytical functions, making them valuable for diagnostics, environmental monitoring, and food safety testing.

Based on the methods of combining hydrophobic substances with paper, the fabrication methods for microfluidic channels can be categorized into three major types: (i) physical barrier/deposition method, (ii) chemical modification method, and (iii) biochemical paper modification. In terms of physical barrier/deposition method, physical barrier involves impregnating hydrophobic substances such as photoresist or Polydimethylsiloxane (PDMS) into the paper to block the gaps in the paper fiber array, achieving a hydrophobic effect, while physical deposition method, on the other hand, involves depositing hydrophobic materials like wax or polystyrene (PS) onto the fiber surface. By lowering the surface energy of the paper, this method makes the liquid more likely to penetrate the hydrophilic channels, thereby controlling the liquid transport. Both processes alter the wettability of the paper through the physical interaction between the hydrophobic substances and the fibers. Chemical modification method, on the other hand, involves using hydrophobic compounds such as alkyl ketene dimer (AKD) or other substances that react with the hydroxyl (–OH) groups on the cellulose molecules. This process introduces hydrophobic groups onto the cellulose molecular chains, creating hydrophobic boundaries on the paper [24]. The above two methods are primarily used to fabricate paper-based microfluidic chips. By creating hydrophobic boundaries on the naturally hydrophilic paper substrate, these methods form patterned contrasts between hydrophilic and hydrophobic areas. This enables the handling and analysis of fluids directly on the paper [23].

Biochemical modification, on the other hand, involves treating the paper substrate through physical modification, chemical covalent bonding, or loading with nano/microparticles. This process enables the immobilization of biomolecules or indicators on the paper, endowing it with biological detection capabilities. This method is commonly used to create bioactive paper.

Currently, the methods used for fabricating paper-based microfluidic chips include the wax printing method [25, 26], photolithography method [27, 28], inkjet printing method [29–31], etching method [32, 33], stamping method [34, 35], drawing method [36, 37], plasma treatment method [38, 39], screen printing method [40–43], flexographic printing method [44], cutting method [45, 46], laser treatment method [47, 48], spray coating method [49], chemical vapor deposition method [50–52], and so on. The specific classifications are shown in Figure 9.3.

Common materials used for creating hydrophobic regions in paper-based microfluidic channels include photoresist, wax, PDMS, PS, and AKD. Additionally, methods for creating bioactive paper mainly involve treating the paper substrate with protein solutions, chitosan solutions, or other substances to improve its detection performance, allowing detection reagents to be immobilized on the paper without hydrophobic boundaries for easy testing. Alternatively, biomolecules with strong adsorption capacity can be transferred to untreated paper substrates through printing, followed by introducing detection reagents to react with them, facilitating the determination of target analytes.

In summary, fabrication methods for paper-based detection chips can be classified into two categories: physical methods and chemical methods for paper-based

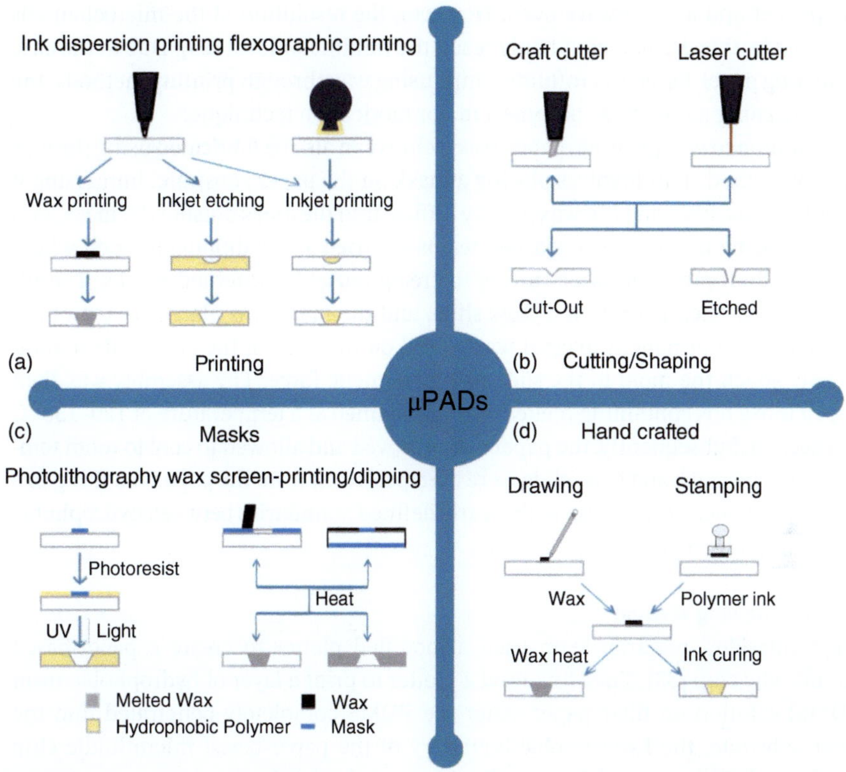

Figure 9.3 Classification of fabrication methods for paper-based microfluidic detection chips.

microfluidic chips, while methods for creating bioactive paper mainly involve paper modification. Below are specific introductions to several commonly used methods.

9.2.1 Physical Barrier/Deposition Method

9.2.1.1 Wax Patterning Method

Wax, as an inexpensive and commonly available hydrophobic material, is widely used in the fabrication of paper-based microfluidic chips. There are several methods for fabricating paper-based microfluidic chips using wax. First, there is the wax printing method, which is further divided into inkjet printing wax and screen printing wax. Both methods involve applying wax to the surface of the paper substrate, then heating to melt the wax and allow it to penetrate into the paper, followed by cooling to solidify the wax onto the paper, thus forming hydrophobic regions.

In 2009, the Whitesides team used a solid wax printer to print wax onto filter paper according to a designed pattern. The printed wax was melted in a microwave oven and permeated into the porous structure of the filter paper, forming clear microchannels on the paper. The entire fabrication process was very simple, requiring only a wax printer and a microwave oven. However, the resolution of the microchannels produced by this method could only reach the millimeter level [25]. Due to the ease of creating paper-based microfluidic chips using wax through printing methods, the wax patterning method has become a major fabrication technique.

In addition to wax printing, paper-based chips can also be fabricated using the wax dipping method. This involves placing a mask on the paper substrate, immersing it in molten wax, allowing the wax to fully diffuse into the areas outside the mask, and then removing the mask. Songjaroen teams reported a wax dipping method where they first used precision laser cutting to create reusable patterned masks. The filter paper was then placed on a glass slide, and the mask was placed on top of the filter paper. A permanent magnet was placed on the back of the glass slide to temporarily attach the mold to the slide using magnetic force. The assembly was then dipped into a box containing melted wax maintained at a temperature of 120–130 °C for 1 second. Subsequently, the paper was removed and allowed to cool to room temperature. The mask and glass slide were then peeled off from the paper, resulting in a paper-based microfluidic chip with clearly defined boundaries between hydrophobic and hydrophilic regions [26].

9.2.1.2 Drawing Method

The Whitesides' research team used a modified plotter to create a paper-based microfluidic chip [53]. They first used a plotter to print a layer of hydrophobic resin (PDMS) solution on filter paper. After the PDMS completely penetrated into the paper substrate, the hydrophobic boundary of the paper-based microfluidic chip was formed. This method is cost-effective, and the fabrication process is simple. In addition to the plotter, using wax or marker pens to draw hydrophobic areas is also a common method for fabricating paper-based microfluidic chips. Among them, the method using marker pens omits the heating process, making it simpler and more convenient. However, although the drawing method for fabricating paper-based microfluidic chips is cost-effective and easy to operate, it is difficult to control the size of the hydrophobic area. Therefore, it is not suitable for making paper-based microfluidic chips with complex patterns, which severely limits the application range of this method.

9.2.1.3 Physical Inkjet Etching Method

Generally speaking, etching is a method used in the fabrication of semiconductor materials. With the development of microfabrication processes, etching has become a method of using ion solutions or mechanical means to remove or eliminate materials. Abe teams applied this method to the fabrication of paper-based microfluidic chips [54]. They first soaked the filter paper in a 1.0% PS/acetone solution for 2 hours. The pretreated filter paper became hydrophobic. Then, they used an inkjet printer to pattern acetone onto the hydrophobic filter paper. The PS printed with acetone was then removed cleanly to create hydrophilic channels. Although this method of inkjet etching produces paper-based microfluidic chips with high resolution, the fabrication process is complex, and the materials used do not meet environmental-friendly requirements.

9.2.1.4 Screen Printing Method

The screen printing method is also commonly used for fabricating paper-based detection chips. It involves creating patterned screens and depositing materials with water-resistant or hydrophobic properties onto the paper substrate through the screen. Subsequent processing forms distinct boundaries between hydrophilic and hydrophobic regions. Dungchai teams reported a method based on screen printing, where solid wax was transferred onto filter paper by friction on the screen. The wax was then melted and permeated into the paper to create paper-based chips. The diffusion of wax on the paper substrate was also investigated [41]. Additionally, Sameenoi utilized the screen printing method to fabricate paper-based chips. However, they substituted a toluene solution of PS for wax. After printing, the toluene on the paper substrate was simply evaporated using a blower [55]. As the most common printing method, screen printing is well-suited for the small-scale fabrication of paper-based detection chips.

9.2.1.5 Flexographic Printing Method

Flexographic printing is a direct printing method. The Olkkonen team used this technique to print PS directly onto the filter paper. The PS permeated the filter paper, forming hydrophobic boundaries, while the areas that did not come into contact with PS remained hydrophilic [38]. This method allows for the large-scale fabrication of paper-based microfluidic chips.

9.2.1.6 Cutting Method

Cutting is also a common, easy-to-operate, and straightforward method for fabricating paper-based microfluidic chips. Fenton teams used a computer-controlled cutter to create microfluidic chips based on nitrocellulose membranes [56]. The fabrication process takes only 60 seconds to complete, making this method cost-effective and highly practical. Additionally, paper cutting and engraving can be performed using cutters, embossers, and punch machines. However, all cutting methods share the same drawback: the precision of the microchannel dimensions is challenging to control. Moreover, this method is only suitable for creating paper-based microfluidic chips with simple patterns.

9.2.2 Chemical Modification Method

9.2.2.1 Photolithography Method

The photolithography method primarily involves using photoresist in combination with masks to create hydrophilic/hydrophobic boundaries. Whitesides teams reported a method for fabricating paper-based microfluidic chips using photolithography [12]. They first soaked chromatography paper in photoresist, then, after complete wetting, they heated the paper in a 95 °C environment for 5 minutes to remove the solvent from the photoresist. Subsequently, the paper was exposed to UV light at a wavelength of 405 nm for 10 seconds. This process controlled the curing crosslinking of the photoresist on the chromatography paper using patterned masks. After UV exposure, the uncured photoresist on the paper was washed away, and finally, the paper was treated in a plasma cleaner to increase the hydrophilicity of the hydrophilic regions on the chromatography paper. Using photoresist allows for the fabrication of high-resolution paper-based microfluidic chips. However, the fabrication process requires large and expensive equipment, limiting its widespread use.

9.2.2.2 Inkjet Printing Method

Inkjet printing does not require predesigned masks, and the operation process is simple. By using an inkjet printer, different types of reagents can be simultaneously added to the paper, offering broad prospects for batch production of paper-based chips. Shen teams printed AKD solution on filter paper using a digital inkjet printer, then heated the filter paper to cure the AKD on the paper fibers. After the paper-based chip dried, the fabrication process was complete [11]. Meanwhile, Wang teams compared the use of siloxane, wax, and AKD as hydrophobic boundary materials using an inkjet printer. The authors found that hydrophobic boundaries made with siloxane had better performance, effectively preventing the leakage of easily diffused solutions. This demonstrated that siloxane is more suitable for the fabrication of paper-based microfluidic chips [57]. Maejima teams also employed inkjet printing to fabricate paper-based microfluidic chips. However, they used hydrophobic UV-curable acrylic instead of AKD. They first printed this special ink on the paper and then cured the paper under UV light for 60 seconds to create hydrophobic boundaries. This method is straightforward, and the fabrication process is very rapid [58]. However, when using modified ink, it is prone to clogging the nozzles of the inkjet printer. Therefore, before achieving large-scale manufacturing, further research is needed to develop more suitable inks.

9.2.2.3 Plasma Treatment Method

Plasma treatment is another common chemical method used to fabricate paper-based microfluidic chips, first reported by Li et al. [38]. The filter paper is first soaked in a glue solution containing AKD and then heated to cure the AKD, rendering the filter paper hydrophobic. Subsequently, the hydrophobic-treated filter paper is sandwiched between patterned metal masks and placed in a vacuum plasma cleaner for 15 minutes. Exposure to the plasma results in hydrophilic areas in the patterned regions. AKD, as a common industrial adhesive, is inexpensive and readily available, but the fabrication process of the patterned metal masks is relatively complex.

9.2.2.4 Chemical Etching Method

Cai teams proposed a method for fabricating paper-based microfluidic chips using selective wet etching of hydrophobic filter paper. This method involves using a patterned paper mask to achieve partial silanization of the filter paper fibers, creating hydrophobic boundaries [32]. First, the filter paper is rendered hydrophobic by treatment with a trimethoxyoctadecylsilane (TMOS) solution. The hydrophobic filter paper is then fixed between a patterned paper mask soaked in NaOH solution and a glass slide, followed by heating. The silanized cellulose exposed to the NaOH solution becomes hydrophilic, while the areas not in contact with the mask remain hydrophobic. This method does not require any expensive equipment or metal masks; however, the resolution of the fabricated paper-based chips is relatively low, and the reproducibility is poor.

9.2.3 Paper Modification Method

The fabrication of bioactive paper primarily involves modifying the paper substrate so that detection reagents can be immobilized on the paper substrate without hydrophobic boundaries. After introducing the detection reagents, they can still react at the fixed positions, enabling the determination of target analytes.

The Yamada team utilized a colorimetric method to achieve semiquantitative detection on paper, with the results displayed in a text-based format [59]. They first pretreated the filter paper with a sodium citrate buffer solution. Then, using an inkjet printer, they printed the reagent tetrabromophenol blue (TBPB) ink onto the paper in the form of character patterns, creating bioactive paper. After drying, they added protein solutions of varying concentrations to the detection areas for colorimetric reactions. The reaction results were then compared with a transparent film printed with different color gradients to read the measurement results. This study utilized inkjet printing to achieve text-based result display without hydrophobic boundary limitations, contributing to the expansion of applications for paper-based detection chips.

The Brennan research group similarly used inkjet printing to create paper-based aptamer chips without the need for hydrophobic boundaries [60]. The continuous DNA aptamers used as detection elements were directly inkjet-printed onto the paper substrate in a patterned manner. After adding the sample, biological detection was achieved. The paper-based chip effectively immobilizes the sample in its original position. This method is simple, efficient, and enables quantitative detection of analytes and text-displayed results without the need for complex hydrophobic barrier fabrication processes, offering new opportunities for developing various paper-based detection chips for real-time testing.

In addition, Zhang teams achieved heavy metal ion detection on paper-based chips without hydrophobic boundaries by using reagents that can react with metal ions to form insoluble precipitates, employing a writing-based approach [61]. Similarly, Liu teams accomplished a series of biodetection assays without the need for hydrophobic boundaries by first treating the paper substrate with chitosan solution to enhance the detection performance of the filter paper, employing a writing-based approach [62].

9.3 Detection Methods of Paper-based Functional Materials for Analysis and Detection

The existing literature reports detection methods for paper-based microfluidic chips including electrochemical method [63], chemiluminescence method [64, 65], electrochemiluminescence (ECL) method [66, 67], fluorescence method [68, 69], and colorimetric method [29, 70]. Among them, the electrochemical method is commonly used to detect glucose, uric acid, lactic acid, and metal ions such as

Pb and Au. Additionally, it has been reported for the detection of ascorbic acid, cholesterol, and total iron content in health diagnostics, as well as for determining ethanol content in food. Chemiluminescence and ECL methods have also been reported for the detection of glucose and uric acid, as well as various samples in biomedical applications, such as nicotinamide adenine dinucleotide. The fluorescence method can be used to detect heavy metal ions such as Mn, Co, Ni, Cu, Zn, and Hg in water, as well as pesticides. Moreover, the colorimetric method is the most widely used in paper-based chip detection method, with significant potential in health diagnostics, biochemical analysis, and environmental monitoring. It is commonly used to analyze glucose, proteins, uric acid, ketones, lactic acid, pH levels, pathogenic bacteria, and nitrite. It also measures enzyme activities such as alkaline phosphatase or peroxidase and quantitatively or semiquantitatively analyzes the content and concentration of heavy metals in the environment. Below, each method will be introduced in detail.

9.3.1 Electrochemical Method

The electrochemical method is based on the electrochemical reactions of analytes occurring on the paper-based working electrode, where the analyte concentration is determined by measuring the magnitude of the oxidation or reduction current produced by the reaction. Due to its relatively high sensitivity, the electrochemical method can detect low concentrations of analytes. Integrating paper-based microfluidic chips with electrochemical detection provides a sensitive and selective platform for biomarker detection. Noiphung teams reported a reusable paper-based microfluidic chip utilizing screen-printed carbon electrodes, modified with a mediator-ferricyanide, for the detection of glucose in whole blood. Glucose present in the plasma is separated and enters the detection zone, where its concentration is measured through current analysis due to the presence of glucose oxidase and peroxidase in the detection area [71].

The application of electrochemical methods for the detection of low-concentration cancer biomarkers has become a focal point in early medical diagnosis and screening. Utilizing a sandwich method for immunoassay of biomarkers, the dynamic signals generated by the immunoreactions are detected electrochemically. Researchers have also focused on drug screening by observing the activities of cultured cancer cells. Su teams reported a method for culturing cancer cells in a three-dimensional paper-based microfluidic chip for drug screening. The paper-based microfluidic chip used in their study could detect cell apoptosis and the production of polysaccharides. Additionally, the paper fibers were coated with gold nanoparticles, allowing peroxidase-labeled aptamers to capture cancer cells on the electrode surface. The catalytic oxidation of hydrogen peroxide produced by the polysaccharides was then detected electrochemically by the electrode [72].

9.3.2 Chemiluminescence Method

Paper-based microfluidic chips utilizing the chemiluminescence method determine the analyte concentration based on the intensity of light produced by chemical reactions. Chemiluminescent reagents are typically inexpensive, and the measurement of light intensity offers high sensitivity. Consequently, chemiluminescence has become a popular, low-cost, high-sensitivity detection method.

Zhou teams demonstrated the application of chemiluminescence in POCT using a paper-based detection chip. They utilized biotinylated single-stranded oligonucleotide probes and utilized the biotinylated DNA target probes to hybridize with a phospho-Streptavidin complex. The conjugated probes were excited to emit light at 980 nm, and a fluorescence microscope was used to analyze and detect the concentration of the target probe [73].

9.3.3 Electrochemiluminescence Method

ECL, which combines electrochemical methods with chemiluminescence, involves a luminescent phenomenon resulting from further reactions of specific substances generated during electrochemical processes. This method determines the concentration of analytes by measuring the emitted light signal.

Delaney teams utilized ECL combined with smartphone technology for analyte detection [74]. They fabricated paper-based detection chips using a combination of screen-printed electrodes and inkjet printing. Traditional photodetectors were employed to detect the concentrations of the target analytes, 2-dibutylaminoethanol (DBAE), and nicotinamide adenine dinucleotide (NADH). This study integrated electrochemistry with chemiluminescence and utilized smartphone detection and analysis of the luminescent phenomena, showcasing significant implications for the development of paper-based chips.

9.3.4 Fluorescence Method

The fluorescence method is a technique for determining the concentration of fluorescent substances by measuring the intensity of light emitted upon excitation. Fluorescein, due to its high sensitivity, has garnered significant attention, and several paper-based detection chips based on fluorescence have been reported. However, there is a drawback to conducting fluorescence reactions on paper, namely, the whiteness of the paper affects fluorescence expression. In the past two years, fluorescence-based paper-based detection chips have been used for the detection of bacteria, biomolecules, and cancer biomarkers and have also been applied in DNA detection.

Scida teams introduced a method for detecting single-stranded DNA through hybridization experiments, achieving a detection limit of 5 nmol/L [75]. This method utilizes a paper-based detection chip composed of four layers, with three of these layers each containing four independent detection zones. Fluorescently labeled single-stranded DNA and a quencher are added to the detection zones, and the layers are then folded to ensure connectivity between the base layers. In each detection zone, the fluorescently labeled single-stranded DNA and the quencher are premixed. The entire chip is secured with a clamp, and the target analyte is added for detection.

9.3.5 Colorimetric Method

Colorimetric analysis, as one of the most common detection methods, is based on the colored reaction resulting from the interaction between the analyte and a reagent or substrate. By comparing the color change before and after the reaction, the type and concentration of the analyte can be determined.

Colorimetric methods can be used to detect common heavy metal ions in water, food additives, and proteins in the human body. By adding the appropriate reagents to the paper substrate and then introducing the analyte to produce a color reaction, this method enables straightforward visual analysis. The colorimetric method is also suitable for detecting enzyme-catalyzed reactions, such as glucose or uric acid, in liquids. The specific catalytic action of enzymes produces intermediates that react with the reagent, resulting in a color change. Colorimetric methods are widely used in immunoassays as well. By using paper as the detection substrate and combining it with enzyme-linked immunosorbent assays (ELISA), low-cost immunodetection can be achieved. Additionally, nanomaterials have been extensively applied in paper-based detection chips. Nanoparticles are favored for their ability to enhance the sensitivity of analytes, such as cancer antigens. Many research groups have improved the detection limits and sensitivity of analytes by using nanoparticle modifications. Currently, nanoparticle modifications are applied to the detection of cancer markers, antibodies, bacteria, proteins, and disease pathogens.

9.4 Applications of Paper-based Functional Materials for Analysis and Detection

Paper-based chips, as a novel analytical platform, have been applied in multiple fields, primarily summarized as medical diagnostics, environmental monitoring, and food safety control.

9.4.1 Application of Paper-based Detection Chips in Medical Diagnosis

The application of paper-based microfluidic chips in medical diagnostics primarily includes the detection and diagnosis of analytes in real samples such as sweat, blood, saliva, and urine. The types of analytes include glucose, uric acid, proteins, cholesterol, nucleic acids, ketones, and lactic acid. Additionally, these chips can be used for blood type identification, cancer marker detection, and the diagnosis of pathogens such as malaria [76–85]. Based on the nature of analytes, medical diagnostics can be divided into four categories: protein analysis, nucleic acid analysis, cell analysis, and the analysis of other small molecules/ions.

9.4.1.1 Analysis of Proteins

Proteins are widely used as biomarkers for disease diagnosis. Common immunoassay techniques include enzyme-linked immunosorbent assay (ELISA), immunofluorescence, immunoblotting, and immunodiffusion [86]. Traditional detection methods are often complex and require expensive instruments. For example, ELISA, one of the most commonly used techniques, can take hours or even overnight to complete and relies on costly equipment like microplate readers. Therefore, there is a need to develop low-cost, simple, rapid, and highly sensitive diagnostic methods. In recent years, many paper-based detection platforms have been developed to analyze protein biomarkers for cancer and infectious diseases.

Whitesides and colleagues developed a simple paper-based detection chip capable of simultaneously detecting glucose and bovine serum albumin (BSA) in urine [12]. Cheng and colleagues invented a 96-well plate paper-based chip for ELISA experiments. This paper-based device features a circular array of test zones (12×8), matching the layout and size of standard plastic 96-well plates, enabling multiple simultaneous enzyme-linked immunosorbent assays. This innovation facilitated the detection of rabbit IgG and HIV-1 envelope antigens [87]. Songjaroen teams developed a paper-based diagnostic device capable of separating plasma from whole blood and quantifying plasma proteins in a single step. This advancement is highly significant for subsequent diagnostic applications [88]. Wang teams reported a paper-based chip capable of performing chemiluminescence enzyme-linked immunosorbent assay (ELISA). Antibodies were immobilized in the hydrophilic region of the paper-based chip through cross-linking with glutaraldehyde, and the detection of antibodies labeled with horseradish peroxidase (HRP) was achieved through a chemiluminescence experiment enhanced by luminol-iodophenol. The concentrations of three tumor markers, alpha-fetoprotein, cancer antigen, and carcinoembryonic antigen, were successfully measured. This study laid the groundwork for further development of sensitive, specific, and low-cost paper-based detection chips by utilizing chitosan-modified paper-based chips to immobilize antibodies [89].

9.4.1.2 Analysis of Nucleic Acids

Paper-based detection chips can also be used for nucleic acid analysis, enabling sensitive identification and characterization of pathogens using amplified microbial DNA or RNA. Typically, the amount of available DNA/RNA is limited, so amplifying the target sequences is a crucial step in nucleic acid analysis to enhance sensitivity.

Rivas teams utilized DNA's screening ability on paper and recombinase polymerase amplification to create a DNA vertical microarray paper-based chip capable of isothermal amplification. By exploring different detection parameters such as hybridization buffer, flow rate, and printing buffer, they optimized the concentration of capture probes for the detection of *Neisseria meningitidis* (the main cause of bacterial meningitis) [90]. Additionally, Ihalainen teams proposed two different supramolecular recognition systems for impedance detection of DNA hybridization on paper-based chips with inkjet-printed gold electrodes [91]. Although paper-based chips can be used for nucleic acid detection, the analysis of DNA hybridization and the nucleic acid amplification process still rely on expensive and bulky equipment. Therefore, integrating the DNA amplification steps into paper-based devices remains a challenge.

9.4.1.3 Analysis of Cells

The development of paper-based cell detection chips has made it possible to analyze, detect, separate, and culture cancer cells and other types of mammalian cells on paper. Various paper-based chips have been developed for cancer research. Wu teams created a paper-based chip based on ECL for the specific capture of cancer cells. They used aptamer-modified gold/paper electrodes as the working electrodes to accurately analyze four types of cancer cells: human breast adenocarcinoma cells (MCF-7), human acute promyelocytic leukemia cells (HL-60), human chronic myelogenous leukemia cells (K562), and human acute lymphoblastic leukemia cells (CCRF-CEM)[92].

9.4.1.4 Analysis of Other Small Molecules/Ions

Paper-based chips are also commonly used for the analysis of glucose, uric acid, lactate, ketones, and nitrite ions in human sweat, urine, or blood. He teams invented a thermoresponsive paper-based detection system that combines a shape-memory polymer-modified textile with a paper-based colorimetric chip for detecting glucose in human sweat [93]. Furthermore, by integrating this system with a smartphone analysis platform, quantitative assessment of glucose concentration was achieved. This study expanded the applications of shape-memory polymers and paper-based chips in analyzing human sweat. Wang teams designed a paper-based chip based on a colorimetric method, which can simultaneously detect uric acid and glucose. They placed the test paper on an LED light, effectively improving the resolution and sensitivity of detection [94].

9.4.2 Application of Paper-based Detection Chips in Environmental Monitoring

With the development of industry and agriculture, environmental pollution is becoming increasingly serious. The degradation of water quality has far exceeded its self-purification capacity. Pollutants, toxic heavy metals, and other harmful substances in water bodies have attracted widespread attention worldwide. Therefore, the timely and effective detection of water quality is crucial for us to take further measures to mitigate or control water pollution. Currently, low-cost paper-based detection chips have been developed for the detection of heavy metal ion types and concentrations in water bodies, as well as inorganic compounds in the environment, such as phosphates [95], nitrate and nitrite [96, 97], iodide [98], bromide [99], chloride [100], organic compounds [101], phenolic compounds [102], and cyanide [103].

Human intake of metals has been identified as one of the factors causing morbidity and mortality. Redox-active metals such as Fe, Cu, Cr, and Co have the ability to generate free radicals that may cause oxidative stress in organisms, while metals like Pb and Cd are well-known neurotoxins. Rattanarat teams developed a paper-based detection chip by combining colorimetric and electrochemical detection on different layers for the detection of Ni, Cu, Fe, Pb, Cr (VI), and Cd. This technology targets pollutants present in particulate matter in the air. Samples are pretreated in four separate channels to achieve colorimetric determination of Cu, Ni, Fe, and Cr, as well as electrochemical detection of Pb and Cd. The multilayer separation detection mode minimizes the possibility of cross-contamination, while different reaction chemistries improve the specificity and sensitivity of the analysis [104]. Lewis teams patterned paper using wax and overlaid the treated layers through lamination. They designed a paper-based sensor to detect lead and mercury in water by timing to obtain the concentration of the analytes [105].

9.4.3 Application of Paper-based Detection Chips in Food Safety Control

Paper-based detection chips also play an important role in the real-time monitoring of food safety. The content of food additives such as glucose [106] , Nitrite [107], as well as pesticide residues [108], can now be detected using paper-based microfluidic chips. These chips provide a simple and effective method for controlling food safety.

9.4.3.1 Detection of Food Additives

In the food and beverage industry, widely used food additives include nitrites, glucose, fructose, and sucrose. Glucose, fructose, and sucrose are specifically used as sweeteners, while other food additives are used to improve or enhance the flavor or color of food or beverages. Although these food additives are generally nontoxic, excessive intake can lead to unhealthy nutrition, and some may even cause poisoning if consumed in large amounts. Therefore, there is a strong demand for portable paper-based chips to perform rapid and highly sensitive analysis of food additives.

Nitrites are commonly added to meat products as food additives, and excessive intake can lead to poisoning. He teams prepared a paper-based chip for detecting nitrites added to food by coupling hydrophobic octadecyltrichlorosilane (OTS) with the fibers of Whatman No. 1 filter paper, followed by UV photolysis of the OTS coating [109]. He chose the Griess reagent as the colorimetric reagent. Nitrites react with sulfanilamide to form a diazonium salt, which then reacts with N-(1-naphthyl) ethylenediamine dihydrochloride in a two-step diazotization reaction to produce a strong pink azo dye, enabling the quantitative analysis of nitrites in food.

9.4.3.2 Detection of Pesticide Residues

Pesticides have been used in agriculture for many years, making significant contributions to maintaining food quality and yield. However, these substances also have adverse effects on human health. Pesticides are well-known toxins found in air, water, soil, food, and feed products. They can enter the human body through ingestion, inhalation, and skin absorption and are associated with neurotoxicity, hepatotoxicity, nephrotoxicity, dermatitis, and cancer. Common separation methods such as gas and high-pressure liquid chromatography are typically used for pesticide-sensitive and selective determination, but the analysis is costly and not suitable for on-site measurements. Therefore, paper-based detection chips provide a practical and cost-effective means for the rapid analysis of pesticide residues in food. Hossain teams employed a method that combines the analytical reagent with sol–gel for the production of paper-based detection chips using piezoelectric inkjet printing. First, the reagent capable of capturing cations, polyethyleneimine, was printed on the paper-based chip. Then, the enzyme was sandwiched between two layers of biocompatible, soluble sol–gel-derived silica. By using the Ellman colorimetric method to determine the residual activity of acetylcholinesterase (AChE) on paper, they detected the residues of organophosphorus pesticides in milk and vegetables at very low concentrations [108].

9.4.3.3 Detection of Bacteria

Pathogenic bacteria such as Staphylococcus aureus, Salmonella typhi, Escherichia coli, and Listeria monocytogenes pose significant threats to human health and safety. Raw materials and any processed food may be contaminated by pathogenic bacteria. According to the WHO, nearly one-tenth of the world's population suffers from foodborne diseases, with 420 000 people being unable to recover. Therefore, rapid and accurate diagnostic techniques, as well as the ability to timely identify pathogens, are crucial in reducing mortality rates and minimizing the spread of pathogens. In a study reported by Jokerst teams, they developed a paper-based detection chip for the microspot detection of Escherichia coli, Listeria monocytogenes, and Salmonella typhimurium in ready-to-eat meat samples [63]. The pathogens were collected from food using a swab sampling technique, then cultured in a medium, and finally added to a paper-based chip containing chromium. The color change of the reaction indicated the presence of enzymes related to the target pathogens, thus enabling detection. This study was able to detect pathogenic bacteria in consumed meat and significantly reduce the time required for bacterial detection and counting.

In the same year, Hossain teams confirmed the use of a multiplex device for the specific detection of nonpathogenic and pathogenic Escherichia coli. They detected nonpathogenic Escherichia coli in the presence of β-glucuronidase through the reaction of sodium salt of 5-bromo-4-chloro-3-indolyl-β-D-glucuronide, which produced highly colored indole species and glucuronic acid [64]. Using the same β-galactosidase chemical structure as mentioned above indicates the total Escherichia coli. In addition, the authors demonstrated the use of immunomagnetic nanoparticles for concentrating Escherichia coli through a magnetic separation step without the need for external cell culture. They tested the effect of other components in artificially contaminated orange juice and milk samples, confirming the specificity of the device for the presence of Bacillus subtilis and Escherichia coli in liquid samples.

9.5 Application Cases of Paper-based Functional Materials for Analysis and Detection

Paper-based sensing chips have great potential for rapid diagnostic testing in low-resource settings and can be widely applied in clinical diagnostics, environmental monitoring, food testing, and other fields. While paper-based chips may not match or surpass advanced large-scale analytical diagnostic equipment in terms of detection sensitivity, they excel in achieving rapid analysis, detection, diagnosis, and mass screening under conditions of relatively accurate and acceptable sensitivity, meeting the ASSURED standards proposed by the WHO. The development of low-cost, rapid detection systems that can be used for mass surveys or self-testing enables the "home-based," "personalized," "on-site," and "ubiquitous" transformation of analytical laboratories, serving as a supplement to the standard testing systems used by large hospitals and testing institutions and possessing significant practical application value. This section will further illustrate examples of the application of paper-based analytical detection materials.

9.5.1 Optimization and Regulation of Blood Typing Paper-chip and Paper-based Structure

Since 2007, Shen, Junfei Tian, and others have been engaged in research related to paper-based detection and diagnostic devices. They have accumulated rich experience in various aspects, including the design of device substrates, optimization of detection techniques, rational construction of detection units, biochemical modification of paper substrates, and expansion of application areas. Their systematic research exploration has progressed from basic theory to product samples [11, 38, 110, 111]. In early 2008, they explored the use of a commonly used paper-sizing agent, AKD, in the industry to print paper surfaces with regularly patterned microfluidic channels with strong hydrophilic–hydrophobic contrast. The materials and techniques involved in this process are derived from the papermaking and printing industries. The research creatively recombines these traditional materials and existing technologies, making it not only easier to achieve large-scale manufacturing of paper-based devices in an economically efficient manner but also creating new development opportunities for the traditional papermaking and printing industry [11]. Subsequently, they "textualized" the hydrophobic boundaries. By modifying antibodies used for blood type detection onto four hydrophilic "letters" regions, they effectively detected surface antigens A, B, and RhD on red blood cells, thus achieving, for the first time internationally, the display of ABO/RhD blood type results in the form of "text" (Figure 9.4). Even laypersons could clearly read the blood type results, thereby reducing one of the major causes of blood typing errors, human error, and revolutionizing rapid blood typing [82]. This research achievement has attracted widespread attention from many world-renowned academic and news media outlets, such as Xinhua News Agency and Xinhua Net in China, the British Broadcasting Corporation (BBC), and the Australian Broadcasting Corporation (ABC), among others. This invention was commented on by Dr. Yetisen from the University of Cambridge and others as the first user-friendly paper-based sensing chip in which results are displayed without the need for additional devices. Its innovation in displaying results using "text" provides a new solution to meet the WHO's requirements for convenience in the use of low-cost diagnostic devices.

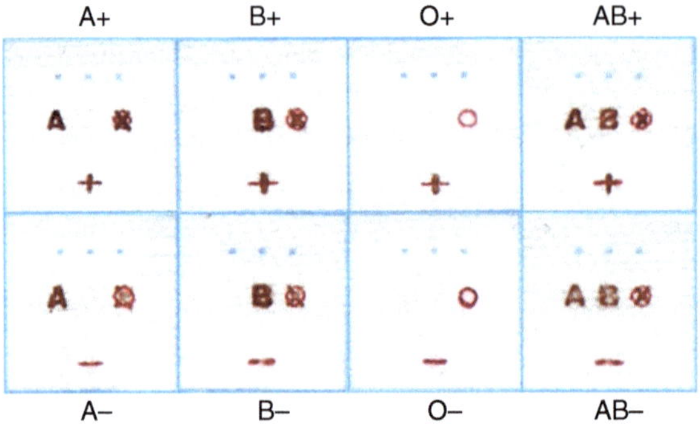

Figure 9.4 Shows the first-generation ABO/RhD blood type diagnostic paper constructed using hydrophobic boundary domain technology *Source:* [82] / with permission of John Wiley & Sons.

Lizi Li, Shen, and others controlled the fiber types and paper quantification to construct paper-based materials with different three-dimensional network structures. They used these paper-based materials to create blood type detection paper-based chips to study the effect of paper fiber network structure on the detection results of blood type detection paper-based chips [112]. In experimental design, the influence of fiber types on the paper network structure was isolated from various other interfering factors by sieving out fine fibers and not adding any papermaking additives. The physical structure characteristics and internal pore-structure diameter distribution of the paper-based materials were characterized using surface thickness, bulk thickness, and mercury intrusion methods. The study of red blood cell flow behavior in the paper was conducted in two ways: one was chromatographic separation parallel to the paper surface, and the other was fluid flushing perpendicular to the paper surface. The research results, as shown in Figure 9.5, indicate that paper made from shorter broadleaf wood fibers has a higher porosity and simpler pore structure compared to paper made from longer coniferous wood fibers. The complexity of the internal pore structure of the paper has the most significant impact on the flow behavior of red blood cells in the paper. Paper made from low quantification of broadleaf wood fibers has a simpler internal pore structure, making it easier for red blood cells to flow within the paper. Test papers for blood typing made from this type of paper have higher resolution, which can enhance the accuracy of the detection results. In contrast, paper made from coniferous wood fibers has a more complex internal pore structure, making it more difficult for red blood cells to flow within the paper. Test papers for blood typing made from this type of paper have lower resolution, thus reducing the accuracy of the detection results. Therefore, it is preferable to use broadleaf wood fibers with low quantification for making paper-based chips for blood type detection.

Junfei Tian, Rong Cao, Miaosi Li, and others studied the filtering effect of paper-based fiber arrays on red blood cells and the mechanism of the "coffee ring" phenomenon, achieving quantitative control over the dynamic migration behavior of red blood cells on paper [113]. During incubation and elution, the individual distribution of red blood cells in the wetting center and their aggregation behavior at the boundaries within the porous paper fiber array result in the formation of the "coffee ring" phenomenon. The study explored factors potentially affecting the red

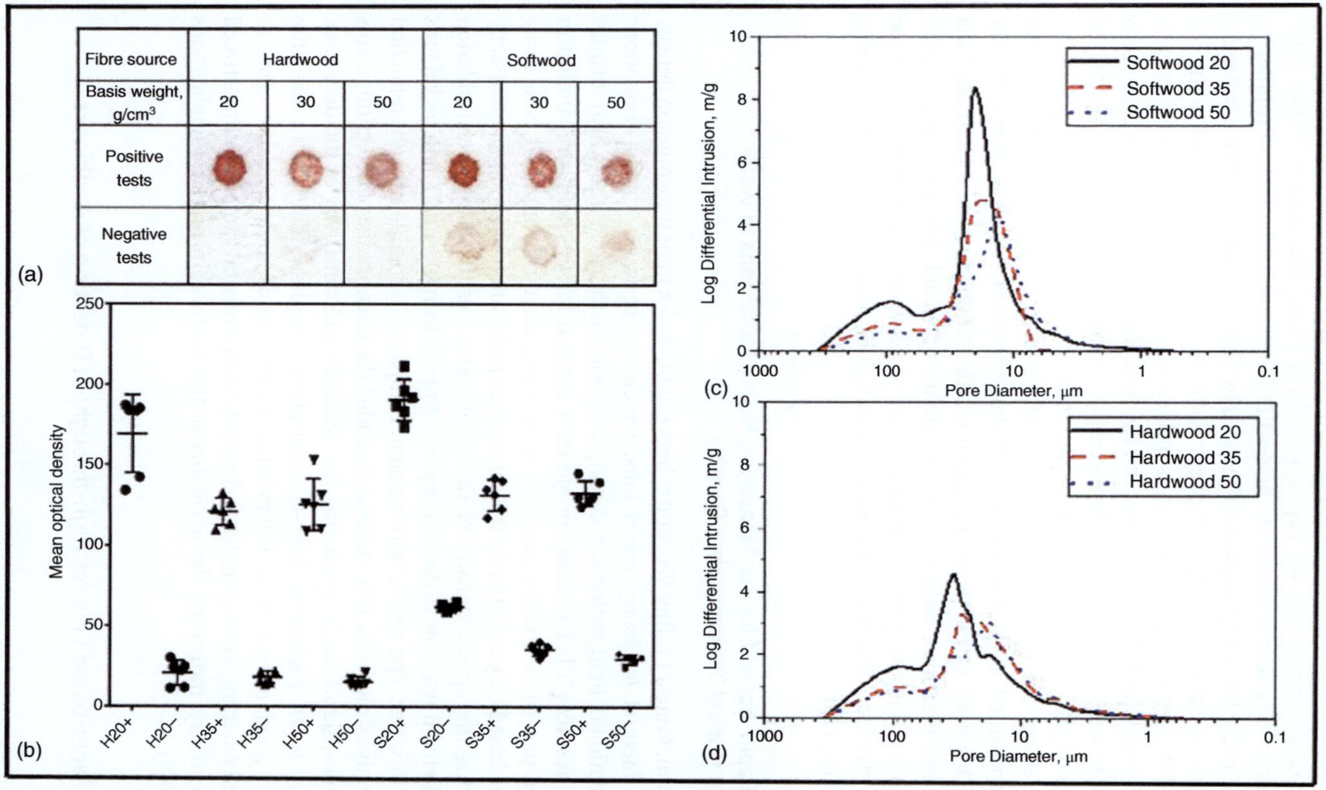

Figure 9.5 Impact of paper-based material structure on blood type detection results. (a) Blood type detection results using fluid flushing method with different structured paper-based materials *Source:* [112] / American Chemical Society; (b) color density analysis of blood type detection results; (c) internal structure characterization results of paper-based materials made from coniferous wood fibers with different quantifications using mercury intrusion method; and (d) internal structure characterization results of paper-based materials made from broadleaf wood fibers with different quantifications using the mercury intrusion method.

blood cell "coffee ring" through orthogonal experiments. By designing the paper structure and surface chemical properties, the study achieved regulation and control of the number of red blood cells deposited and quantified the dynamic migration behavior of red blood cells on paper. The study found that (i) the "coffee ring" phenomenon on paper is primarily the result of the combined effects of capillarity, filtration, and solvent evaporation, influenced by fiber structure, red blood cell incubation time, relative humidity, and paper additives. (ii) The prominence of the red blood cell "coffee ring" phenomenon is influenced by both the filtration efficiency of the fiber array and the affinity between the cells and fibers. Filtration efficiency is the primary factor; the cell-fiber affinity only becomes a major factor affecting red blood cell migration and distribution when there are enough red blood cells on the fiber structure. The experimental results indicated that red blood cells are more easily transmitted in fiber structures with larger pore sizes and weaker cell-fiber affinity. This study can improve the uniform distribution of reactants and products in detection reactions, enhance the sensitivity of visual detection, and reduce the likelihood of false positive or false negative results in diagnostic tests. These findings provide a theoretical basis for the design of paper-based diagnostic chips (Figure 9.6) [113].

9.5.2 Paper-based Chip for Measuring Heavy Metal Ions in Water Based on Optical Transmittance Density

Junfei Tian, Miaosi Li, Jing Wu, and others established a light transmission density analysis method for heavy metal ions, which compensates for the limitations of conventional light reflection density analysis methods in the linear display range of results [70]. The study proposed a method for the rapid preparation of patterned paper-based chips using UV curing and screen printing. By combining the paper-based chip with the light transmission result analysis method and directly quantifying the optical density of the colorimetric results using a transmission densitometer, it was able to detect copper (0.5 ppm), iron (0.5 ppm), and nickel ions (2 ppm) in water. The results showed that, compared to traditional light reflection image analysis methods, this method expanded the linear range of detection results by 10 times, significantly improving the accuracy and sensitivity of the detection, especially when measuring high-concentration samples. Additionally, the chip exhibited good resistance to interference from various ions. Measurements of simulated environmental samples and actual environmental samples demonstrated good specificity, indicating the feasibility of this method in practical applications (Figure 9.7).

9.5.3 Paper-based Chip for High-throughput Detection of Antibiotics in Milk

Antibiotic residues in animal-derived food can have serious adverse effects on human health, such as allergic reactions or antibiotic resistance. Efficient, low-cost, and portable food safety detection methods have garnered widespread attention

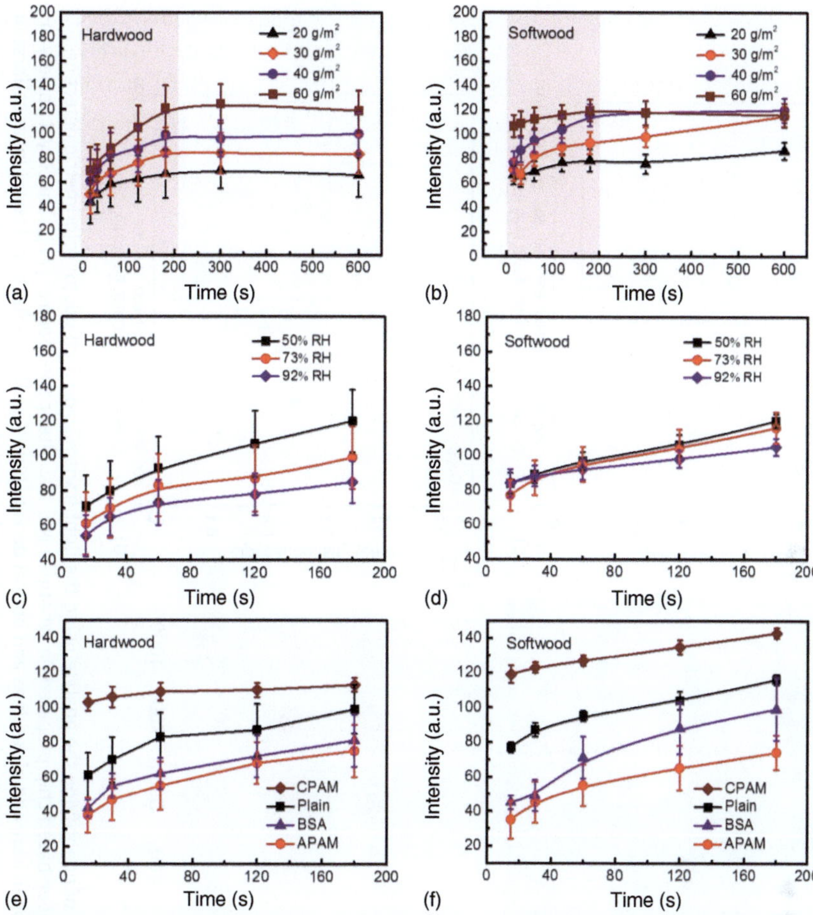

Figure 9.6 Impact of key factors on the formation of red blood cell rings. The ring formation of red blood cells in eucalyptus fiber-based paper over incubation time, influenced by grammage (a), ambient humidity (c), and different additives (e). The ring formation of red blood cells in pine fiber-based paper over incubation time, influenced by grammage (b), ambient humidity (d), and different additives (f). Standard errors are derived from three parallel measurements *Source:* [113] / with permission of Elsevier.

in the global market. Junfei Tian, Miaosi Li, Huimin Lu, and their team developed a paper-based detection chip with the potential for high-throughput and on-site monitoring of representative antibiotics like tetracycline in milk [114]. This study utilized UV-curing and screen printing technology to achieve large-scale and automated printing of patterned paper-based chips. By optimizing detection reagents and the paper substrate, the researchers were able to rapidly detect residual antibiotics in different types of milk through the chelation reaction between phenolic hydroxyl and enol groups in tetracycline molecules and metal ions, which forms colored complexes. The study evaluated the detection effect from both macroscopic and microscopic perspectives. (i) Macroscopic evaluation: This focused on the practical properties and detection effectiveness of the paper.

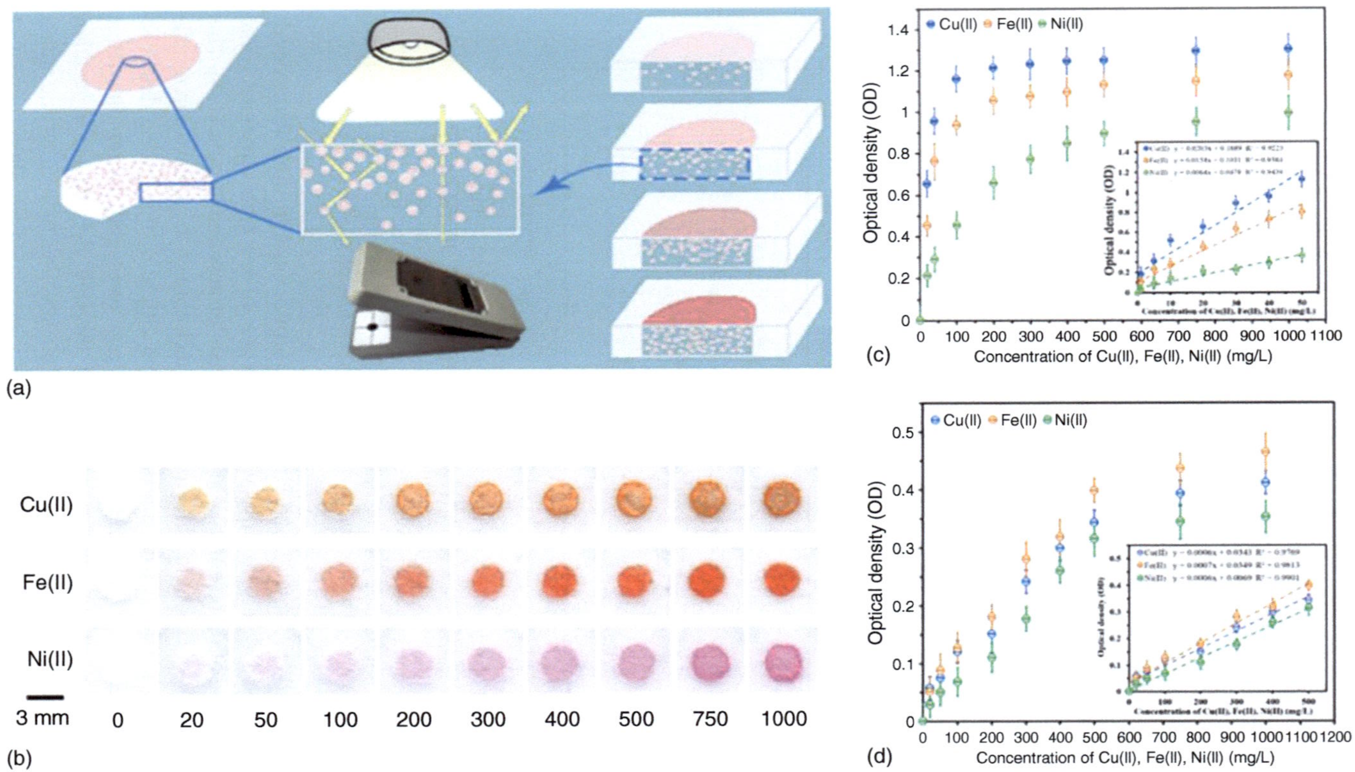

Figure 9.7 (a) Schematic diagram illustrating the principle of quantifying the analyte using the light transmission method; (b) colorimetric detection results of 0–1000 mg/L copper, iron, and nickel ions on a paper-based chip prepared on Whatman No. 102 filter paper *Source:* [72] / The Royal Society of Chemistry; (c) relationship between the concentration of copper, iron, and nickel ions and optical density based on light reflection. Inset: Linear calibration curves for the three metal ions; (d) relationship between the concentration of copper, iron, and nickel ions and optical density based on light transmission. Inset: Linear calibration curves for the three metal ions *Source:* [70] / with permission of Royal Society of Chemistry.

As the paper's basis weight increased, the uniformity of the precipitate distribution and color presentation improved to varying degrees. (ii) Microscopic evaluation: Using scanning electron microscopy, the study examined the influence of paper fiber quantities on the distribution and adsorption of the precipitate. Tests on copper precipitates' distribution and aggregation on papers of different basis weights revealed that low basis weight eucalyptus fiber paper showed significant differences in aggregation and precipitate distribution, whereas high basis weight paper provided better aggregation result discrimination. The porosity of the paper substrate, causing filtration and concentration effects, enhanced color intensity, allowing for ideal detection limits (1 ppm) and linear ranges for tetracycline quantification and sensitivity. Additionally, the study monitored the dynamic degradation behavior of tetracycline using the low-cost, efficient paper-based chip, providing valuable guidance for handling antibiotics in food processing [114]. This simple, rapid, and inexpensive paper-based detection platform holds great potential for application in the field of food safety. It is expected to be widely used for the detection of antibiotics in meat, seafood, dairy products, and other related items (Figure 9.8).

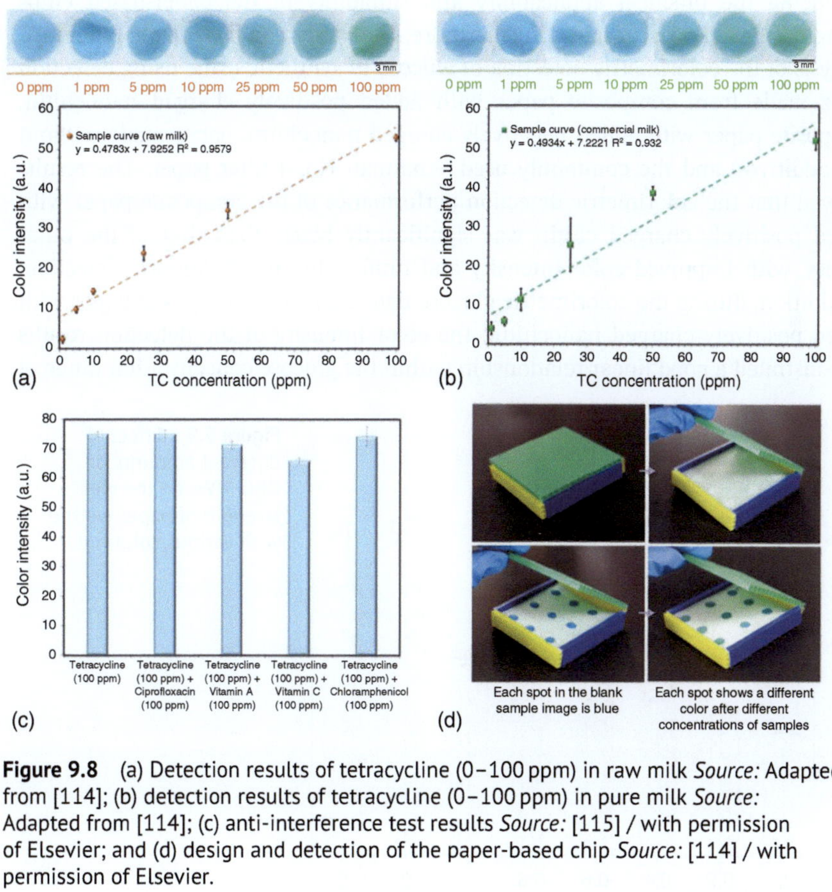

Figure 9.8 (a) Detection results of tetracycline (0–100 ppm) in raw milk *Source:* Adapted from [114]; (b) detection results of tetracycline (0–100 ppm) in pure milk *Source:* Adapted from [114]; (c) anti-interference test results *Source:* [115] / with permission of Elsevier; and (d) design and detection of the paper-based chip *Source:* [114] / with permission of Elsevier.

9.5.4 Construction of Composite Substrate for Paper-based Chips and Optimization of Physical and Detection Performance

The physical performance of paper as the substrate material for paper-based detection chips is crucial for the feasibility of detection. Low-grammage paper has insufficient strength and is prone to damage during the detection process, especially in multistep detection, making it unsuitable for detection requirements. Based on the performance requirements of the substrate material for wet reaction conditions in paper-based detection, Junfei Tian, Hua Tang, and others studied the effects of adding surface-positive and surface-negative nanochitin additives on the performance of the chip substrate, such as tensile strength, tear resistance, and wet strength [115]. The results revealed that adding positively charged nanochitin (DEChNWs) and negatively charged nanochitin (TOChNWs) to the paper improved its tensile and tear resistance, but the enhancing effect of negatively charged chitin was not as significant as that of positively charged chitin. In terms of wet strength, the paper made with negatively charged nanochitin showed almost no difference from the paper made without any additives. However, the paper with added positively charged nanochitin exhibited a significant improvement in wet strength, enhancing the practical applicability and suitability of the paper-based chips for actual detection scenarios. Furthermore, the study conducted a comparative analysis of the colorimetric detection of glucose in synthetic urine using detection chips made from composite paper with added positively charged nanochitin, composite paper with added negatively charged nanochitin, paper made without any additives, and the commonly used Whatman No. 4 filter paper. The results showed that the colorimetric detection performance of the composite paper with added positively charged chitin was significantly better than that of the other papers, with improved color intensity and uniformity in colorimetric detection. In addition, during the colorimetric glucose detection using composite paper with added positively charged nanochitin, the color intensity of the detection results demonstrated a good linear relationship within the glucose concentration range of

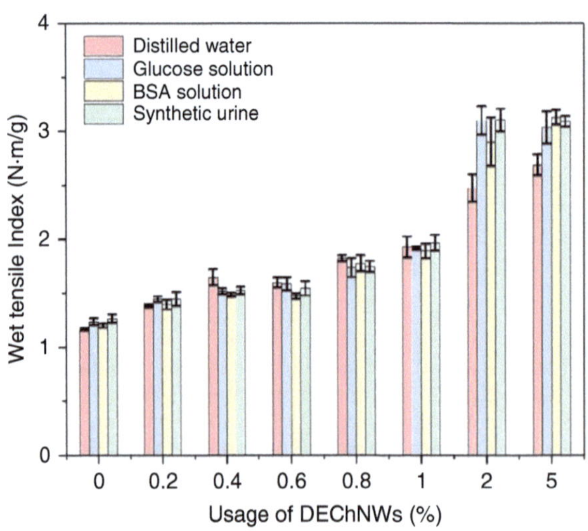

Figure 9.9 Effect of different amounts of DEChNWs on the wet strength of paper wetted with various solutions.

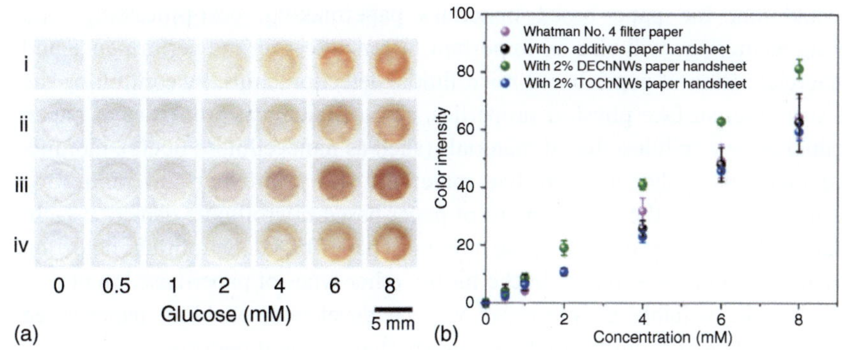

Figure 9.10 (a) The colorimetric detection of glucose using different types of paper: [(i) Whatman NO.4 filter paper, (ii) paper without additives, (iii) paper with 2% DEChNWs, (iv) paper with 2% TOChNWs]. (b) The color intensity graph for glucose detection using paper-based methods *Source:* [115] / with permission of Elsevier.

0, 0.5, 1, 2, 4, 6, and 8 mmol/L. This indicates that the composite paper with added positively charged nanochitin possesses excellent detection performance (Figure 9.9 and Figure 9.10) [115].

9.6 Future Prospects of Paper-based Functional Materials for Analysis and Detection

Paper-based microfluidic detection chips, with their advantages of low cost, high efficiency, low sample consumption, portability, and ease of operation, have attracted great attention and found wide applications. However, as a new technology, there are still many bottlenecks that need to be overcome in this field and warrant further research. For example:

1) Further improve the sensitivity and specificity of detection.
2) Simplify the operation process of paper-based microfluidic chips to make them more suitable for nonprofessional users.
3) Improve the detection result reading technology and explore the possibility of integration with more portable detection instruments.
4) Extend the shelf life of paper-based microfluidic chips and improve their stability.
5) Develop more types of paper-based microfluidic detection chips to expand their application scope.
6) Explore simpler and more affordable fabrication methods.
7) Enhance the biocompatibility of paper-based materials by combining them with new materials.
8) Study the flow mechanism of fluids in paper-based materials to achieve more precise control.
9) Explore the possibility of combining with more detection methods.
10) Minimize the influence of external environmental factors on detection.

In addition, the paper-based material's papermaking, post-processing, and printing technology will also be important factors affecting the performance and commercialization of paper-based microfluidic detection chips. By controlling the pore structure, surface physical properties, surface chemical properties of paper, introducing new cellulose-based materials (such as nanocellulose fibers), biofunctional materials, polymer electrolyte materials, and improving and innovating printing technology, the development of paper-based microfluidic detection technology will be greatly promoted. Up to now, most paper-based microfluidic detection chips are based on filter paper. In the future, other types of paper-based materials that are more suitable as substrates will be developed, enabling paper-based microfluidic detection chips to achieve unexpected excellent performance.

In summary, through continuous optimization, paper-based microfluidic detection chips will no longer be limited to medical diagnosis in underdeveloped areas but can also provide low-cost and convenient detection methods for emergency medical treatment, clinical trials, and healthcare in developed regions. They have good prospects for development in clinical diagnosis, food safety analysis, environmental monitoring, and other fields.

References

1 Eamon, W. (1980). New light on Robert Boyler and the discovery of colour indicators. *Ambix* 27 (3): 204–209.

2 Müller, R.H. and Clegg, D.L. (1949). Automatic paper chromatography. *Analytical Chemistry* 21 (9): 1123–1125.

3 Comer, J.P. (1956). Semiquantitative specific test paper for glucose in urine. *Analytical Chemistry* 28 (11): 1748–1750.

4 Bhattacharya, S., Kumar, S., and Agarwal, A.K. (2019). *Paper Microfluidics: Theory and Applications*. Singapore: Springer Nature Singapore Pte Ltd.

5 Harvey, D.R., Cooper, L.V., Fancourt, R.F. et al. (1976). The use of dextrostix and dextrostix reflectance meters in the diagnosis of neonatal hypoglycemia. *Journal of Perinatal Medicine* 4 (2): 106–110.

6 Chard, T. (1992). Pregnancy tests: a review. *Human Reproduction* 7 (5): 701–710.

7 von Lode, P. (2005). Point-of-care immunotesting: approaching the analytical performance of central laboratory methods. *Clinical Biochemistry* 38 (7): 591–606.

8 Tian, T., Huang, Y., Lin, B. et al. (2015). Development and application of paper-based microfluidic technology. *Journal of Analysis and Testing* 34 (3): 257–267.

9 Wu, J. (2020). *Preparation of Paper-Based Detection Chips and their Application in Colorimetric Detection Methods*. Guangzhou: South China University of Technology.

10 Manz, A., Graber, N., and Widmer, H.M. (1990). Miniaturized total chemical analysis systems: a novel concept for chemical sensing. *Sensors and Actuators B: Chemical* 1 (1): 244–248.

11 Li, X., Tian, J., Garnier, G. et al. (2010). Fabrication of paper-based microfluidic sensors by printing. *Colloids and Surfaces B* 76 (2): 564–570.

12 Martinez, A.W., Phillips, S.T., Butte, M.J. et al. (2007). Patterned paper as a platform for inexpensive, low-volume, portable bioassays. *Angewandte Chemie International Edition* 48 (8): 13401318–13401320.

13 Martinez, A.W., Phillips, S.T., Carrilho, E. et al. (2008). Simple telemedicine for developing regions: camera phones and paper-based microfluidic devices for real-time, off-site diagnosis. *Analytical Chemistry* 80 (10): 3699–3707.

14 Martinez, A.W., Phillips, S.T., and Whitesides, G.M. (2008). Three-dimensional microfluidic devices fabricated in layered paper and tape. *Proceedings of the National Academy of Sciences* 105 (50): 19606–19611.

15 Liu, H. and Crooks, R.M. (2011). Three-dimensional paper microfluidic devices assembled using the principles of origami. *Journal of the American Chemical Society* 133 (44): 17564–17566.

16 Pelton, R. (2009). Bioactive paper provides a low-cost platform for diagnostics. *Trends in Analytical Chemistry* 28 (8): 925–942.

17 Wu, J., Xu, J., Shi, C. et al. (2018). Research Progress and trends in paper-based microfluidic chips. *Journal of China Pulp and Paper* 33 (02): 57–64.

18 Li, L., Tian, J., Ballerini, D. et al. (2013). A study of the transport and immobilisation mechanisms of human red blood cells in a paper-based blood typing device using confocal microscopy. *Analyst* 138 (17): 4933–4940.

19 Li, X., Ballerini, D.R., and Shen, W. (2012). A perspective on paper-based microfluidics: current status and future trends. *Biomicrofluidics* 6 (1): 011301.

20 Yetisen, A.K., Akram, M.S., and Lowe, C.R. (2013). Paper-based microfluidic point-of-care diagnostic devices. *Lab on a Chip* 13 (12): 2210–2251.

21 Yager, P., Edwards, T., Fu, E. et al. (2006). Microfluidic diagnostic technologies for global public health. *Nature* 442 (7101): 412–418.

22 Martinez, A.W., Phillips, S.T., Whitesides, G.M. et al. (2010). Diagnostics for the developing world: microfluidic paper-based analytical devices. *Analytical Chemistry* 82 (1): 3–10.

23 Xia, Y., Si, J., and Li, Z. (2016). Fabrication techniques for microfluidic paper-based analytical devices and their applications for biological testing: a review. *Biosensors and Bioelectronics* 77: 774–789.

24 He, Y., Wu, Y., Fu, J.-Z. et al. (2015). Fabrication of paper-based microfluidic analysis devices: a review. *RSC Advances* 5 (95): 78109–78127.

25 Carrilho, E., Martinez, A.W., and Whitesides, G.M. (2009). Understanding wax printing: a simple micropatterning process for paper-based microfluidics. *Analytical Chemistry* 81 (16): 7091–7095.

26 Songjaroen, T., Dungchai, W., Chailapakul, O. et al. (2011). Novel, simple and low-cost alternative method for fabrication of paper-based microfluidics by wax dipping. *Talanta* 85 (5): 2587–2593.

27 Yu, L. and Shi, Z.Z. (2015). Microfluidic paper-based analytical devices fabricated by low-cost photolithography and embossing of Parafilm®. *Lab on a Chip* 15 (7): 1642–1645.

28 Asano, H. and Shiraishi, Y. (2015). Development of paper-based microfluidic analytical device for iron assay using photomask printed with 3D printer for fabrication of hydrophilic and hydrophobic zones on paper by photolithography. *Analytica Chimica Acta* 883: 55–60.

29 Shibata, H., Hiruta, Y., and Citterio, D. (2019). Fully inkjet-printed distance-based paper microfluidic devices for colorimetric calcium determination using ion-selective optodes. *Analyst* 144 (4): 1178–1186.

30 Alfadhel, A., Ouyang, J., Mahajan, C.G. et al. (2018). Inkjet printed polyethylene glycol as a fugitive ink for the fabrication of flexible microfluidic systems. *Materials & Design* 150: 182–187.

31 Hamad, E.M., Bilatto, S.E.R., Adly, N.Y. et al. (2016). Inkjet printing of UV-curable adhesive and dielectric inks for microfluidic devices. *Lab on a Chip* 16 (1): 70–74.

32 Cai, L., Xu, C., Lin, S. et al. (2014). A simple paper-based sensor fabricated by selective wet etching of silanized filter paper using a paper mask. *Biomicrofluidics* 8 (5): 056504.

33 Chaiyo, S., Siangproh, W., Apilux, A. et al. (2015). Highly selective and sensitive paper-based colorimetric sensor using thiosulfate catalytic etching of silver nanoplates for trace determination of copper ions. *Analytica Chimica Acta* 866: 75–83.

34 de Tarso, G.P., Garcia Cardoso, T.M., Garcia, C.D. et al. (2014). A handheld stamping process to fabricate microfluidic paper-based analytical devices with chemically modified surface for clinical assays. *RSC Advances* 4 (71): 37637–37644.

35 Curto, V.F., Lopez-Ruiz, N., Capitan-Vallvey, L.F. et al. (2013). Fast prototyping of paper-based microfluidic devices by contact stamping using indelible ink. *RSC Advances* 3 (41): 18811–18816.

36 Dossi, N., Toniolo, R., Pizzariello, A. et al. (2013). Pencil-drawn paper supported electrodes as simple electrochemical detectors for paper-based fluidic devices. *ELECTROPHORESIS* 34 (14): 2085–2091.

37 Nuchtavorn, N. and Macka, M. (2016). A novel highly flexible, simple, rapid and low-cost fabrication tool for paper-based microfluidic devices (μPADs) using technical drawing pens and in-house formulated aqueous inks. *Analytica Chimica Acta* 919: 70–77.

38 Li, X., Tian, J., Nguyen, T. et al. (2008). Paper-based microfluidic devices by plasma treatment. *Analytical Chemistry* 80 (23): 9131–9134.

39 Hecht, L., Philipp, J., Mattern, K. et al. (2016). Controlling wettability in paper by atmospheric-pressure microplasma processes to be used in μPAD fabrication. *Microfluidics and Nanofluidics* 20 (1): 25.

40 Namwong, P., Jarujamrus, P., Amatatongchai, M. et al. (2018). Fabricating simple wax screen-Printing paper-Based analytical devices to demonstrate the concept of limiting reagent in acid–Base reactions. *Journal of Chemical Education* 95 (2): 305–309.

41 Dungchai, W., Chailapakul, O., and Henry, C.S. (2011). A low-cost, simple, and rapid fabrication method for paper-based microfluidics using wax screen-printing. *The Analyst* 136 (1): 77–82.

42 Liu, M., Zhang, C., and Liu, F. (2015). Understanding wax screen-printing: a novel patterning process for microfluidic cloth-based analytical devices. *Analytica Chimica Acta* 891: 234–246.

43 Yao, Y. and Zhang, C. (2016). A novel screen-printed microfluidic paper-based electrochemical device for detection of glucose and uric acid in urine. *Biomedical Microdevices* 18 (5): 92.

44 Olkkonen, J., Lehtinen, K., and Erho, T. (2010). Flexographically printed fluidic structures in paper. *Analytical Chemistry* 82 (24): 10246–10250.

45 Thuo, M.M., Martinez, R.V., Lan, W.-J. et al. (2014). Fabrication of low-cost paper-based microfluidic devices by embossing or cut-and-stack methods. *Chemistry of Materials* 26 (14): 4230–4237.

46 Giokas, D.L., Tsogas, G.Z., and Vlessidis, A.G. (2014). Programming fluid transport in paper-based microfluidic devices using razor-crafted open channels. *Analytical Chemistry* 86 (13): 6202–6207.

47 Chitnis, G., Ding, Z., Chang, C.-L. et al. (2011). Laser-treated hydrophobic paper: an inexpensive microfluidic platform. *Lab on a Chip* 11 (6): 1161–1165.

48 Nie, J., Liang, Y., Zhang, Y. et al. (2013). One-step patterning of hollow microstructures in paper by laser cutting to create microfluidic analytical devices. *Analyst* 138 (2): 671–676.

49 Nurak, T., Praphairaksit, N., and Chailapakul, O. (2013). Fabrication of paper-based devices by lacquer spraying method for the determination of nickel (II) ion in waste water. *Talanta* 114: 291–296.

50 Kwong, P. and Gupta, M. (2012). Vapor phase deposition of functional polymers onto paper-based microfluidic devices for advanced unit operations. *Analytical Chemistry* 84 (22): 10129–10135.

51 Demirel, G. and Babur, E. (2014). Vapor-phase deposition of polymers as a simple and versatile technique to generate paper-based microfluidic platforms for bioassay applications. *Analyst* 139 (10): 2326–2331.

52 Haller, P.D., Flowers, C.A., and Gupta, M. (2011). Three-dimensional patterning of porous materials using vapor phase polymerization. *Soft Matter* 7 (6): 2428–2432.

53 Bruzewicz, D.A., Reches, M., and Whitesides, G.M. (2008). Low-cost printing of poly (dimethylsiloxane) barriers to define microchannels in paper. *Analytical Chemistry* 80 (9): 3387–3392.

54 Abe, K., Suzuki, K., and Citterio, D. (2008). Inkjet-printed microfluidic Multianalyte chemical sensing paper. *Analytical Chemistry* 80 (18): 6928–6934.

55 Sameenoi, Y., Nongkai, P.N., Nouanthavong, S. et al. (2014). One-step polymer screen-printing for microfluidic paper-based analytical device (μPAD) fabrication. *Analyst* 139 (24): 6580–6588.

56 Fenton, E.M., Mascarenas, M.R., López, G.P. et al. (2009). Multiplex lateral-flow test strips fabricated by two-dimensional shaping. *ACS Applied Materials & Interfaces* 1 (1): 124–129.

57 Wang, J., Monton, M.R.N., Zhang, X. et al. (2014). Hydrophobic sol-gel channel patterning strategies for paper-based microfluidics. *Lab on a chip* 14 (4): 691–695.

58 Maejima, K., Tomikawa, S., Suzuki, K. et al. (2013). Inkjet printing: an integrated and green chemical approach to microfluidic paper-based analytical devices. *RSC Advances* 3 (24): 9258–9263.

59 Yamada, K., Suzuki, K., and Citterio, D. (2017). Text-displaying colorimetric paper-based analytical device. *ACS Sensors* 2 (8): 1247–1254.

60 Carrasquilla, C., Little, J.R.L., Li, Y. et al. (2015). Patterned paper sensors printed with long-chain DNA aptamers. *Chemistry* 21 (20): 7369–7373.

61 Zhang, L., Guan, L., Lu, Z. et al. (2019). Barrier-free patterned paper sensors for multiplexed heavy metal detection. *Talanta* 196: 408–414.

62 Liu, S., Cao, R., Wu, J. et al. (2019). Directly writing barrier-free patterned biosensors and bioassays on paper for low-cost diagnostics. *Sensors and Actuators B: Chemical* 285: 529–535.

63 Jokerst, J.C., Adkins, J.A., Bisha, B. et al. (2012). Development of a paper-based analytical device for colorimetric Detection of select foodborne pathogens. *Analytical Chemistry* 84 (6): 2900–2907.

64 Hossain, S.M.Z., Ozimok, C., Sicard, C. et al. (2012). Multiplexed paper test strip for quantitative bacterial detection. *Analytical and Bioanalytical Chemistry* 403 (6): 1567–1576.

65 Lamas-Ardisana, P.J., Casuso, P., Fernandez-Gauna, I. et al. (2017). Disposable electrochemical paper-based devices fully fabricated by screen-printing technique. *Electrochemistry Communications* 75: 25–28.

66 Alahmad, W., Uraisin, K., Nacapricha, D. et al. (2016). A miniaturized chemiluminescence detection system for a microfluidic paper-based analytical device and its application to the determination of chromium (iii). *Analytical Methods* 8 (27): 5414–5420.

67 Li, W., Ge, S., Wang, S. et al. (2013). Highly sensitive chemiluminescence immunoassay on chitosan membrane modified paper platform using TiO_2 nanoparticles/multiwalled carbon nanotubes as label. *Luminescence* 28 (4): 496–502.

68 Huang, Y., Li, L., Zhang, Y. et al. (2019). Auto-cleaning paper-based electrochemiluminescence biosensor coupled with binary catalysis of cubic Cu_2O-au and polyethyleneimine for quantification of Ni^{2+} and Hg^{2+}. *Biosensors and Bioelectronics* 126: 339–345.

69 Delaney, J.L. and Hogan, C.F. (2015). Mobile Phone Based Electrochemiluminescence Detection in Paper-Based Microfluidic Sensors. In: *Mobile Health Technologies: Methods and Protocols, Rasooly* (ed. A. Rasooly and K.E. Herold), 277–289. New York: Humana Press.

70 Wu, J., Li, M., Tang, H. et al. (2019). Portable paper sensors for the detection of heavy metals based on light transmission-improved quantification of colorimetric assays. *Analyst* 144 (21): 6382–6390.

71 Noiphung, J., Songjaroen, T., Dungchai, W. et al. (2013). Electrochemical detection of glucose from whole blood using paper-based microfluidic devices. *Analytica Chimica Acta* 788: 39–45.

72 Su, M., Ge, L., Ge, S. et al. (2014). Paper-based electrochemical cyto-device for sensitive detection of cancer cells and in situ anticancer drug screening. *Analytica Chimica Acta* 847: 1–9.

73 Zhou, F., Noor, M.O., and Krull, U.J. (2014). Luminescence resonance energy transfer-based nucleic acid hybridization assay on cellulose paper with Upconverting phosphor as donors. *Analytical Chemistry* 86 (5): 2719–2726.

74 Delaney, J.L., Hogan, C.F., Tian, J. et al. (2011). Electrogenerated Chemiluminescence Detection in paper-based microfluidic sensors. *Analytical Chemistry* 83 (4): 1300–1306.

75 Scida, K., Li, B., Ellington, A.D. et al. (2013). DNA Detection using origami paper analytical devices. *Analytical Chemistry* 85 (20): 9713–9720.

76 Schazmann, B., Morris, D., Slater, C. et al. (2010). A wearable electrochemical sensor for the real-time measurement of sweat sodium concentration. *Analytical Methods* 2 (4): 342–348.

77 Noiphung, J., Nguyen, M.P., Punyadeera, C. et al. (2018). Development of paper-based analytical devices for minimizing the viscosity effect in human saliva. *Theranostics* 8 (14): 3797–3807.

78 Mohammadifar, M., Tahernia, M., and Choi, S. (2019). An equipment-free, paper-based electrochemical sensor for visual monitoring of glucose levels in urine. *SLAS TECHNOLOGY: Translating Life Sciences Innovation* 24 (5): 499–505.

79 Ruecha, N., Rangkupan, R., Rodthongkum, N. et al. (2014). Novel paper-based cholesterol biosensor using graphene/polyvinylpyrrolidone/polyaniline nanocomposite. *Biosensors and Bioelectronics* 52: 13–19.

80 Dong, T., Wang, G.A., and Li, F. (2019). Shaping up field-deployable nucleic acid testing using microfluidic paper-based analytical devices. *Analytical and Bioanalytical Chemistry* 411 (19): 4401–4414.

81 Klasner, S.A., Price, A.K., Hoeman, K.W. et al. (2010). Paper-based microfluidic devices for analysis of clinically relevant analytes present in urine and saliva. *Analytical and Bioanalytical Chemistry* 397 (5): 1821–1829.

82 Li, M.S., Tian, J.F., Al-Tamimi, M. et al. (2012). Paper-based blood typing device that reports patient's blood type "in writing". *Angewandte Chemie International Edition* 124 (22): 5593–5597.

83 Guan, L., Tian, J., Cao, R. et al. (2014). Barcode-like paper sensor for smartphone diagnostics: an application of blood typing. *Analytical Chemistry* 86 (22): 11362–11367.

84 Ratajczak, K. and Stobiecka, M. (2020). High-performance modified cellulose paper-based biosensors for medical diagnostics and early cancer screening: a concise review. *Carbohydrate Polymers* 229: 115463.

85 Reboud, J., Xu, G., Garrett, A. et al. (2019). Paper-based microfluidics for DNA diagnostics of malaria in low resource underserved rural communities. *Proceedings of the National Academy of Sciences* 116 (11): 4834–4842.

86 Dou, M., Sanjay, S.T., Benhabib, M. et al. (2015). Low-cost bioanalysis on paper-based and its hybrid microfluidic platforms. *Talanta* 145: 43–54.

87 Cheng, C.-M., Martinez, A.W., Gong, J. et al. (2010). Paper-based ELISA. *Angewandte Chemie International Edition* 49 (28): 4771–4774.

88 Songjaroen, T., Dungchai, W., Chailapakul, O. et al. (2012). Blood separation on microfluidic paper-based analytical devices. *Lab on a Chip* 12 (18): 3392–3398.

89 Wang, S., Ge, L., Song, X. et al. (2012). Paper-based chemiluminescence ELISA: lab-on-paper based on chitosan modified paper device and wax-screen-printing. *Biosensors and Bioelectronics* 31 (1): 212–218.

90 Rivas, L., Reuterswärd, P., Rasti, R. et al. (2018). A vertical flow paper-microarray assay with isothermal DNA amplification for detection of Neisseria meningitidis. *Talanta* 183: 192–200.

91 Ihalainen, P., Pettersson, F., Pesonen, M. et al. (2014). An impedimetric study of DNA hybridization on paper-supported inkjet-printed gold electrodes. *Nanotechnology* 25 (9): 094009.

92 Wu, L., Ma, C., Ge, L. et al. (2015). Paper-based electrochemiluminescence origami cyto-device for multiple cancer cells detection using porous AuPd alloy as catalytically promoted nanolabels. *Biosensors and Bioelectronics* 63: 450–457.

93 He, J., Xiao, G., Chen, X. et al. (2019). A thermoresponsive microfluidic system integrating a shape memory polymer-modified textile and a paper-based colorimetric sensor for the detection of glucose in human sweat. *RSC Advances* 9 (41): 23957–23963.

94 Wang, X., Li, F., Cai, Z. et al. (2018). Sensitive colorimetric assay for uric acid and glucose detection based on multilayer-modified paper with smartphone as signal readout. *Analytical and Bioanalytical Chemistry* 410 (10): 2647–2655.

95 Jayawardane, B.M., Wongwilai, W., Grudpan, K. et al. (2014). Evaluation and application of a paper-based device for the determination of reactive phosphate in soil solution. *Journal of Environmental Quality* 43 (3): 1081–1085.

96 Bui, M.-P.N., Brockgreitens, J., Ahmed, S. et al. (2016). Dual detection of nitrate and mercury in water using disposable electrochemical sensors. *Biosensors and Bioelectronics* 85: 280–286.

97 Jayawardane, B.M., Wei, S., McKelvie, I.D. et al. (2014). Microfluidic paper-based analytical device for the determination of nitrite and nitrate. *Analytical Chemistry* 86 (15): 7274–7279.

98 Nie, J., Brown, T., and Zhang, Y. (2016). New two-dimensional liquid-phase colorimetric assay based on old iodine–starch complexation for the naked-eye quantitative detection of analytes. *Chemical Communications* 52 (47): 7454–7457.

99 Loh, L.J., Bandara, G.C., Weber, G.L. et al. (2015). Detection of water contamination from hydraulic fracturing wastewater: a μPAD for bromide analysis in natural waters. *Analyst* 140 (16): 5501–5507.

100 Rahbar, M., Paull, B., and Macka, M. (2019). Instrument-free argentometric determination of chloride via trapezoidal distance-based microfluidic paper devices. *Analytica Chimica Acta* 1063: 1–8.

101 Fraiwan, A., Lee, H., and Choi, S. (2016). A paper-based cantilever array sensor: monitoring volatile organic compounds with naked eye. *Talanta* 158: 57–62.

102 Alkasir, R.S.J., Ornatska, M., and Andreescu, S. (2012). Colorimetric paper bioassay for the Detection of phenolic compounds. *Analytical Chemistry* 84 (22): 9729–9737.

103 Saraji, M. and Bagheri, N. (2018). Paper-based headspace extraction combined with digital image analysis for trace determination of cyanide in water samples. *Sensors and Actuators B: Chemical* 270: 28–34.

104 Rattanarat, P., Dungchai, W., Cate, D. et al. (2014). Multilayer paper-based device for colorimetric and electrochemical quantification of metals. *Analytical Chemistry* 86 (7): 3555–3562.

105 Lewis, G.G., Robbins, J.S., and Phillips, S.T. (2014). A prototype point-of-use assay for measuring heavy metal contamination in water using time as a quantitative readout. *Chemical Communications* 50 (40): 5352–5354.

106 Li, Z., Zhu, Y., Zhang, W. et al. (2017). A low-cost and high sensitive paper-based microfluidic device for rapid detection of glucose in fruit. *Food Analytical Methods* 10 (3): 666–674.

107 Trofimchuk, E., Hu, Y., Nilghaz, A. et al. (2020). Development of paper-based microfluidic device for the determination of nitrite in meat. *Food Chemistry* 316: 126396.

108 Hossain, S.M.Z., Luckham, R.E., Smith, A.M. et al. (2009). Development of a bioactive paper sensor for detection of neurotoxins using piezoelectric inkjet printing of sol–gel-derived bioinks. *Analytical Chemistry* 81 (13): 5474–5483.

109 He, Q., Ma, C., Hu, X. et al. (2013). Method for fabrication of paper-based microfluidic devices by Alkylsilane self-assembling and UV/O_3-patterning. *Analytical Chemistry* 85 (3): 1327–1331.

110 Tian, J., Jarujamrus, P., Li, L. et al. (2012). Strategy to enhance the wettability of bioactive paper-based sensors. *ACS Applied Materials & Interfaces* 4 (12): 6573–6578.

111 Tian, J. and Shen, W. (2011). Printing enzymatic reactions. *Chemical Communications* 47 (5): 1583–1585.

112 Li, L., Huang, X., Liu, W. et al. (2014). Control performance of paper-based blood analysis devices through paper structure design. *ACS Applied Materials & Interfaces* 6 (23): 21624–21631.

113 Cao, R., Pan, Z., Tang, H. et al. (2020). Understanding the coffee-ring effect of red blood cells for engineering paper-based blood analysis devices. *Chemical Engineering Journal* 391: 123522.

114 Lu, H., Li, M., Nilghaz, A. et al. Paper-based analytical device for high-throughput monitoring tetracycline residue in milk. *Submitted.*

115 Tang, H., Wu, J., Li, D. et al. (2020). High-strength paper enhanced by chitin nanowhiskers and its potential bioassay applications. *International Journal of Biological Macromolecules* 150: 885–893.

10

Other Paper-based Functional Materials

10.1 Cream Paper-based Functional Materials

10.1.1 Overview

10.1.1.1 Definition of Cream Paper-based Functional Materials

Cream paper, also known as soft moisturizing paper (also called cream moisturizing paper), is a new type of high-end household paper that adds cream to the surface of facial tissue through mechanical methods such as spraying or roller coating. After sealing and standing treatment, the cream completely penetrates into the fibers of the facial tissue, giving the facial tissue softness, smoothness, and moisturizing effects.

What is cream? Cream is a safe, nontoxic, odorless emulsion or microemulsion containing specific components such as smoothing agents, moisturizing agents, preservatives, emulsifiers, and water, and has a specific viscosity. When the amount of cream is low, it can improve the surface softness of facial tissue; when the amount of cream increases to a certain extent, it can give facial tissue some new properties, such as skin moisturizing, that is, improving the smoothness of the facial tissue, reducing the surface friction coefficient (roughness), and showing obvious moisturizing properties [1]. Cream-treated facial tissue, with its moist and soft texture, has been widely used for wiping the nose of rhinitis patients, wiping the mouth and nose of infants and young children, and removing makeup for women [2, 3]. In recent years, cream has gradually attracted widespread attention from papermaking companies due to its unique skin moisturizing properties, and more and more papermaking workers have begun to try to apply it to the production of facial tissue.

10.1.1.2 Characteristics of Cream Paper-based Functional Materials

Safety, nontoxic, odorless, skin-friendly smoothness, moisturizing, antibacterial, nonirritating to the skin, good paper strength, etc., are the main characteristics of cream paper and are also important criteria for consumers to select cream paper and evaluate its performance. Cream paper must strictly comply with cosmetic-grade or food-grade health standards.

Paper-Based Functional Materials: Preparation and Applications, First Edition.
Xianping Yao, Genrong Chen, Zhen Yao, and Liping Zheng.

1) Human Safety: Cream paper is an emerging product in the domestic market in recent years. As a high-end variety of household paper series, safety, nontoxicity, odorlessness, antibacterial properties, and non-irritation to the skin are the most basic health requirements for cream paper in contact with the human body. Since there are no clear national regulations on the safety standards of cream paper, the industry's safety standards for cream paper are mainly based on the safety technical specifications of the cosmetics industry. Specific safety standard testing items include skin irritation tests, antibacterial performance, heavy metal content, oral toxicity tests, preservative effects, ROHS (certain harmful substances in electronic and electrical equipment), etc. The specific implementation standards are shown in Table 10.1.
2) Odorlessness: Odorlessness is the most important indicator in the safety and performance indicators of cream paper. Odor not only affects the subsequent market promotion of cream paper but also involves major quality issues such as quality complaints and customer compensation, and most importantly, it may have adverse effects on human health. Therefore, achieving odorlessness in cream paper is the most basic goal pursued by experts and all scientific researchers in the cream paper industry, and it is the first checkpoint for ensuring the quality of cream paper. Common odors in cream paper include a greasy smell, fishy smell, herbal smell, and foot odor. The sources of these odors are mainly the base paper of the tissue, the cream, and the production process, among which the cream and the base paper are the key influencing factors.

1) Impact of Base Paper on Odor

According to the current production process of base paper for tissue paper, it is necessary to add chemicals such as cylinder release agents, cylinder adhesives, and wet strength agents during the production process. The main components of these chemicals are mineral oil, oil derivatives, amine polymers, and emulsifiers. The refining purity of mineral oil and oil, the completeness of emulsification, and the content of free monomers in amine polymers greatly affect the odor of the base paper for tissue paper.

2) Impact of Water Quality on Odor

During the production process of base paper for household paper, various chemicals accumulate in the closed white water circulation system of the paper machine, which can easily cause bacterial growth in the closed circulation system, leading to the deterioration of the water quality used to make the base paper, resulting in odor. The base paper produced with this water, when coated with cream to make cream paper, can easily develop an odor under humid and sealed packaging conditions.

3) Impact of Cream on Odor

The quality of the cream is also one of the main factors affecting the odor of cream paper. The cream mainly consists of glycerin, moisturizers, smoothers, preservatives, and deionized water. Glycerin and moisturizers are low-molecular substances with multiple hydroxyl or carboxyl groups, which can easily breed

Table 10.1 Standards and technical requirements for safety testing of cream paper.

Safety testing items	Reference standards	Technical requirements
Total Viable Count	"Cosmetic Safety Technical Specifications" (2015 Edition)	≤1000 cfu/mL
Thermotolerant Coliforms	"Cosmetic Safety Technical Specifications" (2015 Edition)	Not detected
Molds and Yeasts	"Cosmetic Safety Technical Specifications" (2015 Edition)	<100 cfu/mL
Staphylococcus aureus	"Cosmetic Safety Technical Specifications" (2015 Edition)	Not detected
Pseudomonas aeruginosa	"Cosmetic Safety Technical Specifications" (2015 Edition)	Not detected
Primary Skin Irritation Test	"Disinfection Technical Specifications" (2002 Edition, Part 2.3.3)	Erythema and edema irritation score = 0
Antibacterial Activity (Streptococcus hemolyticus)	GB 15979—2002 Appendix C4 Antibacterial Performance	Bacteriostatic rate > 99.97%
Antibacterial Activity (*Escherichia coli*)	GB 15979—2002 Appendix C4 Antibacterial Performance	Bacteriostatic rate > 99.97%
Antibacterial Activity (*S. aureus*)	GB 15979 – 2002 Appendix C4 Antibacterial Performance	Bacteriostatic rate > 99.97%
Antibacterial Activity (*Candida albicans*)	GB 15979—2002 Appendix C4 Antibacterial Performance	Bacteriostatic rate > 99.97%
Heavy Metal Content (Arsenic)	"Cosmetic Safety Technical Specifications" (2015 Edition, Chapter 4, 1.6)	≤2 mg/kg
Heavy Metal Content (Cadmium)	"Cosmetic Safety Technical Specifications" (2015 Edition, Chapter 4, 1.6)	≤5 mg/kg
Heavy Metal Content (Lead)	"Cosmetic Safety Technical Specifications" (2015 Edition, Chapter 4, 1.6)	≤10 mg/kg
Heavy Metal Content (Mercury)	"Cosmetic Safety Technical Specifications" (2015 Edition, Chapter 4, 1.6)	≤1 mg/kg
Certain Harmful Substances in Electrical and Electronic Equipment	IEC 62321-6 : 2015, Determination of polybrominated biphenyls and polybrominated diphenyl ethers in polymers and electronic materials; IEC 62321-8 : 2017, Determination of phthalates in polymers	≤1000 mg/kg
Acute Oral Toxicity Test	"Cosmetic Safety Technical Specifications" (2015 Edition)	LD50 > 5000 mg/kg

bacteria and mold under long-term humid and sealed conditions. If the preservative system in the cream product is not well selected or the dosage is insufficient, it will lead to mold and odor in the cream paper. Additionally, different types and structures of smoothers in the cream can result in different odors. Creams made with fatty acid derivatives and silicone-based smoothers, although having a good smoothing effect, have a noticeable amine and oil characteristic odor, leading to a significant odor in the cream paper. Therefore, the selection of raw materials for the cream is crucial.

3) Moisturizing Experience: For daily fast-moving consumer goods like cream paper, the comprehensive experience of softness and moisturizing is the most important quality indicator for consumers when selecting cream paper products. It is also the most obvious feature distinguishing it from ordinary tissue paper.

1) Softness

Softness is an important factor determining the grade of cream paper. The higher the grade of cream paper, the higher the requirement for a soft and skin-friendly feel. There are two main methods to characterize softness: surface softness and bulk softness. The factors affecting surface softness are mainly the elastic modulus and roughness. The elastic modulus can be considered an indicator of the difficulty of elastic deformation of the paper; the larger the value, the greater the stress required to cause elastic deformation of the paper. Roughness can be used to characterize the smoothness and flatness of the paper surface.

For household paper, there are three main methods to characterize softness: the expert group evaluation method, the Handle-O-meter method (TAPPI T498cm-85), and the tissue softness analyzer (TSA) method. Currently, the measurement of cream paper softness mainly uses the expert group evaluation method and the TSA method. The expert group evaluation method is related to the friction, bending stiffness, bulk, and thermal properties of the paper surface. These properties are related to human psychological processes, i.e., sensitivity to smoothness, stiffness, bulk, and thermal properties. The expert group evaluation method involves selecting a certain number of paper samples with different weights and thicknesses, conditioning them at a specific humidity and temperature, then sealing them in moisture-proof bags. Multiple expert group members open the packaging bags under constant temperature and humidity conditions and evaluate the samples. This analysis method can only evaluate overall softness, not surface softness.

The TSA method uses a softness tester to quantitatively test the handfeel value of the paper and calculate the softness of the paper quantitatively. The handfeel value is a comprehensive representation of fiber softness TS750, surface smoothness (roughness) TS7, deformation coefficient (stiffness) D, base paper weight G, and other factors, quantitatively representing fiber softness. It is actually a comprehensive reflection of fiber softness and surface smoothness, i.e., touch feel HF. HF = f (TS7, TS750, D, thickness, G, layers).

Currently, there are many types of cream paper on the market, and the brands with high consumer recognition include Japan's Daio GOO. Nelleair, C&S Lotion,

Nepia, etc. The cream used in these brands of cream paper is produced by Japan's Miyoshi Oil & Fat Co., Ltd. This cream has a market share of over 40% in the domestic cream paper market and is expensive. In recent years, some domestic research institutions and many scientific workers have devoted a lot of effort to the research and development of cream, and the quality of domestic cream has gradually matured. Its smoothness performance has basically reached the quality of Japan's MIYOSI® series products.

The research team led by Yao Xianping at the Hangzhou Chemical Research Institute (National Engineering Research Center for Papermaking Chemicals) has done a lot of effective work in the research of cream and successfully developed the Lotion-HH® series products. Currently, this series of products represents the highest level of smoothness performance of domestic cream paper (see Figure 10.1), comparable to the smoothness performance of Japanese cream paper (see Figure 10.2), with a bright market prospect.

2) Moisturizing

Moisturizing is one of the most important features for consumers to evaluate the experience of cream paper. Good moisturizing performance gives consumers a smooth, soft, and skin-friendly feeling when using it, which is also the main difference in experience between cream paper and ordinary tissue paper. The moisturizing performance of cream paper is reflected by the moisturizer used in the cream. Therefore, the choice of moisturizer directly affects the experience of cream paper.

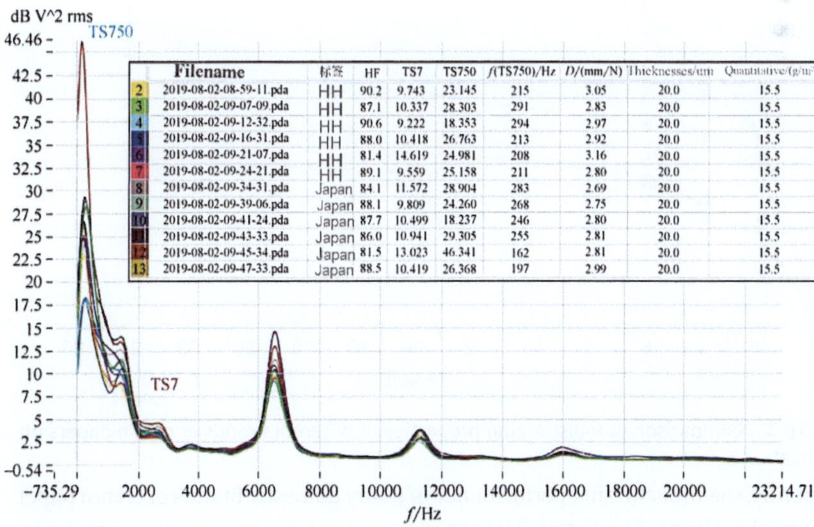

	Filename	标签	HF	TS7	TS750	f(TS750)/Hz	D/(mm/N)	Thicknesses/um	Quantitative/(g/m²)
2	2019-08-02-08-59-11.pda	HH	90.2	9.743	23.145	215	3.05	20.0	15.5
3	2019-08-02-09-07-09.pda	HH	87.1	10.337	28.303	291	2.83	20.0	15.5
4	2019-08-02-09-12-32.pda	HH	90.6	9.222	18.353	294	2.97	20.0	15.5
5	2019-08-02-09-16-31.pda	HH	88.0	10.418	26.763	213	2.92	20.0	15.5
6	2019-08-02-09-21-07.pda	HH	81.4	14.619	24.981	208	3.16	20.0	15.5
7	2019-08-02-09-24-21.pda	HH	89.1	9.559	25.158	211	2.80	20.0	15.5
8	2019-08-02-09-34-31.pda	Japan	84.1	11.572	28.904	283	2.69	20.0	15.5
9	2019-08-02-09-39-06.pda	Japan	88.1	9.809	24.260	268	2.75	20.0	15.5
10	2019-08-02-09-41-24.pda	Japan	87.7	10.499	18.237	246	2.80	20.0	15.5
11	2019-08-02-09-43-33.pda	Japan	86.0	10.941	29.305	255	2.81	20.0	15.5
12	2019-08-02-09-45-34.pda	Japan	81.5	13.023	46.341	162	2.81	20.0	15.5
13	2019-08-02-09-47-33.pda	Japan	88.5	10.419	26.368	197	2.99	20.0	15.5

Figure 10.1 Comparison of smoothness performance of different brands of cream sheets in the market (TSA method).

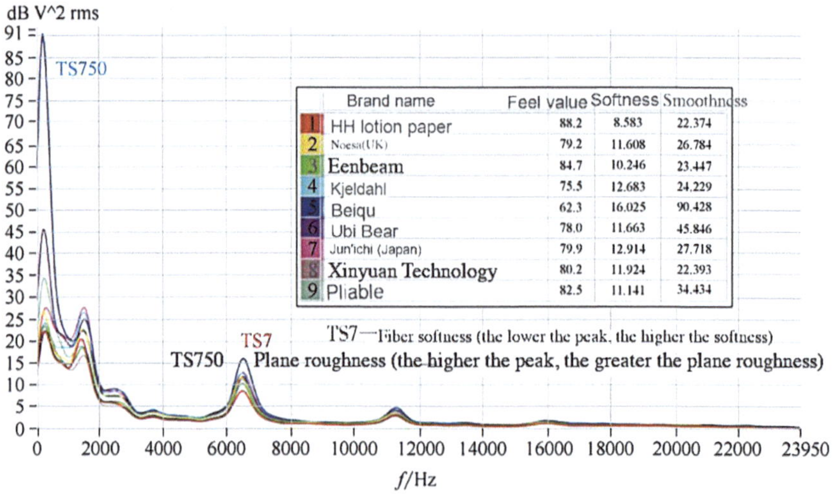

Figure 10.2 Comparison of the smoothing properties of domestic creamer paper and Japanese creamer paper (TSA method).

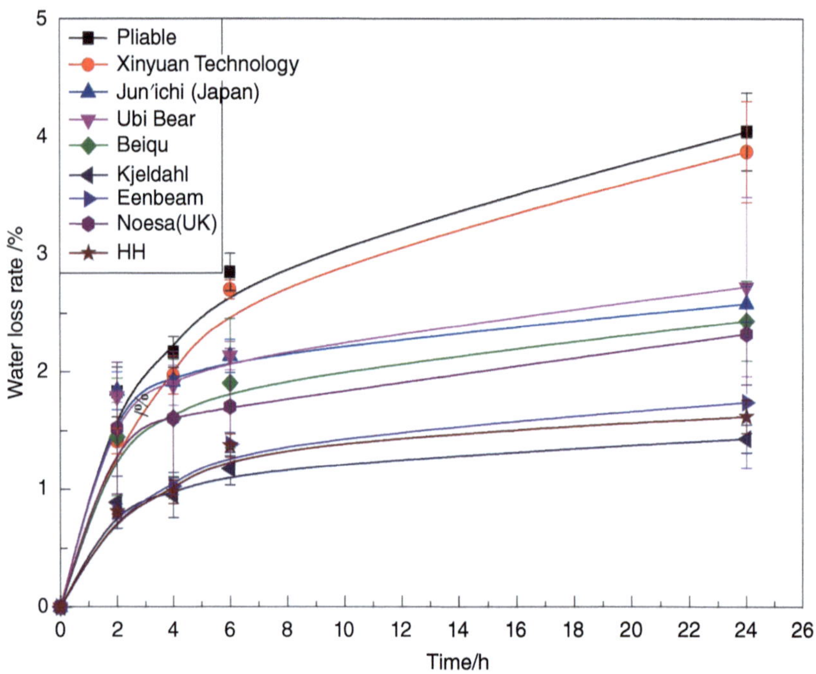

Figure 10.3 Comparison of moisturizing properties of different brands of cream papers on the market.

Currently, the moisturizing performance of many domestic brands of cream paper on the market varies. The Lotion-HH series products have better moisturizing performance (see Figure 10.3), and the moisturizing performance of the cream is better than that of Japanese cream (see Figure 10.4).

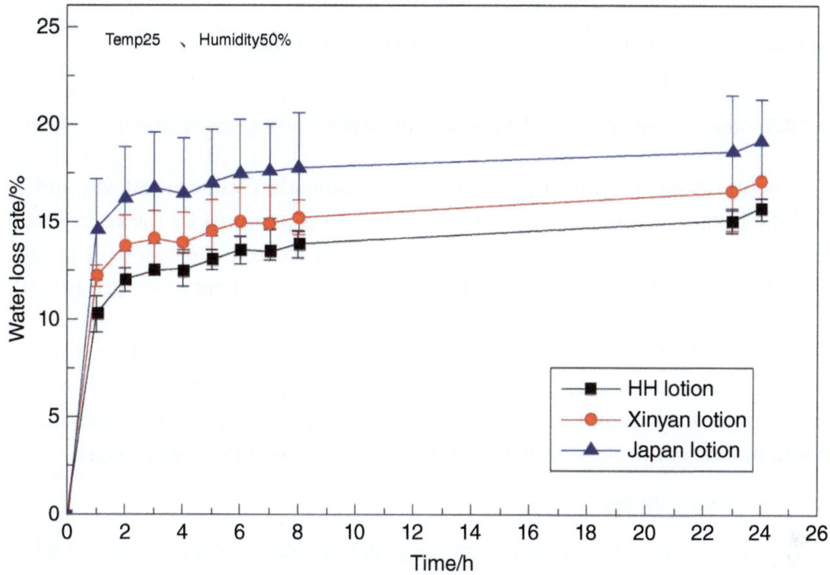

Figure 10.4 Comparison of moisturizing properties of domestic and Japanese creams.

10.1.2 Preparation Technology and Principles of Cream

The preparation technology of cream mainly includes smoothing technology and high moisturizing technology.

10.1.2.1 Smoothing Technology

1) Cationic Quaternary Ammonium Salt Softener

The cationic quaternary ammonium salt softeners used for preparing cream mainly include stearic acid triethanolamine ester quaternary ammonium salt, glycerol ether quaternary ammonium salt, and di (stearoyloxy)-propyl trimethyl ammonium salt. Cationic quaternary ammonium salt softeners carry a positive charge, which can directly combine with negatively charged fibers. The hydrophilic groups adsorb onto the fibers, while the hydrophobic groups arrange outward, forming a low surface energy surface. The fibers are coated with a film of long-chain fatty alkyl groups. Therefore, the fibers of cream paper treated with this type of softener are soft, and the paper surface is also smooth, with a full and smooth hand feel. Additionally, cationic quaternary ammonium salt softeners in cream products cannot be used in combination with anionic moisturizers, as they can easily precipitate and stratify, and they can cause some irritation to human skin.

2) Fatty Acid Derivative Softener

Stearic acid polyoxyethylene ester (also known as softener SME-4) is also one of the common softener components in cream. It is a nonionic softener, soluble in water, ethanol, ether, and toluene, and can be mixed with various surfactants. Its

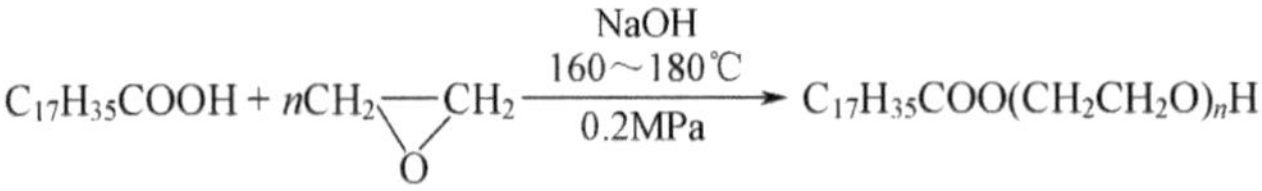

Figure 10.5 Reaction equation for the preparation of polyoxyethylene stearate.

aqueous solution has a neutral pH value, a saponification value of 80–90, and a hydrophilic-lipophilic balance (HLB) value of 10.0. It is prepared by the reaction of stearic acid with ethylene oxide under the catalysis of sodium hydroxide (or potassium hydroxide) (see Figure 10.5), with the number of ethylene oxide groups (n) being 5 to 10.

Softener SME-4 has good permeability and both softening and smoothing effects. When used as a smoothing agent in the production and processing of cream paper and napkins, SME-4 not only allows the cream to quickly penetrate the paper, making it soft, but also improves the hygroscopic performance of the cream paper.

3) Silicone-based Softener

In terms of the smoothing effect of cream paper, the cream paper prepared with silicone-based softeners has a better smoothing effect than the above two types of softeners. The main structure of silicone-based softeners is polysiloxane as the main chain, with organic functional groups such as methyl groups on the silicon atoms. Side groups are introduced into the siloxane molecules to improve the affinity of siloxane to the substrate, giving it both the characteristics of siloxane compounds and the features of branched functional groups. The silicone-based softeners suitable for cream mainly include silicone quaternary ammonium salt emulsion and amino silicone oil, and its emulsion-type softeners.

a) Silicone Quaternary Ammonium Salt Emulsion

Silicone quaternary ammonium salt contains both quaternary ammonium salt and polysiloxane main chain. It has easily hydrolyzable methoxy groups that can hydrolyze into easily cross-linkable silanol groups, which can firmly adhere to the fiber surface, providing a softening effect. At the same time, silicone quaternary ammonium salt also has strong bactericidal and antibacterial abilities. It can penetrate the cell walls of bacteria, causing the proteins in the cell walls to coagulate, thereby achieving sterilization. This is crucial for cream paper that requires good antibacterial properties.

b) Amino Silicone Oil and Its Emulsion

Amino silicone oil is also one of the smoothing components needed for cream preparation. Terminal amino silicone oil and side-chain amino silicone oil are the two commonly used amino silicone oils in cream preparation. The preparation reaction is shown in Figure 10.6.

Amino silicone oil can exhibit cationic properties under acidic conditions. Its smoothing effect on fibers is achieved by the oxygen atoms in the siloxane bonds adsorbing onto the fiber surface, causing the hydrophobic methyl groups to arrange directionally and cover the fiber surface. Due to the ability of the bond angles in the

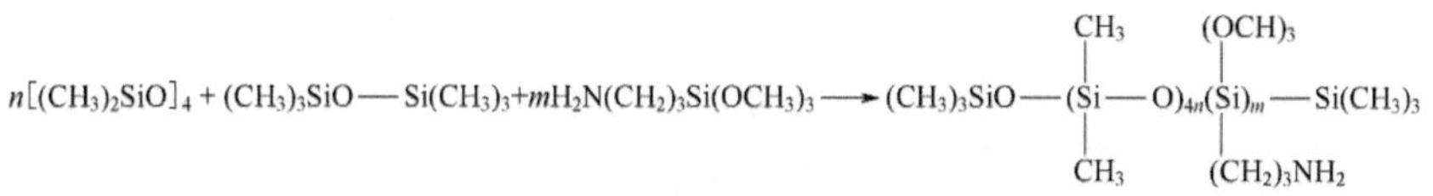

Figure 10.6 Reaction equation for the preparation of amino silicone oils.

siloxane bonds to change and stretch under external force, a tough, water-repellent, and stretchable continuous film is formed on the fiber surface treated with silicone softeners, making the fibers softer. Amino-modified silicone softeners bond very firmly with the fibers and significantly improve the performance of the paper. Among amino-modified silicone softeners, those with aminoethyl and aminopropyl groups perform the best.

4) Polymer Softener

This type of softener is an emulsion made from high-molecular polymers such as polyethylene, polypropylene, silicone resin, and polyurethane resin. As a smoothing component in cream, this type of polymer emulsion has an outstanding softening effect, not only giving the cream paper a good soft and smooth feel but also providing good bulkiness and thickness to the cream paper.

a) Mechanism of Action

Polymer softeners have hydrophobic groups and cationic hydrophilic groups. The hydrophilic groups combine with the fibers, while the hydrophobic groups cover the fibers. The hydrophobic groups break the hydrogen bonds between the pulp fibers, having a very noticeable bond-breaking effect on the paper fibers. The high molecular weight and smooth characteristics of the silicone resin in the polymer give the cream paper excellent smoothness. However, as the hydrogen bonds are broken, the bonding force between the fibers becomes loose (see Figure 10.7).

b) Impact of Particle Size

The particle size of the polymer has a certain impact on the softening effect of the cream. The Lotion-HH series cream developed by Hangzhou Chemical Research Institute uses polymer softeners with a nanoscale particle size (100 to 500 nm) (see Figure 10.8), which can provide the paper with a very good softening effect. The area

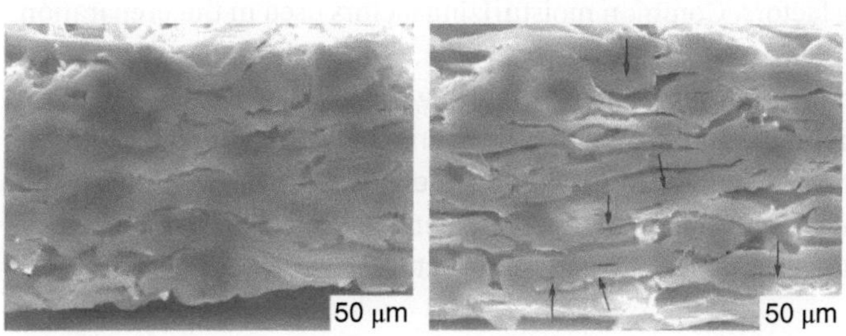

Figure 10.7 Effect of polymer softener on paper fiber structure.

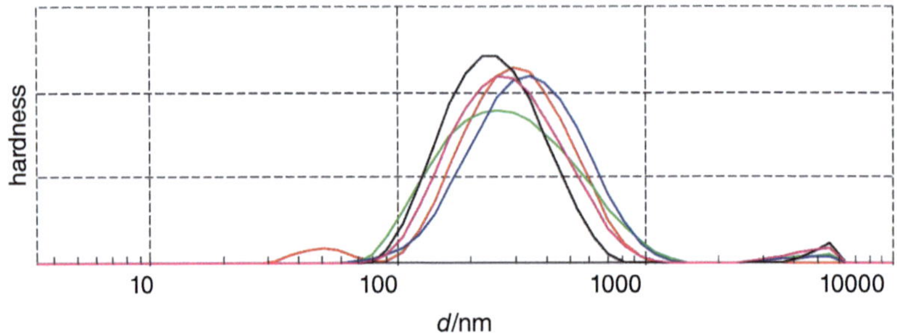

Figure 10.8 Polymer softener nanoparticle size distribution.

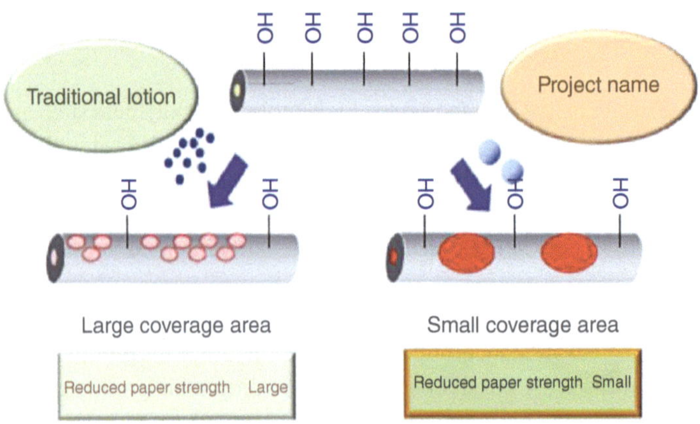

Figure 10.9 Schematic diagram of the relationship between polymer particle size and softening mechanism.

covered by the fibers is smaller, resulting in a smaller decrease in paper strength. Traditional creams use fatty acid softeners that cover a larger area of the fibers, resulting in a greater decrease in paper strength. Nanoscale polymer softeners not only give the cream paper an outstanding smoothing effect but also significantly improve the strength of the cream paper (see Figure 10.9).

10.1.2.2 High Moisturizing Technology

The good moisturizing effect of cream paper is mainly achieved by introducing natural high moisturizing factors. Common moisturizing factors used in the preparation of cream include glycerin, propylene glycol, sorbitol, polyethylene glycol, nanocellulose derivatives, and hyaluronic acid (HA). Among them, glycerin, nanocellulose derivatives, and HA are commonly used high moisturizing factors in cream. Numerous research experiments have shown that among these moisturizers, glycerin is the most economical and effective moisturizing ingredient, while nanocellulose derivatives and HA are relatively efficient moisturizing factors.

1) Nanocellulose Derivatives

Nanocellulose derivatives are natural, high molecular green bio-based materials that are nontoxic and harmless to humans and the environment. This makes them an ideal natural high moisturizing factor for cream paper that requires high moisturizing performance. The nanocellulose derivatives used for the moisturizing function of cream paper mainly include cationic etherified nanocellulose and carboxylated oxidized nanocellulose.

a) Cationic Etherified Nanocellulose

Nanocellulose with moisturizing function and multi-hydroxyl molecular structure is modified by cationic etherification to make it quaternized and positively charged (see Figure 10.10). Introducing cationic etherified nanocellulose components into the cream not only greatly enhances the moisturizing effect of the cream paper but also strengthens the bonding force between nanocellulose and paper fibers, thereby improving the tensile strength of the cream paper.

The research team led by Yao Xianping at the Hangzhou Chemical Research Institute has conducted extensive in-depth research on improving the moisturizing properties of cream using nanocellulose. They found that introducing etherified modified nanocellulose as a high moisturizing factor in the preparation of cream paper resulted in cream paper with both dry and wet tensile strength better than domestic XY-type cream and comparable to Japan's Miyoshi Oil & Fat MIYOSIYY-8 cream (see Table 10.2). For cream paper with smoothing and moisturizing functions, having high paper strength at the same time will provide consumers with a brand new experience.

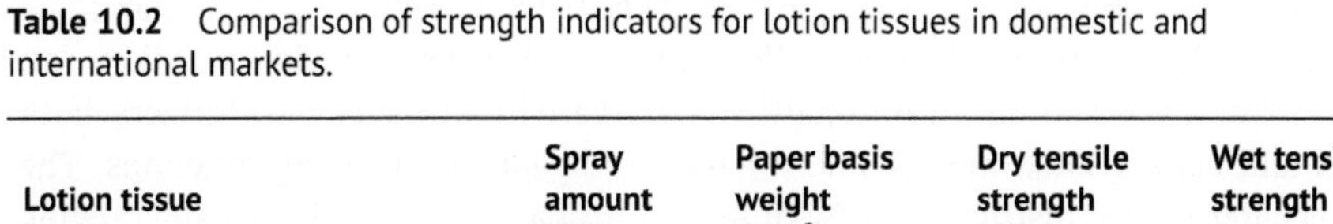

Figure 10.10 Reaction equation for the preparation of cationic etherified nanocellulose.

Table 10.2 Comparison of strength indicators for lotion tissues in domestic and international markets.

Lotion tissue brand	Spray amount (%)	Paper basis weight (g/m²)	Dry tensile strength (N/m)	Wet tensile strength (N/m)
Hanghua Lotion-HH-201	30	15	12.20	4.89
Domestic XY-Type Lotion	30	15	12.01	4.45
Japanese MIYOSIYY-8 Lotion	30	15	12.23	4.66

b) Carboxylated Nanocellulose

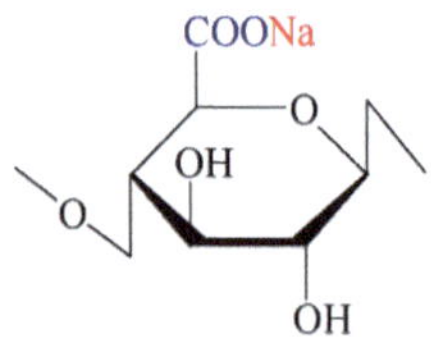

Figure 10.11 TEMPO oxidized nanocellulose.

Carboxylated nanocellulose mainly includes two structures: 2,2,6,6-tetramethylpiperidine-1-oxyl (TEMPO) oxidized nanocellulose and carboxymethylated nanocellulose. TEMPO oxidized nanocellulose is an effective method for the oxidative pretreatment of lignocellulose fibers invented by Professor Ariko Isogai from the University of Tokyo, Japan. TEMPO has weak oxidizing properties and, in the presence of a TEMPO co-oxidant system, can selectively oxidize primary alcohol groups in cellulose molecules without affecting secondary alcohol groups. The TEMPO/NaBr/NaClO oxidation system oxidizes nanocellulose under alkaline conditions with a pH of 8.0–10.5, introducing carboxyl groups into the nanocellulose (see Figure 10.11), which enhances its hydrophilicity and provides strong water-locking and moisturizing functions.

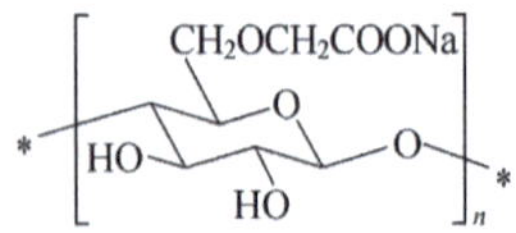

Figure 10.12 Carboxymethylated nanocellulose.

Carboxymethylated nanocellulose is prepared by etherifying nanocellulose with sodium chloroacetate, linking carboxymethyl groups to the cellulose molecules to produce carboxymethylated nanocellulose (see Figure 10.12). Carboxymethylated nanocellulose also has excellent water-locking and moisturizing functions. When combined with glycerin, it forms the moisturizing system of the cream, significantly enhancing the moisturizing feel of the cream paper.

Using TEMPO oxidation and carboxymethylation methods to prepare modified nanocellulose results in the highest moisture absorption at low relative humidity and the lowest moisture absorption at high relative humidity. This unique property meets the moisturizing requirements of cream paper in different seasons and environmental humidity conditions (such as dry winter and humid summer).

1) Hyaluronic Acid

HA, also known as hyaluronan, is a high-molecular-weight polysaccharide composed of repeating disaccharide units of N-acetylglucosamine and D-glucuronic acid (see Figure 10.13). Commercially available HA is generally in its sodium salt form, known as sodium hyaluronate. Compared with ordinary moisturizers, both HA and carboxylated nanocellulose have intelligent moisturizing functions. The molecular weight distribution of sodium hyaluronate is very wide, usually ranging

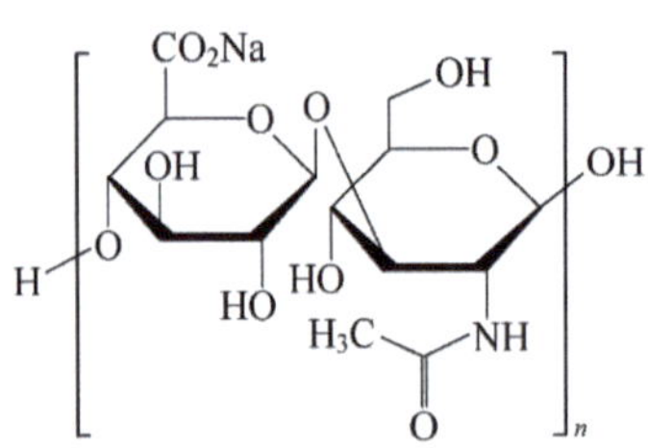

Figure 10.13 Molecular structure of hyaluronic acid.

from 1 kDa to 3 MDa. The properties and application fields of HA with different molecular weights vary greatly.

For cream moisturizing performance, the molecular weight of HA used is 100 kDa to 1 MDa. HA within this molecular weight range also has good lubricity and film-forming properties, which enhance the smoothness of cream paper. However, the stability of HA components in the cream system is currently not ideal. This is mainly due to the coexistence of cationic softeners and anionic HA in the cream system, which affects the stability of the cream product.

10.1.3 Preparation of Cream Paper

In the processing and production of cream paper, methods such as spraying, roller coating, or extrusion coating are used to apply the cream to the surface of the tissue paper. The finished product is then sealed and left to stand for a period of time, resulting in a product with good moisturizing effects on the skin and excellent cleaning effects, suitable for makeup removal and daily use by women [4].

The correct coating method has a profound impact on the application of softeners and cream products. The main processing methods include online spraying, rotary spraying, gravure roller coating, and high-precision spraying [5].

10.1.3.1 Online Spraying Method

The technology used in tissue paper spraying equipment is commonly used in the automotive spraying industry. The nozzles on the equipment can apply fine spray liquid to the tissue paper, completing the spraying of one or both sides of the tissue paper without easily polluting the environment. However, this method can lead to uneven coating, mist generation, and waste of coating material. Additionally, the viscosity of the coating material is limited to a certain range.

10.1.3.2 Rotary Spraying Method

This spraying technology uses a set of high-speed rotating spray discs supplied with liquid from a central reservoir. The centrifugal force of the spray discs generates a flat arc-shaped spray that evenly distributes the coating material onto the passing paper web. This method does not contact the paper web. The equipment is suitable for tissue paper coatings with an optimal viscosity of less than 200 mPa·s. Due to the use of high centrifugal force, the coating process is stable and does not easily cause interlayer separation, but the possibility of bubbles must be considered.

10.1.3.3 Gravure Roller Coating Method

Gravure roller coating technology uses a roller to apply a certain amount of liquid to the tissue paper. When the paper web passes between the gravure printing roller and the offset printing pressure roller, the coating material is applied to the paper web. When the offset printing pressure roller contacts the paper web, each part of the paper web receives an even coating. This method can provide a uniform coating on one or both sides of the tissue paper, regardless of the physical state of the coating material. This system does not produce mist and does not require the use of a vacuum on the paper web. The possibility of coating loss and contamination is very small. However, compared to spraying equipment, this equipment is relatively expensive. In addition, this equipment needs to contact the paper web, which may affect the embossing and creping properties of the paper.

10.1.3.4 High-Precision Spraying Method

This method is also noncontact with the tissue paper and is suitable for continuous or intermittent coating methods. Adjusting the spray diameter of the cream can make certain surfaces of the paper web free of coating. This method is important for changing the absorbency of certain paper and tissue paper products. High-precision spraying equipment can use both liquids and solids and does not contact the tissue paper. This system can be directly integrated into the tissue paper processing production line, and the amount of waste is very small. However, like the gravure roller coating method, this equipment is relatively expensive.

10.1.4 Outlook

With the continuous improvement of people's living standards, the future market demand for cream paper will continue to rise, and higher requirements will inevitably be put forward for the environmental protection, smoothness, moisturizing, and antibacterial properties of cream paper. This requires experts and scientific researchers in the industry to have a deeper understanding of the development and application performance of creams and to propose creative ideas and solutions closely related to human health and safety. Based on the feedback on the user experience of cream paper in the current market and the future industry development trends, odorless paper, antibacterial properties, ultrasoft feel, high strength, and high moisturizing properties will be the focus of future development in the cream paper industry. At the same time, the continuous introduction of pure green natural smooth and moisturizing factors will also be the direction of future research and development in the cream paper industry. More environmentally friendly natural smooth factors will inevitably replace the current synthetic softener components, and green materials such as modified nanocellulose will also have broader development prospects in the leapfrog development process of the cream paper industry.

10.2 Photocatalytic Paper-based Functional Materials

10.2.1 Overview of Photocatalytic Materials

10.2.1.1 Background of Photocatalytic Materials Research

Entering the 21st century, energy and the environment have become two important questions facing the sustainable development of human society.

Semiconductor photocatalytic materials have attracted increasing attention from domestic and foreign scholars due to their potential to address environmental pollution and energy shortages, becoming a research frontier and hotspot in the current international fields of chemistry, environment, energy, materials, and so on.

In 1972, Fujishima and Honda discovered the photocatalytic decomposition of water on N-type semiconductor titanium dioxide (TiO_2) electrodes, opening a new era of heterogeneous catalysis research [6]. Since the 1980s, TiO_2 photocatalytic technology has made significant progress in the photocatalytic removal of organic and inorganic pollutants in water and gas phases in the field of environmental protection. Long-term studies have shown that photocatalytic methods can completely mineralize various organic pollutants, providing a promising environmental pollution purification technology for various organic pollutants and reducible inorganic pollutants, especially toxic and harmful substances that are biologically difficult to degrade. Since the 1990s, with the rapid development of nanoscience and technology, it has provided excellent opportunities for the widespread application of photocatalytic technology. In order to improve the catalytic efficiency of photocatalytic materials, researchers have focused on aspects such as photochemical synthesis, conversion and storage of solar energy, and the reaction mechanism of heterogeneous photocatalysis, further advancing photocatalytic technology.

Photocatalytic technology is a photooxidation method based on the band theory of N-type semiconductors, using N-type semiconductors as sensitizers, and utilizing abundant solar energy as an energy source to act on target substances, ultimately achieving energy conversion or pollution control purposes. Photocatalysis has become an ideal environmental pollution control technology and clean energy production technology due to its characteristics, such as room temperature deep reaction and the ability to directly use solar energy as a light source to drive reactions.

10.2.1.2 Classification of Photocatalytic Materials

Photocatalytic materials refer to catalysts required for a series of photochemical reactions under the action of light. Semiconductors commonly used for photocatalysis are mostly metal oxides and sulfides. Commonly used N-type semiconductors include: TiO_2, tin oxide (SnO), zinc oxide (ZnO), copper oxide (CuO), cadmium sulfide (CdS), etc. [7]. Among them, TiO_2 is the most researched and widely applied photocatalytic material, known for its advantages such as easy availability of raw materials, low cost, good stability, strong oxidizing ability, nontoxicity, and absence of secondary pollution.

10.2.1.3 Application Areas of Photocatalytic Materials

1) Energy is a necessary factor for human survival and social development. With the development of industry and the improvement of material and spiritual life, energy consumption has been increasing day by day. The energy consumption in the world over the past 25 years is equivalent to that of the past hundred years [8]. The primary energy sources on Earth, such as petroleum, coal, and natural gas, are rapidly depleting, and the discovered mineral energy will be exhausted within a limited period. At the same time, the rapid growth in energy consumption will

lead to the deterioration of the human ecological environment. Therefore, the problems faced by humans are not only the scarcity of energy but also the pressure of survival environment and sustainable development.

Among these, hydrogen energy, as a highly promising future energy system, has become a consensus in many countries around the world. Currently, the interest in hydrogen energy among countries worldwide is continuously rising due to its enormous economic and social benefits. Utilizing solar energy to produce hydrogen is the most economical and ideal hydrogen circulation system, which will permanently solve the energy problem for humanity under the premise of zero environmental pollution and low cost. Once the technology matures and reaches practical levels, it will greatly promote the utilization of hydrogen energy. Currently, the main approaches for using solar energy to decompose water and produce hydrogen include: visible light photocatalytic water splitting, photoelectrolysis of water using organometallic complexes, and semiconductor photocatalytic water splitting. Among these, the method of using semiconductor photocatalysis for water splitting to produce hydrogen is the most economical, clean, and practical, making it the most promising method for development.

Ye et al. synthesized a series of NiM_2O_6 (M = Nb, Ta) type visible light-responsive photocatalysts using the solid-phase method. These catalysts can decompose pure water and generate hydrogen gas under visible light irradiation ($\lambda > 420$ nm) without the need for any co-catalysts [9]. Zou et al. developed a series of novel visible light-responsive oxide photocatalysts, such as $InMO_4$ (M = Ta, Nb, V). Without any sacrificial agents, these catalysts can achieve complete water splitting to release hydrogen and oxygen. By substituting a small amount of nickel (10%) for indium in the catalyst, its structure undergoes minimal changes, yet the photocatalytic activity is significantly enhanced [10].

2) Environmental Governance. A report released by the World Health Organization and the United Nations Environment Programme pointed out that air and water pollution have become an unavoidable reality in the lives of urban residents worldwide. While industrial civilization and urban development have created immense wealth for humanity, they have also discharged billions of tons of waste gases and wastewater into the atmosphere and rivers, causing severe damage to the environment upon which human survival depends. Therefore, when harmful gases in the atmosphere and toxic pollutants in water reach certain concentrations, they can bring about enormous disasters to both humanity and the environment.

With the long-term high-speed development of China's economy and the significant progress in social civilization, people's living standards are continuously improving. However, at the same time, environmental pollution issues are becoming increasingly serious and have aroused widespread concern. Among them, air pollution (haze), water pollution, and indoor environmental pollution are the main types of pollution today and are closely related to people's daily lives and health [11–13]. Therefore, how to efficiently and quickly address these environmental pollution problems will directly affect people's vital interests and health safety.

Currently, some traditional pollution control technologies are no longer able to meet the current societal needs, prompting researchers to seek more effective and convenient solutions for environmental governance.

As an advanced oxidation treatment technology, photocatalysis offers several advantages, such as mild preparation conditions, high activity, relatively low cost, safety with no toxic effects, chemical stability, and resistance to acid and alkali. With these advantages, photocatalysis has promising application prospects, becoming a hot topic of research. It has been extensively studied in the fields of water pollution control, air pollution control, and pollution remediation.

3) Sterilization and Antibacterial Effects. In recent years, nanomaterial photocatalytic technology has experienced rapid development in areas such as photocatalytic sterilization and disinfection. Especially, the severe acute respiratory syndrome (SARS) outbreak in the spring and summer of 2003 prompted further research into environmental health issues caused by microorganisms such as bacteria and viruses. This led to the advancement of research on nanophotocatalytic antibacterial and antiviral technologies. Various photocatalytic antibacterial products have emerged and experienced rapid development as a result.

The sterilization effect of semiconductor nanoscale photocatalytic particles can be categorized into two scenarios: (i) Direct Interaction with Cellular Components: Photogenerated electrons and holes directly interact with various components of cells, cell walls, or cell membranes. (ii) Reactive Oxygen Species (ROS) and Hydrogen Peroxide (H_2O_2) Reacting with Biomolecules: Free radicals and H_2O_2 can react with biomolecules such as proteins, enzymes, lipids, and nucleic acids, causing damaging effects on the structure of biological cells through direct action or a series of oxidative chain reactions [14].

10.2.2 Nano TiO_2 Photocatalytic Material

10.2.2.1 Structure of Nano TiO_2 Photocatalytic Material

Titanium dioxide (TiO_2) is a polycrystalline wide bandgap semiconductor material, with its main crystal structures including rutile, anatase, and brookite types [15], as shown in Figure 10.14.

The structure of brookite-type titanium dioxide is a special form formed during the weathering process of ilmenite ore. Its structure is unstable, and it exhibits poor

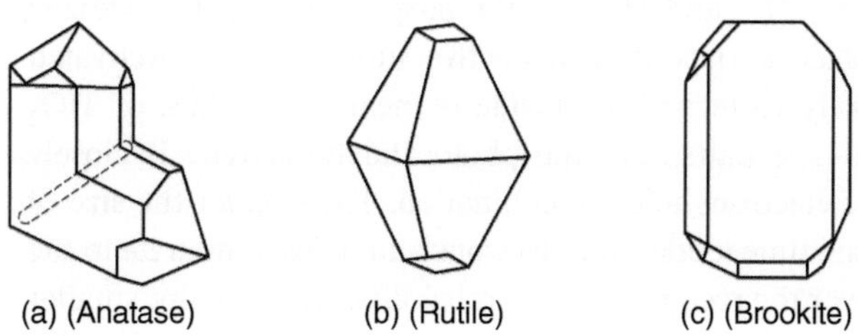

Figure 10.14 Three crystal structures of TiO_2.

thermal stability and photocatalytic activity. Therefore, research and applications regarding the brookite structure are relatively limited. The structure of rutile-type titanium dioxide consists of TiO_6 octahedra sharing vertices, where each octahedron is connected to eight surrounding octahedra. Rutile phase is a thermodynamically metastable phase that can exist stably at room temperature. However, under calcination at 700 °C, an irreversible phase transition occurs from the rutile phase to the anatase phase. In the anatase structure of titanium dioxide, each octahedron is connected to 10 surrounding octahedra. The anatase phase is a thermally stable structure that does not change with temperature fluctuations.

10.2.2.2 Nano TiO_2 Photocatalysis Principle

TiO_2 is an electron-conducting semiconductor, and its photocatalytic reaction mechanism can be explained using semiconductor band theory. According to the band theory of semiconductors, the basic band structure of nano TiO_2 semiconductor consists of a series of filled bands, with the topmost filled band called the valence band (VB), and a series of empty bands, with the bottommost empty band called the conduction band (CB). The region between the VB and the CB is called the bandgap. When it is irradiated with light energy equal to or greater than the bandgap width (Eg), the electrons in the VB of the semiconductor can be excited to transition to the CB, simultaneously generating corresponding holes in the VB. This process creates electron–hole pairs (e^--h^+) within the semiconductor.

For rutile-type TiO_2, with a bandgap width of 3.2 eV, when it absorbs photons with wavelengths less than or equal to 387.5 nm, electrons in the VB are excited to the CB, forming highly active negative charge carriers (e^-), while positive charge carriers (h^+) are simultaneously generated in the VB. Due to the discontinuity of the semiconductor band structure, electrons and holes have relatively long lifetimes. Under the influence of an electric field, electrons and holes are separated and migrate to different locations on the surface of the particles. They can move under the influence of the electric field or through diffusion, react with substances adsorbed on the surface of the semiconductor catalyst particles through oxidation or reduction reactions, be captured by surface lattice defects, or directly recombine.

The electron and hole photocatalytic reaction mechanisms within or on the surface of the semiconductor TiO_2 catalyst particles are illustrated in Figure 10.15.

During the photocatalytic reaction process of TiO_2, hydroxyl radicals (·OH) generated are highly reactive and play a crucial role throughout the photocatalytic oxidation process. O_2, H_2O, OH^-, and HO^- in the suspension can be oxidized by holes adsorbed on the TiO_2 surface into this active substance [16]. Activated electrons and holes are likely to recombine inside or near the surface of TiO_2 particles, forming electron–hole pairs, and the photocatalytic activity is closely related to the probability of electron–hole recombination. The smaller the size of TiO_2 particles, the shorter the time it takes for electrons and holes to migrate to the particle surface, and the lower the recombination probability. Additionally, smaller TiO_2 particles have larger specific surface areas, stronger adsorption capacities for substrates, and more active surface sites, resulting in higher photocatalytic reaction

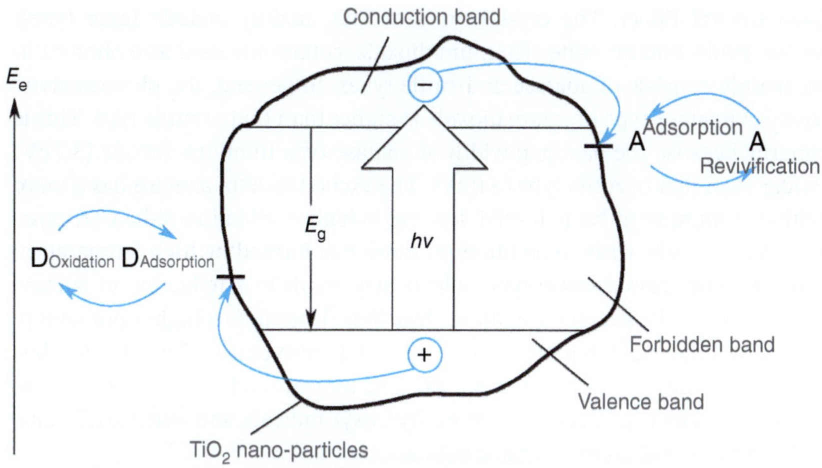

Figure 10.15 Photocatalytic reaction mechanism of TiO_2.

rates [17]. In recent years, researchers have focused on semiconductor particles at the nanometer scale and even quantum-level semiconductors.

10.2.2.3 Factors Affecting the Photocatalytic Activity of Nano TiO_2

The factors influencing the photocatalytic activity of TiO_2 mainly include crystal size, dopant effects, and crystal phase defects.

1) Crystal Size: Crystal size is an important factor influencing photocatalytic activity. When TiO_2 particle size reaches the nanometer scale, the quantum size effect is significantly enhanced. Discrete energy levels are formed between the VB and the CB, widening the bandgap. This leads to higher energy levels for photogenerated electrons and holes, resulting in stronger oxidation-reduction capabilities. Additionally, quantum size effects increase the charge transfer rate of the semiconductor, significantly reducing the probability of hole–electron recombination, thus enhancing photocatalytic activity. Furthermore, smaller TiO_2 particle sizes correspond to larger surface areas. Due to the increased number of surface atoms and unsaturated surface coordination, there is a notable increase in dangling bonds and unsaturated bonds, enhancing the adsorption capacity of TiO_2 particles for reactants and consequently increasing the rate of photocatalytic reactions.

Sun Fengyu obtained nanosized titanium dioxide with different grain sizes under different preparation conditions and investigated the relationship between grain size and photocatalytic activity. The experiments showed that the nanosized titanium dioxide exhibited significant size quantum effects on photocatalytic activity. When the grain size of TiO_2 was less than 16 nm, nanosized titanium dioxide exhibited a pronounced size quantum effect, greatly reducing the recombination of photogenerated electron–hole pairs, thereby facilitating their better separation. Consequently, the photocatalytic degradation activity of organic benzene pollutants by titanium dioxide was correspondingly enhanced [18].

2) Mixed Crystal Effect: The crystal types of TiO_2 mainly include three types: anatase, rutile, and brookite. Titanium dioxide, commonly used as a photocatalyst, mainly consists of anatase and rutile types. In general, the photocatalytic activity of anatase-type titanium dioxide is higher than that of rutile type. This is primarily because the bandgap width of anatase-type titanium dioxide (3.2 eV) is wider than that of rutile type (3.0 eV). The excited state of anatase has a more positive or more negative potential, leading to higher oxidation-reduction capability. Additionally, rutile-type titanium dioxide is formed by high-temperature transformation from anatase type, which may result in a reduction of surface active groups and a smaller specific surface area, leading to a higher probability of electron-hole recombination. Moreover, anatase-type titanium dioxide has a stronger ability to adsorb hydroxyl groups, oxygen, and water molecules on the surface, which can generate more hydroxyl radicals and superoxide ions, facilitating the oxidation of organic substances.
3) Crystal Phase Defects On the surface of TiO_2: There are mainly three types of oxygen defects: lattice vacancies, single-bridge vacancies, and double-bridge vacancies. The presence of oxygen defects can increase the reactivity of surface hydroxyl groups, thereby enhancing the rate of photocatalytic reactions. The more surface defects there are, the more electrons are captured, resulting in a lower probability of recombination between electrons and holes, thus leading to higher photocatalytic activity. However, excessive defects may become recombination centers for electrons and holes, ultimately reducing photocatalytic activity.

Additionally, factors such as surface charge, porosity, average pore size, and heat treatment of TiO_2 photocatalysts can also affect their photocatalytic activity.

10.2.3 Photocatalytic Paper-based Functional Materials Application Research

Nano TiO_2 can be incorporated into paper-based functional materials with photocatalytic activity through methods such as wet-end addition, coating, and impregnation. Robert et al. defined photocatalytic paper-based functional materials as paper sheets or nonwoven fabrics with photocatalytic activity capable of degrading organic pollutants and killing pathogens. In these materials, nano TiO_2 primarily exhibits photocatalytic activity, capable of degrading organic pollutants under near-ultraviolet (UV) light exposure. Other components of the paper, such as plant fibers, ceramic fibers, and zeolite fillers, serve as carriers or adsorbents within the photocatalytic paper.

In photocatalytic paper-based functional materials, nano TiO_2 can be added through methods such as wet-end addition, coating, or impregnation.

Among them, wet-end addition refers to the use of appropriate retention aids to retain nano TiO_2 in the three-dimensional structure of the paper page. Its advantages include the special porous structure of the paper, allowing toxic gases to diffuse inside the paper and providing a higher contact area with pollutants. Additionally, the preparation process is simple and cost-effective. However, it also has disadvantages such as easy loss and agglomeration of nano TiO_2.

Surface coating with the adhesive method involves preparing a coating containing nano TiO_2 using certain base materials and adhesives, and then applying the coating to the surface of the paper through surface coating or impregnation. Although this method can address the retention issue, the formulation of the coating is complex (including pigments, adhesives, etc.), and nano TiO_2 is prone to being encapsulated, resulting in the loss of photocatalytic activity.

Currently, there are three companies worldwide offering commercial photocatalytic paper. Nippon Paper Group in Japan incorporates TiO_2 into pulp to produce newsprint with photocatalytic activity capable of purifying air, and has established partnerships with major domestic news publishers such as Yomiuri. Ahlstrom in Finland uses colloidal silica as an adhesive to produce photocatalytic paper capable of degrading Rhodamine B dye. However, research has found that the concentration of ions in the dye significantly affects the photocatalytic results. Ein Co. Ltd in Japan has developed a pulp with photocatalytic activity using environmental technology, which can be used to produce various functional materials.

Currently, photocatalytic paper-based functional materials are mainly applied in the following aspects:

10.2.3.1 Air Pollution Degradation

With the rapid development of the economy and the increase in industrialization and urbanization, human civilization has continued to advance, marking the onset of the third pollution era characterized by "air pollution" [19]. Indoor air pollution is particularly severe. This is mainly because modern humans spend about 80–90% of their time indoors, with the vast majority of the air they breathe being indoor air [20]. Additionally, due to the relatively enclosed indoor environment, indoor air pollutant concentrations are often higher than outdoor levels [21]. Compared to outdoor air pollution, indoor air pollution poses greater and longer-lasting health risks to humans [22, 23]. Research has shown that indoor air pollution has become one of the most important threats to human health [24].

Indoor air pollution refers to the presence of harmful substances such as formaldehyde, benzene, volatile organic compounds (VOCs), and pathogenic microorganisms mixed into indoor air, posing health hazards to humans [25, 26]. Studies have shown that over 300 pollutants can be detected indoors, and 68% of human diseases are related to indoor air pollution [27]. The main sources of indoor air pollution include indoor decoration materials, smoke, daily necessities, bacteria and viruses carried by humans, and outdoor air [28–30]. These pollutants enter the body through respiration, accumulate over time, and seriously harm human health. With the increasing demand for the aesthetics and quality of living spaces, various types of wooden furniture, marble furniture, and wall coatings are widely used in indoor decoration. These materials slowly release formaldehyde, benzene, ammonia, and VOCs into the environment, posing significant health risks to people living in such environments over the long term.

Compared with ceramics and glass, which can only react with organic substances adsorbed on the surface, when photocatalysts are used in paper, the resulting photocatalytic paper-based functional materials can achieve higher decomposition capabilities for major pollutants in the air. Paper also has the property of adsorbing suspended particles in the air into its interior, serving as an air filter. Utilizing the porous structure of paper, photocatalytic paper can achieve high air permeability and gas permeability. Making photocatalytic paper into a honeycomb structure not only effectively degrades organic pollutants and bacteria in the air but also has low air resistance, low cost, simple process, and easy replacement.

Ichiura mainly studied the composite of zeolite and TiO_2, and prepared TiO_2/zeolite composite paper materials using papermaking technology [31–34]. Using ceramic fibers and plant fibers as carriers, and employing polydiallyldimethylammonium chloride (PDADMAC) and anionic polyacrylamide (APAM) as a dual retention system, the retention problems of both titanium dioxide and zeolite components were solved. After impregnating the well-beaten paper in the sol, it was calcined at a high temperature of 700 °C to remove the plant fibers in the paper, forming a fully inorganic component paper. The mixing of plant fibers can give the pre-impregnated paper a certain initial strength, while the final removal solves the problem of plant fibers as a degradable carrier for photocatalysis.

Wang Yu developed "storage-oxidation" catalysts specifically targeting typical indoor air pollutants such as formaldehyde and benzene, effectively addressing the problem of competitive adsorption of pollutants in the air. This has significant implications for the treatment of other pollutants in the air.

The research team, led by Yao Xianping from the Hangzhou Institute of Chemical Technology (National Engineering Research Center for Papermaking Chemicals), conducted in-depth studies on traditional photocatalytic materials. They utilized core technologies such as homogeneous hydrolysis preparation, electron doping modification, and multicarrier composite modification to develop paper-based photocatalytic functional materials with excellent purification performance and low cost. Creatively using various natural fiber paper substrates as bases, they compounded the above photocatalytic functional materials for enhanced efficiency and successfully developed various home purification products. Among them, the multifunctional indoor air quality assurance paper, formaldehyde-removing wallpaper, formaldehyde-removing and odor-eliminating decorative paintings (see Figure 10.16) have a wide audience, high formaldehyde removal efficiency, and have achieved good economic and social benefits.

In particular, the multifunctional indoor air quality assurance paper product fills the market gap for similar products. The surface of this paper-based functional material consists of a reaction layer, adsorption layer, and catalytic layer. The catalytic layer (outer layer) can continuously decompose indoor air pollutants and has functions such as mildew resistance, sterilization, negative ion release, and odor removal. Due to its small footprint and foldable design, it is suitable for various small spaces (such as drawers, wardrobes, etc.), and can be discarded after use to avoid secondary pollution indoors. This product has been professionally tested by

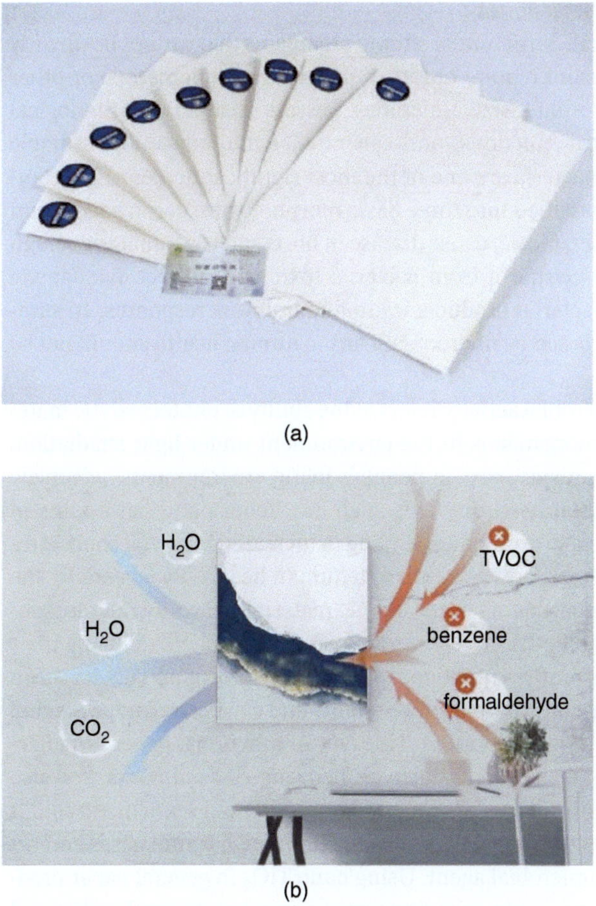

Figure 10.16 Multi-effect indoor air quality assurance paper (a) and aldehyde-removing decorative paintings (b).

a third party: formaldehyde removal rate reaches 95.4%, formaldehyde purification persistence rate reaches 89.8%, and antibacterial rate reaches 99% (test bacteria: *Escherichia coli*, *Staphylococcus aureus*). Meanwhile, this product has successfully passed the Zhejiang Province Industrial New Product Appraisal and obtained one national invention patent.

Additionally, the aforementioned formaldehyde-removing wallpapers and odor-removing decorative paintings are also products developed by the Hangzhou Institute of Chemical Technology that combine both formaldehyde removal and indoor aesthetic decoration functions. These products can continuously and efficiently degrade harmful pollutants such as formaldehyde, benzene, and total VOCs (TVOC) in indoor air, leading the way in various performance indicators among similar products, effectively addressing indoor air pollution issues and greatly safeguarding human health.

10.2.3.2 Bactericidal and Bacteriostatic

Microbes are a group of small, structurally simple organisms that cannot be directly observed by the naked eye and require observation through a microscope or other equipment. Despite their small size, microbes possess distinct morphological structures and physiological functions, and can rapidly multiply under suitable conditions. Among them, bacteria are one of the most significant types of microorganisms. Bacteria can be classified into three basic morphologies: cocci, bacilli, and spirilla. Pathogenic bacteria typically cause disease in hosts such as humans through two main mechanisms: direct harm from bacterial toxins or indirect damage via host allergic reactions to bacterial products leading to immune responses. In summary, the potential threats posed by microorganisms to human health should not be underestimated.

Antibacterial and antimicrobial activity refers to the ability of photocatalytic materials to kill or inhibit microorganisms in the environment under light irradiation. Various harmful microorganisms exist in people's living environments, adversely affecting human life. In damp environments such as kitchens and bathrooms in homes, microorganisms tend to proliferate, leading to increased bacterial concentrations in the air and on surfaces, posing a threat to human health. By leveraging the excellent photocatalytic properties of photocatalytic materials, harmful microorganisms in the environment can be effectively suppressed or eliminated, reducing their threat to human health. Currently, a functional paper containing nano TiO_2 has been developed, which exhibits photocatalytic functionality and can be used as packaging paper and wallpaper for interior decoration. This type of functional paper can effectively absorb harmful gases such as formaldehyde, benzene, and ammonia, and also has the ability to kill bacteria such as *Staphylococcus aureus* and *Escherichia coli*. It can be placed on furniture, indoors, in cars, etc., to eliminate harmful gases and act as an antibacterial and antimicrobial agent. Using nano TiO_2 in general paper products, such as toilet paper and food packaging paper, can exert its sterilization and disinfection effects, which has significant practical implications.

In addition to its significant applications in photocatalytic purification and sterilization, nano TiO_2 also has wide applications in other properties, such as superhydrophobicity, aramid paper, and flexible supercapacitors.

10.2.3.3 Superhydrophobic

In recent years, there has been increasing research on the preparation of superhydrophobic materials, which are increasingly valued for their high water resistance, self-cleaning, and dustproof properties. Among them, superhydrophobic paper-based materials have also received great attention and development. Paper and paperboard are three-dimensional network structure materials composed of a large number of plant fibers. The chemical composition of plant fibers contains a large number of hydrophilic hydroxyl groups, and the cell cavities of the fibers themselves have a microcapillary structure. Moreover, the three-dimensional network structure formed between fibers has a capillary effect, making paper materials inherently hydrophilic. After absorbing water, the physical strength of paper decreases significantly, affecting its performance and shelf life. Internal sizing and surface sizing technologies can impart paper with certain water resistance, ensuring that paper materials can be used as normally as possible when exposed to water.

Huang Lianghui used silane coupling agent-grafted nano TiO_2 and plant fibers to prepare opaque superhydrophobic surface paper, with water contact angles ranging from 125° to 154° and rolling angles less than 3°.

Lai and his team used a hydrothermal method to grow flower-shaped TiO_2 particles on the surface of cotton fibers. After modification with fluoroalkylsilane, the contact angle of the structure exceeded 160° and the rolling angle was less than 10°. This structure can be applied in UV shielding, self-cleaning, and oil-water separation.

10.2.3.4 Application in Aramid Paper

Aramid fiber, as a new type of high-tech and highly valuable organic high-performance fiber, possesses high specific strength and modulus, insulation properties, as well as excellent chemical stability, high-temperature resistance, abrasion resistance, and low density. By using high-performance aramid fibers as raw materials and adopting modern wet papermaking technology, paper sheets are formed, which are then processed through resin curing or hot pressing to prepare sheet-like composite materials called aramid paper substrates [35, 36].

Currently, the most practical types of aramid fibers in the market are mainly divided into two varieties: meta-aramid fibers and para-aramid fibers, corresponding to which are meta-aramid paper and para-aramid paper. Aramid paper inherits the excellent properties of aramid fibers. Due to its outstanding mechanical properties and low density, it can be widely used as lightweight honeycomb structural materials for rigid structural components in aircraft, trains, and other applications. Its stable chemical properties and high dielectric performance make it an important insulating material in transformers, generators, battery separators, and other applications. Additionally, its products are also widely used in individual protection applications such as bulletproof vests and firefighting clothing. Therefore, aramid paper substrates, as a new type of structural and functional material, are important basic materials in modern industries such as aerospace, national defense, military, and electronics, with broad application prospects and significant research and development significance.

Schütz et al. prepared films from nanofibrillated cellulose (NFC) and nano TiO_2. The transparency and mechanical strength were tested using spectrophotometry and nanoindentation, respectively. The results showed that when the nano TiO_2 content was 16%, the modulus and hardness reached 44 GPa and 3.4 GPa, respectively, with optical transmittance exceeding 80%.

Lv Cheng improved the insulation performance of insulating paper with nano TiO_2. The results showed that when the nano TiO_2 content was 3%, the insulating paper exhibited a strong ability to suppress and dissipate space charges, which could reduce the internal field distortion of the insulating paper. Lv Cheng combined the analysis of the multicore interface structure model to explain the mechanism by which nano TiO_2 improves charge accumulation. The results indicated that a strong interaction interface was formed between cellulose and nano TiO_2, improving the structure of the contact surface, which facilitated the formation of more traps. The increase in trap quantity and the change in energy levels enhanced the breakdown strength of the insulating paper. Zhou You et al. modified transformer oil using nano TiO particles. The results showed that nano TiO particles could improve the dissipation rate of charges in transformer oil, to some extent, suppress the distortion of the electric field in transformer oil, thereby enhancing the insulation performance of transformer oil. Wang Binhua et al. found that nano TiO particles had strong absorption and scattering capabilities for UV light. Additionally, nano TiO_2 particles significantly improved the aging resistance of polymer materials. Ban Yan used the sol–gel method to attach nano TiO_2 to aramid fibers. The results showed that nano TiO could absorb UV light energy, slowing down the photodegradation rate of aramid fibers.

10.2.3.5 Flexible Supercapacitors

With the rapid development of flexible, portable, and wearable electronic products in recent years, flexible supercapacitors have become a hot research direction in the field of new energy. Compared to traditional nonflexible supercapacitors, the electrodes, separators, electrolytes, current collectors, and encapsulation shells used in flexible supercapacitors are all made of flexible materials, allowing the shape of flexible supercapacitors to vary greatly. From the current research status, selecting high-performance flexible electrodes is the key to preparing flexible supercapacitors. To date, research on flexible supercapacitors has formed a complex academic system, and the prepared flexible supercapacitors exhibit diverse physical forms and various functions.

In 2015, a fully solid-state flexible supercapacitor was reported, prepared using high surface area activated carbon material as the substrate and composite nanofibers containing TiO_2. This method involves electrospinning to produce TiO_2 nanofibers, followed by calcination of the nanofibers, and finally mixing with activated carbon to prepare flexible and transparent supercapacitors. Polyvinyl alcohol and phosphoric acid are used as the electrolyte. At a current density of 0.3 A/g, the specific capacitance is 310 F/g, and the energy density at a power density of 2.798 kW/kg is 43.05 Wh/kg. After 5000 charge-discharge cycles, the specific capacitance still maintains 98.38%.

Wang Jinguang utilized electrospinning technology, hydrothermal synthesis, and other methods to prepare three-dimensional composite materials based on carbon nanofibers (CNF) with special structures and investigated their supercapacitor performance, exploring the influence of electrode material structure on supercapacitor performance. First, polyacrylonitrile nanofibers were prepared using electrospinning technology, and CNF was obtained after pre-oxidation and high-temperature carbonization. Then, TiO_2 nanorod arrays were "planted" on the fiber surface using CNF as the substrate to obtain CNF/TiO_2 materials. Subsequently, secondary hydrothermal synthesis was conducted to grow nanosheet-like Ni $(OH)_2$ between the titanium dioxide nanorod arrays, resulting in the three-dimensional composite material CNF/TiO_2@Ni$(OH)_2$.

10.2.4 Prospect

Given the broad application prospects and significant socioeconomic benefits of photocatalytic materials, especially photocatalytic paper-based functional materials, in environmental protection, clean energy (such as solar energy conversion to hydrogen energy), national defense, military affairs, medical health, building materials, automotive industry, home appliance industry, textile industry, and many other fields, paper-based functional materials are considered as an important direction for the transformation and upgrading of the traditional papermaking industry. Their structure and performance are completely different from traditional paper, with flexible and designable structural and mechanical, optical, electrical, magnetic, thermal, and acoustic properties. They can overcome the harsh conditions of high impact, high temperature, high humidity, and high corrosion that single natural plant fiber materials cannot adapt to, making them one of the strategic materials.

By using different types of plant fibers as carriers for nano photocatalysts, and utilizing the unique porous three-dimensional structure characteristics of paper, combined with papermaking and photocatalysis technologies, high-performance photocatalytic paper with different functions can be developed. In this process, it is necessary to pay attention to and explore the influence of various environmental factors on photocatalytic paper-based functional materials during application, as well as how to load photocatalysts on suitable paper-based fiber carriers, and how to improve the light utilization efficiency and photocatalytic performance of photocatalysts. It is believed that with continued research and application, photocatalytic paper-based functional materials will further expand their application fields and greatly improve their photocatalytic performance, benefiting society and people more effectively.

10.3 Basic Functional Materials for Fruit and Vegetable Preservation Paper

10.3.1 Overview

10.3.1.1 Definition of Basic Functional Materials for Fruit and Vegetable Preservation Paper

The basic functional materials for fruit and vegetable preservation paper, also known as "fruit and vegetable preservation paper" or "fruit and vegetable antiseptic paper", is a special type of paper that is biodegradable. It uses specific processing methods such as external coating (roller coating or spray coating), impregnation, or internal addition to combine one or more components with specific preservation effects such as antibacterial, ethylene absorption, or humidity regulation. This results in a special composite paper that has preservation functions for fruits and vegetables. The basic functional materials for fruit and vegetable preservation paper have a two-layer structure: a dense layer and a loose layer. The dense layer can inhibit the respiration of fruits and vegetables and prevent the carbon dioxide and water decomposed by fruits and vegetables from leaking out; the loose layer can allow the air outside the bag to penetrate into the bag to keep the fruits and vegetables fresh.

10.3.1.2 Current Status of the Development of Fruit and Vegetable Preservation Materials

There are many types of fruit and vegetable preservation materials at home and abroad. These materials are mainly prepared based on the preservation principles of antibacterial, ethylene removal, and gas adjustment packaging. Table 10.3 lists some representative fruit and vegetable preservation materials at home and abroad.

It's Fresh! Filters product developed by the American Preservation Company (It's Fresh Co, Ltd) is a preservation filter that coats a layer of transition metal-modified porous material on the surface of a paper-based/polymer-based film to improve the adsorption efficiency of ethylene, thereby effectively delaying the rotting speed of fruits and extending the shelf life of fruits by 1-3 days. It is reported that this technology was first applied by Marks & Spencer in the UK in November 2011, with significant effects on strawberry preservation. Currently, this technology has been adopted by many chain supermarkets worldwide. The American Vegetable and Fruit Preservation Specialty Chemicals Company (AgroFresh Solutions, Inc.) uses 1-Methylcyclopropene (1-MCP) technology to maintain the freshness of various fruits, including apples, pears, kiwis, avocados, and bananas, and extend the shelf life of fresh agricultural products. A product called "Blue Apple" in the United States places the ethylene adsorption material in a mold shaped like a blue apple. The bottom of the mold has small gaps for gas to pass through. The ethylene adsorption material inside the "Blue Apple" will adsorb and degrade the ethylene produced by fruits and vegetables, thereby achieving the effect of fruit and vegetable preservation. The Apeel Sciences company in the United States has introduced an edible preservation material made of nontoxic organic compounds of lipids and glycerides, which come from the unnecessary peels, seeds, and pulp of various vegetables and fruits. The material forms a barrier on the surface of fruits and vegetables, preventing water from evaporating from fruits/vegetables, while minimizing the amount of oxygen that may enter, which can extend the shelf life of the product by 2–3 times. The Indian Preservation Company (Keep-it-fresh Co, Ltd) is a company that produces preservation materials. It currently produces preservation materials, including preservation bags, preservation gas adjustment bags, grape preservation pads, and moisture absorption pads, mainly based on reducing ethylene content as the preservation standard. The South African Grape Preservation Pad Company (Grapetek Co, Ltd) has developed UVASYS series grape preservation pads and preservation papers for grape preservation. The product uses a two-stage release technology of sulfur dioxide gas to effectively sterilize grapes and achieves the purpose of grape preservation by inhibiting the growth of bacteria. UVASYS preservation materials are mainly used for grape preservation, especially suitable for long-term storage and transportation of grapes, with a preservation time of 3 weeks to 4 months. The Israeli Phresh company has prepared a type of microcapsule lining paper made from essential oils. For daily use, just place it

Table 10.3 Representative fruit and vegetable preservation materials at home and abroad.

Material name	Producing company	Material form	Application effectiveness
SmartFresh	AgroFreshSolutions, Inc.	Stickers/Boxes	Significantly extends storage life of fruits
–	Apeel Sciences	Edible Film	Extends shelf life 2-3 times
It's Fresh! Filters	It's Fresh Co..., Ltd.	Filter Patches	Effectively extends storage life of fruits
–	Keep-it-fresh Co., Ltd.	Pad Paper	Reduces ethylene content
UVASYS	Grapetek Co..., Ltd.	Pad Paper	Preserves grapes for up to 4 months
FreshPaper	Fenugreen Co., Ltd.	Cardboard	Extends storage time 2-4 times
Fruit Companion	Phresh	Powder/Box Liner	Extends storage time 2-3 times
MS Nano-Preservation Paper	Hubei Zhige Packaging Printing Co., Ltd.	Boxes	Extends storage life 1-2 times
Nano-Preservation Bags	Jiangsu Wanguo Peach Preservation Technology Co., Ltd.	Preservation Plastic Bags	Extends storage life up to 15 days
Dr. Fresh	Xi Qin Biotechnology Co., Ltd.	Bagged Powder	Extends shelf life 1-3 times

next to the food, and it can kill most of the bacteria that cannot be eliminated by conventional washing, and it can extend the preservation time of food by 2–3 times.

The development of domestic fruit preservation materials is also ongoing. Hubei Zhihe Packaging Printing Co., Ltd. has launched an MS nano fruit and vegetable preservation cardboard box. The nano-preservation material in the cardboard box is a natural mineral material treated at the nano level. It is nontoxic, tasteless, and environmentally friendly. It can effectively adsorb the ethylene, water vapor, and other harmful gases volatilized by fruits and vegetables during respiration in the cardboard box, inhibit the respiration and ripening process of fruits and vegetables, achieve the purpose of maintaining the oxygen content inside the cardboard box, and adjust the temperature inside the cardboard box, thereby slowing down the rotting process of fruits and vegetables. This product has good preservation, antibacterial, and moisture-proof effects on fresh agricultural products such as vegetables, fruits, and flowers, can effectively extend the preservation period of fruits and vegetables during transportation, and reduce the rot and loss of fruits and vegetables. Jiangsu Wanguo Peach Preservation Technology Co., Ltd. has launched a peach preservation bag, which mainly uses antibacterial, ethylene adsorption inhibition, and other principles to extend the preservation period of peaches. At the same time, this preservation bag uses nano technology and has the function of automatically adjusting the oxygen, carbon dioxide, and water vapor concentration in the bag, which can inhibit the respiration of peaches and slow down the ripening process of peaches. The normal temperature preservation period of peaches can be extended from 2–3 days to a maximum of 15 days. Xianyang Xi Qin Biotechnology Co., Ltd. has launched a fruit and vegetable preservative using 1-MCP, which can effectively extend the post-harvest storage period and shelf life of fruits, vegetables, and flowers, especially for ethylene-sensitive fruits, vegetables, and flowers. The effect is more significant. Combined with China's national conditions, simple gas-adjusted storage (MA storage) has obvious effects and low investment at this stage, and has been fully successful based on refrigeration, such as the storage of Red Fuji apples in the northwest region, the storage of garlic sprouts in Shandong, the storage of grapes in Liaoning, and the packaging and transportation of bananas in the south and the single fruit packaging storage of citrus [37].

The Hangzhou Chemical Research Institute has done a lot of in-depth research on fruit and vegetable antibacterial preservation and has successfully developed a series of antibacterial preservation papers and powder antibacterial preservation agents that have antibacterial preservation effects on strawberries, small tomatoes, and peaches. The types of materials mainly include microcapsules, spray emulsions, wrapping papers, sandwich papers, and coated cardboard boxes. Among them, the antibacterial preservation paper is a fruit and vegetable preservation material with ethylene adsorption, antibacterial, and humidity regulation functions, which is prepared by using environmentally friendly degradable high-absorption paper as the base material and porous inorganic materials, natural plant functional factors, etc., as raw materials, using special methods. The base material of the antibacterial preservation paper can extend the preservation time of peaches to 6–9 days at room temperature, especially the rot rate of strawberries can be reduced by more than

50% under room temperature storage conditions, and the preservation time of small tomatoes can be extended by 2–3 times. Related products have entered the promotion stage of the farmer's market.

At present, fruit and vegetable antibacterial preservation paper has not been widely commercialized. The main reasons are two aspects: one is that the antibacterial preservation components are complex, the mechanism of action is not very clear, and the sources of medicine, content, and stability of the paper base itself are not easy to solve [38]; the second is that some antibacterial substances have a special smell when used, which will affect the edibility of fruits and vegetables. Natural plant antibacterial substances are harmless to the human body and the environment, and have good preservation performance. In the long run, plant antibacterial fruit and vegetable preservation paper should develop in the direction of natural, safe, and efficient, and has considerable development space.

At present, the following four are the main problems in the fruit and vegetable preservation industry:

a) The popularization rate is low, the storage and preservation capacity is relatively insufficient, and it is not suitable for the multi-variety, industrialized, and export-oriented situation of fruit and vegetable production. Due to the seasonality and regionality of fruit and vegetable production, the perishability of fruits and vegetables themselves, and the diversity of people's demand for fruits and vegetables, fruit and vegetable products that are mainly sold fresh in the market often experience seasonal surpluses once they encounter abnormal weather or market changes, leading to a large amount of fruit and vegetable products rotting and deteriorating, causing serious losses. The phenomenon of increasing production without increasing income or even increasing production and reducing income occurs from time to time, seriously affecting the stability of the fruit and vegetable market and the development of the fruit and vegetable industry.
b) The technology is backward. Traditional preservation methods are still the main means of fruit and vegetable storage and preservation, the utilization rate of storage and preservation facilities is low, and the stored fruit and vegetable products are single, affecting efficiency.
c) The existing preservation facility layout is unreasonable, and there is a lack of storage and preservation infrastructure, and precooling treatment is not timely.
d) Supporting measures cannot keep up, and the harvest is not refined. There are many mechanical injuries during the picking and transportation process, the sorting, grading, and packaging after harvesting are not standardized, and the level of mechanical equipment matching is low, etc. These problems have led to the slow development of China's fruit and vegetable preservation industry.

Facing the new market situation, especially with the rapid increase in fruit and vegetable production, how to increase the storage of fruits and vegetables and improve the comprehensive technical level of fruit and vegetable storage and preservation in various production areas must attract the attention of all departments. It can be predicted that with the globalization of the economy, the global circulation of fruits and vegetables will become a trend. Therefore, the research and application of the functions of different fruit and vegetable preservation materials and the research on the economic multifunctionality are still important development directions for the preservation industry.

10.3.2 Types and Characteristics

After harvesting, fruits and vegetables are prone to spoilage, such as mold and rot, due to their vigorous tissue metabolism, microbial activity, and water loss. In recent years, scholars at home and abroad have conducted a lot of research on the life activities during the preservation and storage of fruits and vegetables, aiming to slow down the respiration of fruits and vegetables, reduce ethylene concentration, inhibit microbial growth, inhibit enzyme activity, and use packaging to reduce mechanical damage through physical, chemical, and biological methods to achieve the purpose of preservation. According to different preservation mechanisms, the types of fruit and vegetable preservation materials are also different. At present, the main fruit and vegetable preservation materials on the market include traditional preservation functional materials and antibacterial preservation paper-based materials.

10.3.2.1 Traditional Preservation Functional Materials

1) Ethylene Removal Preservation Materials

Ethylene is an important plant hormone, generally produced in very small amounts, not exceeding 0.1 μg/g, but it increases sharply during the developmental stages of fruit germination, maturation, and senescence. As a maturation and senescence hormone, ethylene plays an important regulatory role in the maturation and senescence of fruits after harvesting, affecting physiological functions such as nucleic acid metabolism, enzyme activity, hormone levels, and respiration rate, thereby inducing changes in fruit color. Therefore, controlling ethylene concentration can effectively delay the maturation and senescence process of postharvest fruits and extend storage time. There are generally two methods to control ethylene concentration: one is to combine cyclopropene inhibitors with ethylene receptors to reduce ethylene, the most effective being 1-MCP [39]; the other method is to use ethylene absorbers to remove ethylene [40].

For 1-MCP, due to its excellent preservation performance, many experts and scholars have studied its application on paper. Wu Bin et al. uniformly placed 1-MCP inclusion compounds, water absorbents, and dispersants in breathable and moisture-permeable plastic paper, and made preservation paper by heat-sealing. Bananas and tomatoes were placed in the baskets and stored at 20 °C. The respiration peak of bananas was delayed by 6 days, and the respiration rate of tomatoes remained low for the first 3 days, delaying the ethylene release time by 2 days and slowing down the aging process of the fruits [41]. Ma Xiujue et al. used absorbent paper as the core paper, covered both sides with surface paper, bonded with hot melt adhesive, uniformly punched holes in the core paper, loaded a quantitative amount of 1-MCP agent, and made preservation paper. The product had a significant inhibitory effect on the weight loss rate, respiration, and soluble solid content of nectarines [42]. In addition, AgroFresh, Xiqin Biotechnology Co., Ltd., and Litong Co., Ltd. have also developed fruit preservation materials based on 1-MCP.

A large number of studies have shown that 1-MCP has a significant inhibitory effect on the maturation and senescence of many climacteric fruits (such as apples, bananas, pears, kiwis, peaches, citrus, plums, dates, and apricots), delaying fruit softening and skin yellowing, and extending the shelf life of fruits. However, the inhibitory effect of 1-MCP on ethylene production varies with the maturity stage of climacteric fruits. Sun Xisheng et al. treated three different maturity stages of apples with 1-MCP fumigation and found that 1-MCP had no significant inhibitory effect on early-maturing varieties, as the fruits had already reached the respiration peak at harvest. This indicates that for fruits with high ethylene production, especially those that have entered the respiration peak at harvest, the response to 1-MCP is poor, i.e., the inhibitory effect of 1-MCP on ethylene in the late maturity stage may not be significant [43]. However, there are few reports on this aspect at home and abroad, and further research is needed.

For ethylene absorbers, there are usually two types: one is porous materials such as zeolite, diatomaceous earth, and white silica, which rely on strong adsorption to absorb ethylene; the other is strong oxidizing materials such as potassium permanganate, which oxidize ethylene through chemical reactions to control ethylene concentration and extend the preservation period of fruits and vegetables [44]. Chinese patents CN204489574U and CN204489638U introduce a nano-preservation corrugated paper material. The adhesive layer of the corrugated paper is made by mixing multiple nano zeolite powders, nano clinoptilolite powders, nano-silica powders, and glue, which can effectively extend the storage of fruits by 7 to 15 days. However, the porous materials mixed with the adhesive in the inner layer of the corrugated paper have their pores blocked by the adhesive, greatly affecting the adsorption performance of the material [45]. Canadian patent CA2982417A1 introduces a coated fruit preservation paper, which achieves an ethylene adsorption capacity of about 900 $\mu L/m^2$ by controlling the particle size, amount, and type of adhesive of the modified porous material. This patented product has been industrialized, and its application market has expanded from North America to the UK, South Africa, and China. The fruit and vegetable preservation paper made by combining ethylene absorbers with paper has the characteristics of low production cost, easy use, and good preservation effect, making it particularly suitable for rural areas where preservation technology development and popularity are not high, with great market development potential.

2) Modified Atmosphere Packaging Materials

Modified atmosphere packaging materials are used to adjust the gas composition inside the packaging materials of fruits and vegetables, widely used for the preservation of fruit and vegetable products. Modified atmosphere preservation achieves better preservation effects by reducing the oxygen concentration in the storage environment, increasing the carbon dioxide concentration, inhibiting the respiration of fruits and vegetables, and delaying their aging and deterioration.

Modified atmosphere packaging materials mainly include active modified atmosphere preservation and passive modified atmosphere preservation. Active modified atmosphere packaging materials accelerate the modification of gas composition by expelling the gas inside the packaging and refilling it with a specific proportion of mixed gas, avoiding the spoilage of fruit and vegetable products exposed to high concentrations of oxygen for a long time. Passive modified atmosphere packaging materials control the concentration of carbon dioxide and oxygen in the packaging materials through the permeability of the film materials to different gases. Due to the different permeation rates of oxygen and carbon dioxide through the packaging film, the gas in the modified atmosphere packaging materials will reach a balance between the respiration rate of the fruit and vegetable products and the permeability of the film after a certain period.

Paper-based modified atmosphere packaging materials are generally made by combining paper-based materials with polymer film materials, achieving the preservation effect of fruits and vegetables by controlling the types and contents of gases inside the materials. Usually, a low-oxygen and high-carbon dioxide environment can significantly reduce the respiration rate, ethylene biosynthesis, water loss, phenolic oxidation, and microbial growth of fruit and vegetable products, thereby reducing the deterioration of product quality.

At present, the related technologies based on the above mechanisms have not been applied to paper. The high permeability of paper can be reduced by coating the paper surface with a high gas barrier coating. Bowen Yi et al. blended cellulose/chitosan to make a film with a light transmittance of about 80%, and its water contact angle increased from 70° of the cellulose film to 100°, not only improving the hydrophobicity and antibacterial properties of the cellulose film but also having an oxygen permeability coefficient lower than the ideal ethylene-vinyl alcohol (EVA) copolymer oxygen barrier material on the market [46]. The oxygen barrier performance of paper can be improved by coating or impregnating the paper with a cellulose/chitosan blend solution, achieving modified atmosphere preservation of paper-based packaging.

3) Edible Coating Materials

Edible coating materials, as edible materials that act as gas barriers, show preservation effects similar to modified atmosphere packaging. Edible coating materials need to be evenly applied to the surface of fruits and vegetables, and must have sufficient adhesion and durability after drying [47]. When applied to the preservation of fruit and vegetable products, they need to meet some basic requirements, such as suitable barrier properties, good mechanical strength, reasonable microbial, biochemical, and biological stability, safety, low cost, and ease of operation. The preservation effect of edible coating materials mainly depends on the wettability of the coating solution, as the wettability of the coating material directly affects the thickness of the coating.

Edible coatings are usually based on biological materials such as polysaccharides, lipids, and proteins. At present, the polysaccharides used for the preservation of fruits and vegetables mainly include starch and its derivatives, cellulose derivatives, pectin, and alginates. The commonly used proteins mainly include animal proteins (such as casein and whey protein) and plant proteins (such as wheat gluten and soy protein). Among them, whey protein has been widely used to extend the shelf life of fruits and vegetables. In addition, edible coatings can also be used as carriers to load antibacterial compounds to improve the preservation period of fruit and vegetable products. Antibacterial edible coatings can inhibit the bacteria that cause spoilage and pathogenicity by maintaining a certain concentration of effective compounds on the surface of fruit and vegetable products. Although antibacterial edible coatings can play an important role in reducing pathogen contamination and extending the shelf life of fruits and vegetables, they also have certain defects and need further research.

4) Active Packaging Materials

Active packaging technology for fruit and vegetable preservation refers to packaging technology that extends the shelf life of fruits and vegetables by maintaining or changing the internal conditions of the packaging. Active packaging technology for fruit and vegetable preservation uses active materials to absorb, release, or migrate gases or substances (such as absorbing or releasing oxygen, carbon dioxide, and ethylene, or releasing antibacterial substances) to change the storage conditions of fruits and vegetables, thereby maintaining fruit and vegetable quality and extending shelf life. The common methods of active packaging technology for fruit and vegetable preservation include placing active materials in small bags or composite paper sheets inside the packaging bag, spraying them onto the inner surface of the packaging film, or embedding them into the packaging film in some way. According to the functions of active materials, food active packaging can be divided into oxygen absorption/release packaging, carbon dioxide absorption/release packaging, ethylene absorption/release packaging, moisture control/anti-fog packaging, and antibacterial packaging.

10.3.2.2 Antibacterial Preservation Paper-based Materials

The growth of microorganisms is an important cause of spoilage in fruit and vegetable products. During transportation and storage, fruit and vegetable products may come into contact with pathogens. Inhibiting the growth of microorganisms is an important means to extend the preservation period of fruits and vegetables. Antibacterial preservation paper-based materials are used to inhibit the microorganisms produced during the growth, harvesting, storage, and sale of fruits and vegetables. This is mainly achieved by adding antibacterial agents to the paper-based substrate through methods such as impregnation, spraying, and roller coating, then packaging the fruits to create a microenvironment that inhibits the growth and reproduction of spoilage bacteria, thereby delaying the spoilage of the fruits.

Currently, the antibacterial components in antibacterial fruit preservation paper mainly include slow-release sulfur dioxide, slow-release chlorine dioxide, nano-silver ions, antibiotics, organic acids (acetic acid, benzoic acid, lactic acid, acetone, and sorbic acid, etc.), fatty acid esters, nitrites, sulfites, peptides (lysozyme, peroxidase, and lactoferrin, etc.), and plant essential oils. Plant essential oils have attracted widespread attention due to their strong antibacterial ability, good biocompatibility, and low cost, making them one of the most studied antibacterial agents for fruit and vegetable preservation.

Wang Jianqing et al. used a slow-release sulfur dioxide preservative to conduct preservation experiments on cherries. The results showed that sulfur dioxide can significantly reduce the spoilage rate of fruits and extend the storage time [48]. Zheng Yonghua et al. studied the effects of different sulfur dioxide release doses on the quality, polyphenol oxidase activity, respiration intensity, and spoilage rate of loquat fruits. The results showed that the slow-release sulfur dioxide agent can significantly inhibit polyphenol oxidase activity and respiration intensity, slow down the decline of titratable acid and soluble solid content, prevent core browning, and reduce the spoilage rate of fruits, thereby extending the storage time [49]. However, the use of slow-release sulfur dioxide agents often leaves some residue, which can bleach and poison the fruits during storage, affecting the quality of the fruits and having certain limitations.

With the rapid development of chemistry and other disciplines, research on antibacterial substances and their antibacterial effects in plants has become more extensive and in-depth. Many antibacterial substances have been isolated and identified from plants, and more and more experts and scholars have turned their attention to using plant antibacterial substances for fruit and vegetable preservation. Yue Shuli et al. prepared cinnamon essential oil microcapsules using β-cyclodextrin embedding technology and mixed them with polyvinyl alcohol to prepare an antibacterial additive. This antibacterial additive has a good inhibitory effect on *Escherichia coli* and *Staphylococcus aureus*. When applied to paper coating to make antibacterial paper, it has a good preservation effect on cherry tomatoes, improving their sensory evaluation, reducing weight loss rate and respiration intensity, and extending the shelf life of cherry tomatoes [50]. US patent US 6372220B1 introduces the technology of making preservation paper using fenugreek antibacterial components. This technology has been industrialized, and the products are sold on Amazon [51].

The research team led by Yao Xianping at the Hangzhou Chemical Research Institute has successfully developed fruit and vegetable antibacterial preservation paper-based functional materials with essential oil slow-release effects for strawberries after years of in-depth research. Under storage conditions at room temperature (25 °C) and 0 °C, they conducted systematic and in-depth studies on the storage time and apparent characteristics of strawberries (weight loss rate, spoilage index, and surface color), nutritional components (organic acids, soluble solids, total phenols, malondialdehyde [MDA], and anthocyanins), enzyme activity (superoxide dismutase (SOD) enzyme activity, catalase (CAT) enzyme activity, and polyphenol oxidase (PPO) enzyme activity), surface microorganisms (total colony count), and epidermal structure (epidermal cell scanning electron microscopy (SEM) images).

1) Effects of Antibacterial Preservation Paper on the Apparent Characteristics of Strawberries

Weight loss rate, spoilage index, and surface color are important indicators for evaluating the apparent characteristics of strawberries. Table 10.4 lists the effects

Table 10.4 Effects of antibacterial preservation paper on the apparent characteristics of strawberries under different temperature storage conditions.

Experimental days	Apparent characteristics									
					Surface color					
	Weight loss rate/%		Decay index		L		a		b	
	CK	Preservation paper	CK	Preservation paper	CK	Preservation paper	CK	Preservation paper	CK	Preservation paper
D0	0	0	0	0	33.40		36.06		14.83	
D1	0.04	0.04	0	0	33.41	34.49	37.69	39.08	15.74	14.19
D2	0.19	0.31	0	0	30.79	31.97	39.68	40.59	14.28	13.31
D3	0.24	0.49	0	0	29.65	28.33	40.58	36.19	14.13	11.44
D4	1.79	1.96	0	0	31.27	31.66	42.08	37.11	13.85	10.87
D5	2.96	3.03	0	0	30.10	31.99	43.41	41.19	13.67	15.69
D6	3.52	3.53	2.67	0	32.24	32.43	43.11	39.32	13.84	12.63
D6*	1.59	0.38	0	0	0	0	0	0	0	0
D7	4.06	3.93	9.00	0	32.34	32.60	43.35	39.66	13.86	12.33
D8	4.76	4.02	18.67	0	29.94	32.35	44.68	38.18	14.72	12.07
D12*	1.66	1.12								
D18*	2.83	1.89								
D24*	3.63	2.38								

Note: D0, D1 to D8 represent the days before the experiment and the 1st to 8th days of the experiment under room temperature storage conditions at 25 °C. CK represents the control group experiment. D6*, D12*, D18*, and D24* represent the 6th, 12th, 18th, and 24th days of the experiment under 0 °C storage conditions.

of antibacterial preservation paper on the weight loss rate, spoilage index, and surface color of strawberries. Water evaporation is an important cause of fruit weight loss and wilting. The weight loss rate of fruits not only affects their sensory characteristics but also their taste. As fruits mature, respiration and transpiration cause them to gradually lose water, leading to surface wilting and a decrease in glossiness. Therefore, the weight loss rate is an important indicator for judging the preservation effect.

Table 10.4 illustrates that the weight loss rate of strawberries stored at room temperature (25 °C) shows a continuous upward trend. In the early storage period, the weight loss rate of strawberries treated with preservation paper is higher than that of the control group, which may be related to the preservation paper's higher absorption of strawberry moisture. Additionally, the weight loss rate of strawberries shows a significant increase starting from the 4th day. Throughout the storage period, the preservation paper treatment can slow down the increase in the weight loss rate of strawberries, effectively reducing the moisture exchange between strawberries and the external environment, maintaining a slower weight loss rate, reducing dry consumption during storage, and keeping a fresh taste.

Color changes directly reflect the maturity of fruits and vegetables. By observing the color of the fruit, the maturity of the fruit can be intuitively judged, which directly affects the consumer's purchasing psychology. In Table 10.4, the L value reflects the brightness of the color; the higher the L value, the higher the glossiness of the fruit. The a value indicates the color change from green to red, and the b value indicates the color change from blue to yellow. From the experimental results in Table 10.4, it can be seen that the surface color of strawberries changes to a certain extent during room temperature storage. The L value gradually decreases, the a value gradually increases, and the b value does not change significantly. This indicates that the surface red color of strawberries deepens during storage, changing from bright red to dark red. After treatment with antibacterial preservation paper, the decrease in the L value and the increase in the a value can be inhibited, maintaining a brighter surface. This indirectly reflects that antibacterial preservation paper has a certain effect on delaying the aging and browning of strawberries.

The decay rate is a key indicator for measuring the freshness and commercial value of fruits. The spoilage of strawberries not only affects the edible quality of fresh fruits but also reduces the commercial value of the fruits. Figures 10.17 and 10.18 show the changes in the decay rate of strawberries over time under room temperature (25 °C) and low temperature (0 °C) storage conditions, respectively. The experimental results found that under room temperature conditions, the surface of the control group strawberries began to show signs of decay from the fourth day, while the strawberries treated with antibacterial preservation paper did not show obvious signs of decay until the eighth day. This indicates that antibacterial preservation paper can extend the preservation period of strawberries by more than 4 days at room temperature. The results in Table 10.4 show that the decay index of the control group strawberries shows an upward trend throughout the room temperature storage period, while the decay index of the strawberries treated with preservation paper remains at 0 throughout the process, significantly lower than that of the control

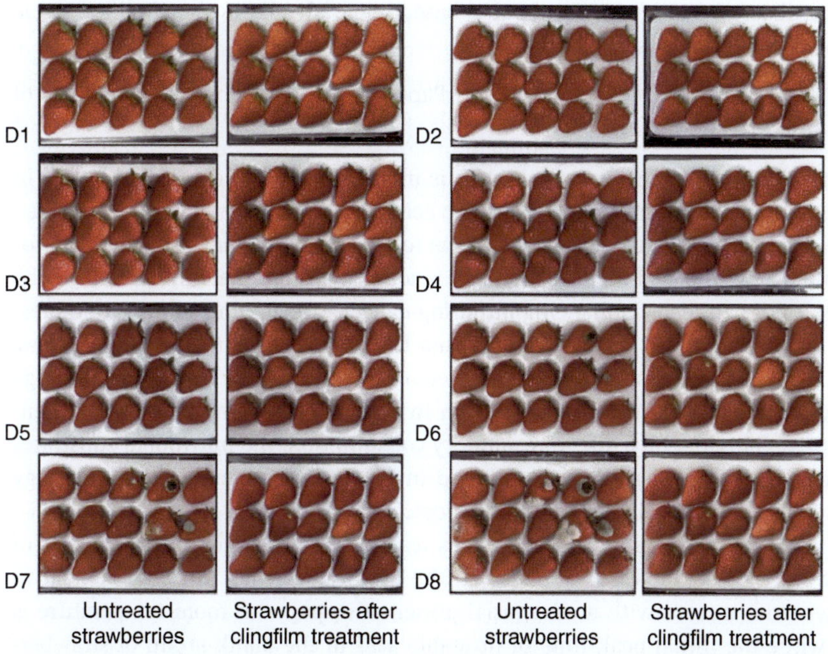

Figure 10.17 Appearance changes of strawberries treated with antibacterial preservation paper under room temperature storage conditions.

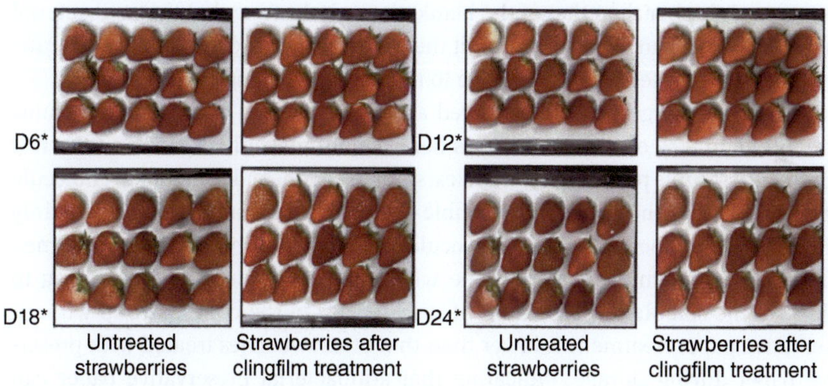

Figure 10.18 Appearance changes of strawberries treated with antibacterial preservation paper under 0 °C storage conditions.

group ($P < 0.05$). Therefore, antibacterial preservation paper can effectively inhibit the decay of strawberries.

Additionally, storage temperature has a significant impact on the degree of spoilage in strawberries. As shown in Figure 10.18, the surface freshness of the control group strawberries stored at 0 °C significantly decreased by the 12th day, while the strawberries treated with antibacterial preservation paper still maintained good freshness after 24 days of storage at 0 °C. This indicates that low temperature and antibacterial preservation of paper-based materials play a very significant role in the preservation of strawberries.

2) Effects of Antibacterial Preservation Paper on the Nutritional Components of Strawberries

The acid composition in strawberries is mainly organic acids. Organic acids in fruits are primarily produced in the citric acid cycle metabolism and are converted from sugars. Some of these acids continue to participate in energy metabolism for essential physiological and biochemical reactions, while others are stored in the vacuoles of fruit cells, enhancing buffering capacity, regulating cell osmotic pressure, maintaining cell texture, and helping fruits cope with environmental stress. The significant increase in acidity observed in strawberries during the early storage period may be a self-defense mechanism in response to postharvest environmental stress. Subsequently, due to respiratory consumption, the nutritional substrates are depleted, necessitating the continued involvement of organic acids in energy metabolism to maintain essential physiological activities.

The research results of Yao Xianping's team on the nutritional components of strawberries (see Table 10.5) show that the initial peak time of titratable acid in strawberries treated with antibacterial preservative paper at room temperature is D2, while the initial peak time of titratable acid in the blank group of strawberries without preservative paper treatment is D1. This indicates that the antibacterial preservative paper delays the appearance of the acidity peak. Moreover, at the end of storage, the titratable acid content of strawberries treated with antibacterial preservative paper is higher than that of the blank group, indicating that the antibacterial preservative paper can effectively inhibit the consumption of organic acids in strawberries during the later stages of storage to maintain physiological activities.

Soluble solids are generally considered as representatives of nutrients in fruits, with more than 80% of the components being small-molecule sugars, and the rest mainly acids, soluble pigments in vacuoles, soluble pectin, and some small-molecule tannins. Therefore, the decrease in soluble solids content during storage is mainly attributed to the reduction of small-molecule sugars, which are primarily consumed in fruit respiration metabolism to cope with environmental stress. According to Table 10.5, the soluble solids content of strawberries in the control group without preservative paper treatment is lower than that of strawberries treated with preservative paper during storage, indicating that antibacterial preservative paper can effectively inhibit the decrease in soluble solids content in strawberries and reduce the impact of environmental stress.

Total phenols are antioxidant active components in strawberry fruits, which are major nutrients and important indicators for evaluating the quality of strawberry fruits. According to Table 10.5, the total phenol content in strawberries tends to decrease with the increase in storage time, indicating that antibacterial preservative paper treatment can slow down the reduction of total phenols in strawberries to a certain extent. Studies have shown that the increase in MDA is a manifestation of aggravated membrane lipid peroxidation and membrane damage, which can reflect the degree of cell membrane peroxidation and is one of the important reasons for

Table 10.5 Effects of antibacterial preservative paper on nutritional components of strawberries under different temperature storage conditions.

	Nutrient content									
	Organic acid (g/kg)		Soluble solids (%)		Total phenols (%)		MDA (%)		Anthocyanins (%)	
Experimental days	CK	Preservation paper	CK	Preservation paper	CK	Preservation paper	CK	Preservation paper	CK	Preservation paper
D0	2.85	2.85	10.60	10.60	113.34	113.84	9.19	9.39	3.03	2.75
D0*	2.85	2.85	12.00	11.17	113.34	113.84	8.64	7.67	8.64	7.67
D1	2.90	2.57	10.97	10.53	112.17	108.47	9.29	9.98	5.31	6.44
D2	2.79	3.96	10.93	11.00	101.86	100.57	14.70	10.61	7.02	5.05
D3	2.74	3.23	10.93	11.80	116.47	109.80	13.65	11.96	10.07	5.41
D4	3.24	3.32	10.97	11.40	112.69	113.83	13.21	12.16	8.29	8.15
D5	3.31	3.31	10.93	11.23	116.18	114.92	17.08	16.51	10.07	9.98
D6	2.66	3.04	11.07	11.37	94.66	100.71	19.22	10.88	12.00	9.60
D6*	2.39	3.13	10.93	11.00	141.77	130.13	16.27	14.72	5.60	3.50
D7	2.59	2.79	10.03	12.10	108.81	111.19	16.27	14.72	10.22	8.98
D8	2.39	2.75	10.93	11.13	109.18	112.83	16.99	12.66	3.03	2.75
D12*	3.29	3.74	10.07	10.43	105.57	113.12	10.45	9.90	5.70	5.98
D18*	2.70	3.04	11.03	10.93	95.35	97.51	9.65	8.87	4.56	5.36
D24*	3.24	2.90	10.87	12.07	111.79	113.59	10.79	6.28	5.66	4.80

Note: D0, D1–D8 represent the days before the experiment and the first to eighth days of the experiment under room temperature storage conditions at 25 °C. CK represents the blank group experiment. D0*, D6*, D12*, D18*, and D24* represent the days before the experiment and the 6th, 12th, 18th, and 24th days of the experiment under 0 °C storage conditions.

fruit aging. According to Table 10.5, the overall MDA content in strawberries shows an upward trend during storage, and it can be seen that the MDA content in the control group strawberries is higher than that in strawberries treated with antibacterial preservative paper throughout the storage period. This indicates that antibacterial preservative paper can delay the process of cell membrane peroxidation and reduce the degree of strawberry aging.

Anthocyanins are the main source of strawberry color, mostly existing in the form of anthocyanins in the plant cell sap. According to Table 10.5, the anthocyanin content shows an upward trend during the normal temperature storage of strawberries, and it can be seen that the anthocyanin content in the control group strawberries is significantly higher than that in strawberries treated with antibacterial preservative paper, which is consistent with the changes in color indicators during strawberry storage. As the anthocyanin content increases, the surface color of strawberries changes from light red to bright red and dark red. It is inferred that antibacterial preservative paper delays the rapid synthesis of anthocyanins during the storage period.

3) Effects of antibacterial preservative paper on the enzyme activity characteristics of strawberries

SOD and CAT are enzymes involved in oxygen metabolism in organisms, with the main function of converting superoxide anions produced in fruits and vegetables into less harmful hydrogen peroxide. Therefore, these enzymes are highly correlated with the aging and stress resistance of fruits and vegetables and are one of the protective enzymes in fruits and vegetables. The higher their activity, the slower the aging process of fruits and vegetables. PPO is the main enzyme causing enzymatic browning in fruits and vegetables. PPO catalyzes the oxidation of endogenous polyphenolic substances in fruits and vegetables to produce melanin, which seriously affects the nutrition, flavor, and appearance quality of fruits and vegetables.

According to Table 10.6, the enzyme activities of SOD and CAT in strawberries gradually increase during the entire storage period, and antibacterial preservative paper has a significant effect on improving the enzyme activities of SOD and CAT, indicating that preservative paper can enhance the stress resistance of strawberries and delay their aging.

During the entire storage period, the PPO enzyme activity in strawberries without preservative paper treatment gradually increased, but in strawberries treated with antibacterial preservative paper, the PPO enzyme activity first increased and then decreased. This indicates that at the end of the storage period, antibacterial preservative paper can inhibit PPO enzyme activity, which is consistent with the results of chromaticity and anthocyanin measurements, indicating that preservative paper can inhibit fruit browning.

4) Effects of Antibacterial Preservative Paper on the Surface Microorganisms of Strawberries

The key pathogenic microorganisms causing fruit spoilage mainly include Botrytis, Penicillium, Aspergillus, Alternaria, Metschnikowia, etc. Since the internal

Table 10.6 Effects of antibacterial preservative paper on enzyme activity characteristics under normal temperature storage conditions.

	Enzyme activity indicators					
	SOD enzyme activity (U/g FW)		CAT enzyme activity (U/g FW)		PPO enzyme activity (U/g FW)	
Experimental days	CK	Preservation paper	CK	Preservation paper	CK	Preservation paper
D0	225.63	225.73	5.42	8.38	19.11	17.07
D2	280.31	311.86	7.45	9.34	20.37	26.94
D4	427.89	412.17	9.06	8.99	27.20	25.43
D6	442.30	474.39	3.89	10.37	36.58	17.19
D8	445.79	459.55	3.34	3.68	31.89	18.28

Note: D0, D2–D8 represent the days before the experiment and the second to eighth days of the experiment under normal temperature storage conditions at 25 °C. CK represents the control group experiment; FW represents fresh weight.

Table 10.7 Effects of antibacterial preservative paper on the total number of surface microorganisms of strawberries.

	Total colony count (lg CFU/g)	
Experimental days	CK	Preservation paper
D0	3.5	3.2
D3	5.4	4.4
D6	7.3	5.3
D12	11.4	7.8
D18	16.2	10.9

environment of most fruits is acidic, the spoilage pathogens causing fruit deterioration are mainly molds [52]. Strawberries have a high water content, and the reproduction speed of surface microorganisms is fast, so it is very important to effectively control the number of surface microorganisms on strawberries. Table 10.7 and Figure 10.19 show the changes in the total number of surface microorganisms on strawberries before and after treatment with antibacterial preservative paper during the storage period.

It can be observed that under normal temperature storage conditions at 25 °C, antibacterial preservative paper can inhibit the growth of surface microorganisms on strawberries after treatment, which is one of the important reasons why antibacterial preservative paper can significantly reduce the decay rate of strawberries.

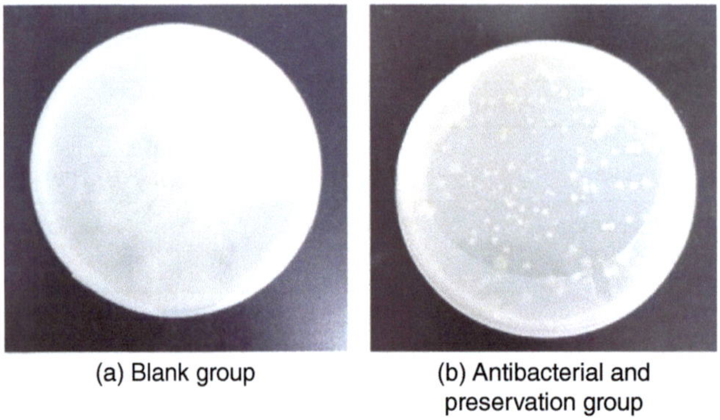

Figure 10.19 Changes in the total number of surface microorganisms on strawberries on the Sixth day under normal temperature storage conditions.

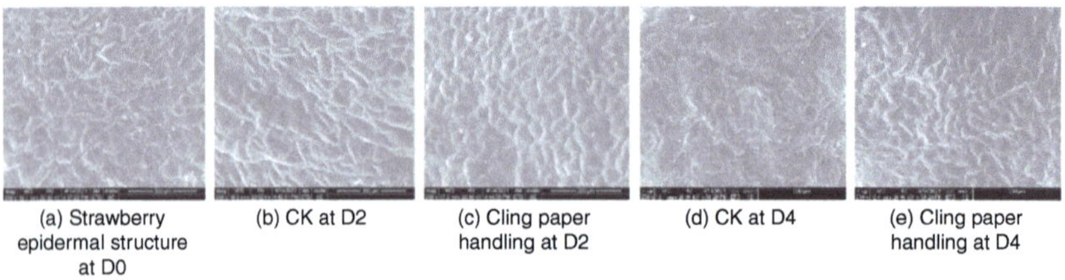

Figure 10.20 Effects of antibacterial preservative paper on the epidermal structure of strawberries (SEM Images).

5) Effects of Antibacterial Preservative Paper on the Epidermal Structure of Strawberries

The epidermal structure of strawberries can directly reflect the impact of antibacterial preservative paper on the epidermal cells of strawberries. As shown in Figure 10.20, at D0, the epidermal cells of strawberries are uniform and plump. After treatment, at D2, the epidermal structure of strawberries treated with antibacterial preservative paper showed slight changes, with the epidermis being full and uniform, without shrinkage or rupture; at this time, the epidermal structure of strawberries in the control group showed slight shrinkage, which may be related to the beginning of water loss and weight loss. At D4 after treatment, the epidermis of strawberries treated with antibacterial preservative paper showed shrinkage, but the epidermal cells of the control group had already shown severe rupture and depression, which may be related to microbial infection and epidermal cell rupture. This indicates that antibacterial preservative paper can inhibit the damage to the epidermal cells of strawberries and maintain the integrity of the epidermal cells.

10.3.3 Preparation Technology

Currently, the main fresh-keeping materials for fruits and vegetables on the market are plastic films, nonwoven fabrics, and paper-based materials. Plastic films have strong chemical stability and are difficult to biodegrade. Although they have a certain fresh-keeping effect on fruits and vegetables, most plastic films have poor air permeability. The moisture produced by the respiration of fruits and vegetables cannot be discharged, leading to an increase in the humidity of the fresh-keeping environment, thereby accelerating the growth of microorganisms. Paper-based fresh-keeping materials are simple to make, relatively low cost, and have relatively good air permeability, which can avoid the problem of fruit and vegetable decay caused by excessive environmental humidity, making them suitable for long-distance transportation. Moreover, paper-based fresh-keeping materials are green, environmentally friendly, and healthy, without causing harm to the environment and the human body.

This section focuses on the preparation technology of several paper-based functional materials for fruit and vegetable preservation, mainly divided into five types: papermaking combined preparation technology, infiltration combined treatment technology, coating combined treatment technology, sandwich combined design technology, and other forms.

10.3.3.1 Papermaking Combined Preparation Technology

Papermaking combined preparation technology refers to the preparation of paper-based materials with fresh-keeping functions by adding raw materials or additives with fresh-keeping functions to the original papermaking raw materials. The added materials have certain size requirements to prevent them from being easily lost during papermaking. Generally, different functional fresh-keeping materials can be mixed to achieve composite fresh-keeping purposes.

Papermaking combined preparation technology is one of the common methods for preparing fruit and vegetable fresh-keeping paper. Usually, hydrophilic resins, antifungal agents, and ethylene absorbers are used as effective fresh-keeping components and added to the pulp. The fruit and vegetable fresh-keeping paper prepared through papermaking technology has stable properties, good ethylene removal, and antibacterial and antifungal functions, making it suitable for fruit and vegetable packaging and preservation, effectively extending the fresh-keeping period of fruits and vegetables. The American company Fenugreen produces a fresh-keeping paper that is made by combining paper-based materials with organic antibacterial materials. It has a maple scent, can inhibit microbial growth, and can extend the storage period of fruits and vegetables by two to four times. This fresh-keeping paper preparation technology can also be used for the preparation of fresh-keeping cartons. For example, a paper layer made of wood pulp, porous fillers, starch-based polyvinyl alcohol, ethylene-vinyl acetate copolymer glue powder, zeolite, sodium benzoate, tea polyphenols, clove powder, chitosan, hexagonal stone particles, and loess particles can effectively improve the fresh-keeping effect of fruits and vegetables, with significant improvements in antibacterial, antifungal, moisture-proof, fresh-keeping, and weather resistance. In addition, the hexagonal stone and loess particles in the paper layer can not only absorb the ethylene released by fruits and vegetables but also release negative oxygen ions, reducing the respiration intensity and aging speed of fruits and vegetables [53]. The Japan Food Distribution System Association has also used Revival stone as an additive to prepare fresh-keeping cartons with good gas adsorption properties, suitable for long-distance storage and transportation.

10.3.3.2 Infiltration Combined Treatment Technology

Paper-based material infiltration combined treatment technology utilizes the water absorption characteristics of paper materials to infiltrate liquid components with different fresh-keeping effects into the paper-based materials. The fresh-keeping components infiltrated into the internal gaps of the paper-based materials gradually release during use or can form a film with the paper-based materials to achieve continuous fresh-keeping and moisturizing purposes. Suitable fresh-keeping materials should be water-soluble or can form a certain emulsion when mixed with water or have a certain volatility. For example, phenolic, aldehyde, and phenolic substances in essential oils, as well as natural polymers such as chitosan and methylcellulose that can form films. This technology is generally only suitable for preparing fresh-keeping paper. For fresh-keeping cartons, the infiltration of liquid fresh-keeping components may reduce the strength and original performance of the cartons to a certain extent.

10.3.3.3 Coating Combined Treatment Technology

Paper-based material surface coating combined treatment technology mainly refers to laminating or surface spraying on the paper surface. Surface coating combined treatment technology generally uses natural polymers such as sodium alginate, oxidized starch, carrageenan, chitosan, gelatin, gum arabic, HA, or basic materials or adhesives such as linear low-density polyethylene, polypropylene, polystyrene, polycaprolactam (nylon 6), porous acrylic resin copolymer, IPN structure polyurethane resin, polyacrylimide, polyvinyl alcohol, ethylene-vinyl acetate, polycarbonate, polyethylene terephthalate, polyvinyl chloride, and their mixtures. These polymer materials can exist in the form of water dispersions, coated or sprayed on the surface of paper-based materials. They form a polymer film on the surface of the paper-based materials during the drying process or directly exist in the form of a polymer film, combined with the paper-based materials using a laminating machine. The resulting cured composite material layer can effectively block the movement of gas molecules, maintaining the internal environment of the carton with "low oxygen concentration and high carbon dioxide concentration," thereby reducing the respiration intensity of the packaged fruits and vegetables to a certain extent and achieving fresh-keeping effects.

In most cases, these polymer materials are used as substrates or adhesives in combination with other materials to achieve composite fresh-keeping purposes. For example, an emulsion made of starch, acrylate, nano-silica, and nano-titanium dioxide can be sprayed on the surface of bamboo fiber paper to make fresh-keeping paper. This fresh-keeping paper can improve air cleanliness, reduce ethylene content, and has excellent fresh-keeping effects, and can be reused at a low cost [54]. Korean scholars have found that compared with ordinary corrugated cardboard, spraying functional materials that can release far-infrared rays on the cardboard increases the far-infrared emission intensity and emissivity of the cardboard as the coating amount increases, and the corrugated cartons with a large coating amount have a relatively better fresh-keeping effect on citrus fruits [55]. This preparation method is simple to operate and has good fresh-keeping effects, making it a common method for preparing fruit and vegetable fresh-keeping paper, cardboard, paper bags, and cartons in laboratories and factories.

10.3.3.4 Sandwich Combined Design Technology

Paper-based material sandwich combined design technology involves adding a fresh-keeping layer inside the sandwich of paper-based materials. This preparation method prevents the fresh-keeping layer from directly contacting the contents, making it safer, greener, and more environmentally friendly. It is commonly used in the preparation of fresh-keeping cardboard, insurance pads, and fresh-keeping cartons.

For fresh-keeping paper, the fresh-keeping agent is generally porous inorganic materials, essential oil microcapsules, polymer films, and other micro-nano particles or flaky substances, which can be well encapsulated inside without leakage problems. The Xinjiang Institute of Physics, Chinese Academy of Sciences, has developed a laminated grape fresh-keeping paper. The paper consists of a fresh-keeping agent and a substrate. The fresh-keeping agent is made by mixing pyrosulfite or bisulfite with Chinese herbal antibacterial agents and slow-release agents, then placing equal amounts into each bag-shaped grid of the substrate, covering it with one or two layers of substrate, and heat-sealing it with a packaging machine, or it can be made manually. This fresh-keeping paper can preserve grapes for 11 days in the laboratory at a high temperature of 32 °C in summer, with a grape integrity rate of over 96%, and maintain the original flavor. Grapes can be preserved for 3 months at 0-5 °C, and only one sheet of fresh-keeping paper is needed for every 5 kg of grapes [56].

When multiple paper-based materials are combined with fresh-keeping materials to form a thicker sandwich paper, the sandwich paper can also be called fresh-keeping cardboard. Beijing Agel Technology Biotechnology Co., Ltd. has developed a cardboard that contains fresh-keeping materials such as 1-MCP microcapsule powder, wetting agents, and stabilizers. It specifically contains two waterproof and airtight layers, a surface layer, and a water-permeable and breathable layer. The stability of 1-MCP in this fresh-keeping cardboard is greatly improved, and after fresh-keeping strawberries, the edibility of the strawberries in the experimental group is significantly higher than that of the control group [57]. For cartons, fresh-keeping films can also be sandwiched between two layers of fresh-keeping paper to control the gas content inside the box.

For cartons, the fresh-keeping film can be sandwiched between two layers of fresh-keeping paper to control the gas content inside the box. Researchers have invented a carton with multiple corrugated structures, each corrugation consisting of two layers of kraft paper with different grammages sandwiching a polymer film layer, and the corrugated carton contains a plastic product lid. This carton can not only control the gas content inside the box but also adjust the moisture content inside the box, maintaining the humidity between 75% and 85% [58]. In addition, the internal corrugated paper can be replaced with other fresh-keeping materials, such as foamed resin and aerogel, which have excellent thermal insulation properties, preventing the loss of nutrients and moisture in fruits and vegetables during storage or transportation due to the increase in temperature inside the box.

10.3.4 Outlook

There has been extensive research on preservatives, fresh-keeping films, and fresh-keeping packaging in the field of fruit and vegetable preservation both domestically and internationally. The research directions are gradually expanding into various fields such as materials science, food chemistry, organic chemistry, genetic biology, and mechanical engineering. Fresh-keeping methods are evolving from single-principle research to composite research and application directions, including refrigeration, modified atmosphere packaging, green preservatives, low-dose radiation pretreatment preservation, UV preservation, and genetic engineering, which are also the international trends in fresh-keeping. In future work, more attention will be paid to in-depth research on the quality of fruit and vegetable flavor, fresh taste, and nutritional components, thereby establishing a comprehensive quality assurance system for evaluating the freshness, maturity, flavor, taste, color, and safety of stored fruits and vegetables. It is believed that with the continuous development of science and technology, through the persistent efforts of researchers, it will be entirely possible to provide fresh, safe, high-quality, and diverse fruits and vegetables all year round in the near future.

10.4 Electromagnetic Shielding Paper-based Functional Materials

10.4.1 Overview

With the rapid development of the electronic information industry, electromagnetic waves have been widely used in various fields such as electronics, automobiles, household appliances, communication tools, computer equipment, information devices, electric tools, aerospace, and medical care. However, as electronic and electrical equipment as well as communication devices evolve toward increased sensitivity, density, high frequency, and diversity, the electromagnetic environment within limited spaces has deteriorated significantly, leading to a series of increasingly serious social and environmental issues such as electromagnetic interference (EMI), electromagnetic information leaks, and harm to the human body [59, 60]. Currently, electromagnetic shielding is the most commonly used protective measure against electromagnetic radiation pollution. Its purpose is to limit electromagnetic waves through technological means and control the radiation intensity below normal thresholds. Electromagnetic shielding materials can effectively address EMI and electromagnetic compatibility issues caused by electromagnetic waves, thus being widely used in communication equipment, computers, mobile phone terminals, automotive electronics, household appliances, defense and military applications, among others. Typically, shielding materials used in electromagnetic shielding bodies are mainly metallic materials. However, their disadvantages, such as heavy weight, large volume, high density, high cost, difficulty in folding, and processing challenges, have restricted their application, particularly in aerospace and military fields where high material requirements are imposed. Compared to metallic materials, electromagnetic shielding paper-based functional materials offer advantages such as lightness, flexibility, foldability, and low cost, making them highly suitable for modern shielding materials that demand "thinness, lightness, breadth, and strength." Shielding materials are crucial in addressing EMI and electromagnetic compatibility issues caused by electromagnetic waves, thus carrying significant implications for social life, economic construction, and national defense.

10.4.2 Shielding Mechanism of Electromagnetic Shielding Materials

Electromagnetic shielding refers to the use of electromagnetic shielding materials to block or attenuate electromagnetic waves generated by electromagnetic field sources, preventing them from entering the shielded area. Currently, there are various explanations for the mechanism of electromagnetic shielding, including the eddy current effect method, electromagnetic field theory method, transmission line theory method, and others. Among these methods, the transmission line theory method [61] has been widely accepted by scholars due to its advantages of simple calculation, high accuracy, and ease of understanding. The mechanism of this method is illustrated in Figure 10.21.

The transmission line theory approach treats the propagation of electromagnetic waves through shielding materials as a transmission process. When electromagnetic waves generated by an electromagnetic field source pass through the shielding

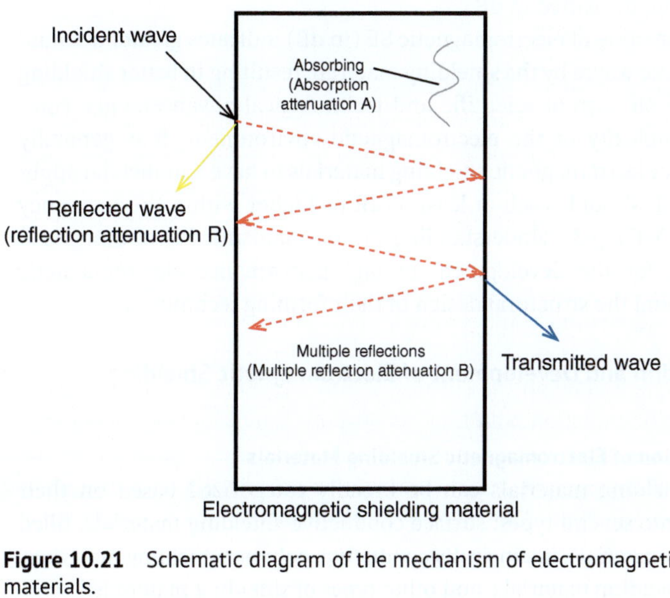

Figure 10.21 Schematic diagram of the mechanism of electromagnetic shielding materials.

material, a portion of the electromagnetic waves is reflected off the outer surface of the shielding material, preventing them from passing through. The remaining electromagnetic waves enter the shielding material and propagate forward. During this transmission process, the shielding material absorbs a certain amount of the electromagnetic waves, further attenuating their energy. After surface reflection and absorption, the remaining electromagnetic waves bounce back and forth between the two internal interfaces of the shielding material, further reducing the electromagnetic wave energy. Therefore, the shielding effect of the shielding material on electromagnetic waves is primarily composed of three components: reflection attenuation (R) of electromagnetic waves on the surface of the shielding material, absorption attenuation (A) of electromagnetic waves by the shielding material, and multiple reflection attenuation (B) of electromagnetic waves within the shielding material.

In general, the shielding characteristics of a shielding material for electromagnetic waves can be characterized by shielding effectiveness (SE). SE is defined as the ratio of magnetic field intensity, electric field intensity, or electromagnetic power before and after the placement of the shielding material.

According to the Schelkunoff formula [62], the SE of electromagnetic waves passing through a shielding material can be calculated using the following equation:

$$\mathrm{SE} = \mathrm{R} + \mathrm{A} + \mathrm{B}$$

In the equation, SE represents electromagnetic SE, measured in decibels (dB);

R represents the reflection attenuation of electromagnetic waves on the surface of the shielding material, measured in dB;

A represents the absorption attenuation of electromagnetic waves by the shielding material, measured in dB;

B represents the multiple reflection attenuation of electromagnetic waves within the shielding material, measured in dB.

Generally, a higher value of electromagnetic SE (in dB) indicates greater attenuation of electromagnetic waves by the shielding material, resulting in better shielding performance. Based on current scientific and technological advancements, combined with the complexity of the electromagnetic environment, it is generally believed that for new electromagnetic shielding materials to have commercial application value, their SE should reach at least 30 dB or higher within the frequency range of 30 to 1000 MHz [63]. Understanding the mechanism of electromagnetic shielding is crucial for the development of high-performance electromagnetic shielding materials and the structural design of their forming techniques.

10.4.3 Classification and Development of Electromagnetic Shielding Materials

10.4.3.1 Classification of Electromagnetic Shielding Materials

Electromagnetic shielding materials can be broadly categorized based on their material structure into several types: surface conductive shielding materials, filled composite shielding materials, intrinsically conductive polymer shielding materials, conductive fabric shielding materials, and other types of shielding materials.

1) Surface Conductive Shielding Materials

Surface conductive shielding materials are those that employ a conductive shielding layer applied to the surface of substrates such as engineering plastics, rubber, and wood through methods like spraying, electroplating, and chemical plating. Their primary purpose is to achieve electromagnetic shielding. These materials are commonly used due to their low cost, excellent SE, and wide applicability.

2) Filled Composite Shielding Materials

Filled composite shielding materials are electromagnetic shielding materials composed of electrically insulating resins and rubbers mixed with fillers that exhibit high conductivity and magnetic properties. Common conductive fillers include metals (such as nickel, copper, silver), carbon-based materials (like carbon nanotubes, graphene, and graphite), and conductive polymers (like polyaniline and polypyrrole). These materials offer ease of processing and scalability for mass production.

3) Intrinsically Conductive Polymer Shielding Materials

Intrinsically conductive polymer shielding materials primarily include polyaniline, polypyrrole, polythiophene, and polyacetylene. Their unique electrical conductivity, corrosion resistance, low density, high strength, and ease of processing have attracted extensive research from scholars globally.

4) Conductive Fabric Shielding Materials

Conductive fabrics typically take two forms: those made by interweaving metal fibers with ordinary fibers, and those where a metal conductive layer is coated onto ordinary fabrics. Conductive fabrics offer excellent electromagnetic shielding capabilities while retaining the inherent flexibility and fold resistance of textiles, making them a commonly used electromagnetic shielding material.

5) Other Types of Shielding Materials

In addition to the four common types of electromagnetic shielding materials mentioned above, there are several others, such as electromagnetic shielding paper-based materials, foam metal shielding materials, and transparent electromagnetic shielding films.

10.4.3.2 Development History of Electromagnetic Shielding Materials

In the 1940s, researchers began to introduce the concept of electromagnetic compatibility and embarked on related research efforts. During this period, metal conductors and ferromagnetic materials such as permalloy (iron–nickel alloy), pure iron, and silicon steel were intensively studied and applied.

By the 1960s, the electromagnetic shielding component industry experienced rapid growth driven by advancements in information technology, testing equipment, digital computers, telecommunications, semiconductors, and other technologies. Surface conductive shielding materials started to be widely used during this time.

The 1980s saw unprecedented attention on filled composite shielding materials, particularly in Europe and the United States, due to the rapid development of emerging fields such as electronics, communications, and automation. These novel materials were created by filling plastic or rubber molding materials with conductive fillers, commonly metal fibers, carbon fibers, ultrafine carbon black, metal alloy powders, and more.

In recent years, the rapid development of nanotechnology has led to the emergence of shielding materials with novel structures such as carbon nanotubes and graphene. Electromagnetic shielding materials have shifted from single-material systems to lighter, thinner, and more effective composite materials. Examples include graphene-based conductive foam materials and carbon nanofiber/nanotube composites. These materials offer superior shielding performance while maintaining a reduced overall weight and thickness, making them increasingly popular in various applications.

Overall, the evolution of electromagnetic shielding materials has been a continuous process of innovation and improvement, driven by technological advancements and the need for more efficient and lightweight shielding solutions. With the ongoing research and development in nanotechnology and other related fields, it is expected that electromagnetic shielding materials will continue to evolve and provide better performance in the future.

10.4.4 Research Status of Electromagnetic Shielding Paper-based Functional Materials at Home and Abroad

The research on electromagnetic shielding paper-based functional materials has been conducted more deeply in developed countries in Europe and the United States. Some products have already been industrialized and widely used in the fields of military and civilian precision instruments, computers, and other equipment. For instance, in the early years, the UK developed shielding wallpaper based on nickel-plated carbon fiber, which can effectively block electromagnetic wave activities and prevent precision instruments from being interfered with. In 2012, the Grenoble Institute of Technology in France and the French Paper Technology Center jointly developed a wallpaper that can shield Wi-Fi signals. This kind of wallpaper does not block the electromagnetic signals of mobile phones but selectively shields Wi-Fi frequencies, allowing mobile phones and other radio equipment to operate normally. Researchers at the Aalto University in Finland successfully prepared a graphene nanocellulose composite paper with good mechanical and electrical properties by combining reduced graphene oxide and modified nanocellulose [64]. Researchers at the Chungnam National University in South Korea prepared cellulose paper coated with silver nanofibers through coating technology. The research shows that the structure of cellulose paper can effectively form a shielding network. Meanwhile, the material is characterized by high toughness and low density, having tremendous application prospects in related shielding and anti-interference fields [65].

The research on electromagnetic shielding paper-based functional materials in China has a relatively short history and is not yet in-depth. However, domestic units and research institutions have gradually recognized the significant social and practical significance of research in this field, showing a trend of diverse development. For example, researchers at Shaanxi University of Science and Technology prepared carbon fiber paper shielding materials by combining carbon fiber as the shielding core material with paper manufacturing technology, and coated them with silver-coated copper powder coatings. Test results indicate that within the frequency range of 100 kHz to 1500 MHz, its SE can reach up to 48 dB, demonstrating ideal electromagnetic shielding performance [66]. Researchers at Beijing Institute of Technology have developed lightweight and ultrathin graphene shielding paper materials using graphene materials. This material exhibits excellent shielding effects: when the thickness is 0.1 mm, the SE reaches 19 dB, and when the thickness is 0.3 mm, the SE can reach 46 dB. The preparation method of this graphene shielding paper material is relatively simple and is expected to be widely used in portable electronic devices [67]. Researchers at Beijing Forestry University prepared ultrathin and highly flexible MXene/nanocellulose composite electromagnetic shielding paper with a shell-like structure using a suction filtration self-assembly process. This new composite electromagnetic shielding paper possesses excellent properties such as light weight and foldability, completely retaining the characteristics of traditional paper. Benefiting from the addition of nanocellulose, the composite shielding paper exhibits excellent mechanical properties. At the same time, the SE of the developed shielding composite paper can reach 25.8 dB at 12.4GHz, meeting the requirements of current commercial shielding. Due to its excellent mechanical properties and electromagnetic shielding performance, this new composite electromagnetic shielding paper-based material is expected to be widely used in fields such as flexible electronic devices, weapons and equipment, and artificial intelligence [68].

The research team led by Yao Xianping from the National Engineering and Technology Research Center for Papermaking Chemicals at Hangzhou Chemical Research Institute proposed an innovative design concept for lightweight, flexible, and efficient composite electromagnetic shielding materials based on the disadvantages of traditional electromagnetic shielding materials at home and abroad and the requirements of modern electromagnetic shielding materials for being "thin, lightweight, wide-range, and strong." They have also developed corresponding electromagnetic shielding paper-based composite materials. This new material is expected to provide a better choice for future electromagnetic shielding applications.

10.4.5 Preparation Methods and Functional Characteristics of Electromagnetic Shielding Paper-based Functional Materials

Electromagnetic shielding paper-based functional materials are a new type of lightweight composite shielding material with integrated structure and functionality. Currently, the main preparation methods include surface coating and pulp addition.

Surface coating refers to the application of electromagnetic shielding coatings on the surface of paper-based materials, thereby endowing the paper-based material with electromagnetic shielding functionality. The preparation process of this type of electromagnetic shielding paper-based functional material is relatively simple and cost-effective. However, it has high requirements for the paper-based material, such as the need for a low quantitative weight (grammage) while maintaining high strength. Additionally, there are risks of peeling and detachment of the electromagnetic shielding coating on the surface.

Pulp addition involves the preparation of electromagnetic shielding paper-based functional materials through the combination of conductive fibers, powders, and other electromagnetic shielding materials with plant fibers using wet-laid paper machine technology. This method produces electromagnetic shielding paper-based functional materials that retain the lightweight, flexible, foldable, and high-strength characteristics of ordinary paper while also possessing excellent electromagnetic shielding properties. Furthermore, due to the advantages of large-scale production on paper machines, this type of shielding material also offers low cost, scalability, and customizability, making it promising for a broad market.

10.4.5.1 Preparation of Basic Shielding Materials for Electromagnetic Shielding Paper-based Functional Materials

The lightweight and flexible nature of electromagnetic shielding paper-based functional materials necessitates the use of lightweight and efficient basic shielding materials. Carbon-based materials, such as carbon fibers and graphene, possess high conductivity, high strength, high modulus, high-temperature resistance, corrosion resistance, low density, and low thermal expansion coefficients. These materials can significantly enhance the mechanical properties of composite materials. Novel carbon-based materials, with their unique nanostructures, high conductivity, and dielectric constants, have significant application potential in electromagnetic shielding composites.

Through extensive literature research, the research team led by Yao Xianping selected five carbon-based materials: graphene, carbon fibers, carbon nanotubes, graphite, and carbon black. Using the preparation method for electromagnetic shielding paper-based functional materials, they developed corresponding electromagnetic shielding paper-based functional materials and conducted SE tests on the samples. Additionally, based on this foundation, the team further enhanced the performance of carbon fiber materials using surface chemical modification techniques, resulting in lightweight and efficient carbon-based electromagnetic shielding materials. These materials were characterized using SEM, as shown in Figure 10.22.

10.4.5.2 Forming Techniques and Functional Characteristics of Electromagnetic Shielding Paper-based Functional Materials

Drawing on the shielding mechanisms of electromagnetic shielding materials and combining them with the preparation methods for electromagnetic shielding paper-based functional materials, the research team led by Yao Xianping

(a) Non-nickel plated carbon fiber (b) Nickel-plated carbon fiber

Figure 10.22 SEM images of shielding materials before and after modification.

innovatively developed a new type of electromagnetic shielding paper-based functional material with a multilayer shielding interface and a combination of reflection and absorption effects using various technologies such as paper machine forming techniques, surface coating techniques, and composite forming techniques. This novel material is composed of three different types of electromagnetic shielding materials, creating a "sandwich-like" shielding structure with multiple shielding interfaces. This significantly reduces the risk of electromagnetic wave leakage and enhances the shielding performance of the electromagnetic shielding paper-based functional material. The SEM characterization of the relevant composite materials is shown in Figure 10.23.

The electromagnetic shielding paper-based functional material developed by the research team led by Yao Xianping has undergone rigorous testing at the China Shanghai Testing Center. The results indicate that its SE exceeds 100 dB across the measurable frequency range of 15 MHz to 10 GHz, with a peak performance of 126.1 dB. Remarkably, despite its exceptional shielding capabilities, the material boasts a unit area mass of only 133.3 g/m^2, which is a fraction of one-twentieth of that of metals with the same volume. This exceptional combination of properties

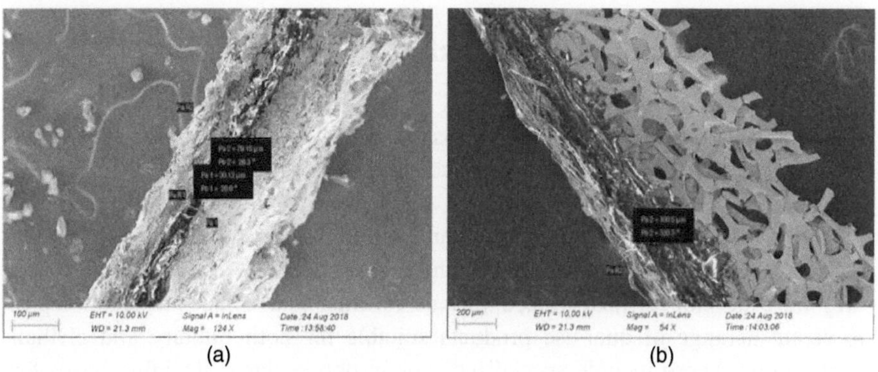

(a) (b)

Figure 10.23 Cross-sectional morphology of different types of electromagnetic shielding paper-based functional materials.

Table 10.8 Comparison of material lightweighting (taking 0.4 mm thick material as an example).

Material	Mass per Unit Area
Steel plate	≈ 3160
Copper plate	≈ 3560
Nickel plate	≈ 3560
Electromagnetic shielding paper-based functional material (developed by Yao Xianping research team)	≈ 133.3

Table 10.9 Actual test data of electromagnetic shielding effectiveness.

Frequency/MHz	Shielding effectiveness/dB	Frequency/MHz	Shielding effectiveness/dB
15	105.2	1500	114.2
30	108.5	1800	116.2
80	107.5	2450	117.9
100	107.1	3000	119.2
300	110.5	5000	121.2
450	111.2	8000	123.5
915	112.5	10000	126.1
1000	113.5		

fully meets the modern requirements for electromagnetic shielding materials, namely being "thin, lightweight, broadband, and strong" (see Tables 10.8 and 10.9 for a detailed comparison).

10.4.6 Applications of Electromagnetic Shielding Paper-based Functional Materials

The research team led by Yao Xianping has successfully developed lightweight, flexible, and highly effective electromagnetic shielding paper-based functional materials, as illustrated in Figure 10.24.

Based on the diverse requirements of military and civilian products, the team, in collaboration with the electromagnetic shielding paper-based functional materials they have developed, has utilized various forming and composite technologies to create a range of dual-use products suitable for both military and civilian applications. These include specialized mobile phone signal shielding boxes for conferences, mobile phone signal shielding bags, and laptop radiation protection bags, as depicted in Figures 10.25 and 10.26.

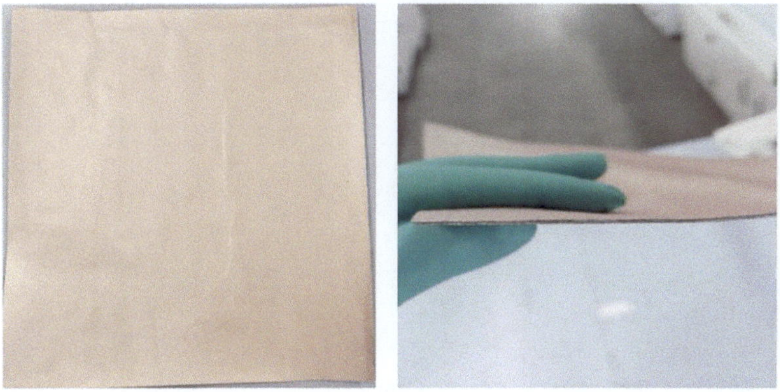

(a) Front view (b) Side view

Figure 10.24 Physical image of electromagnetic shielding paper-based functional material developed by the Yao Xianping research team.

Figure 10.25 Special mobile phone signal shielding box for conferences.

Among them, the conference-specific mobile phone signal shielding box is developed using natural biomass materials and high-performance flexible shielding composite materials as the base materials, integrated with 3D printing technology and composite forming technology. The product is characterized by its small size, portability, simple operation, and environmental friendliness. It can effectively block communication signals such as mobile phones, Wi-Fi (wireless communication technology), and Bluetooth, rendering the signal transmission and reception functions of mobile phones and other communication devices placed in the shielding box inactive, unable to make or receive calls. This achieves the purpose

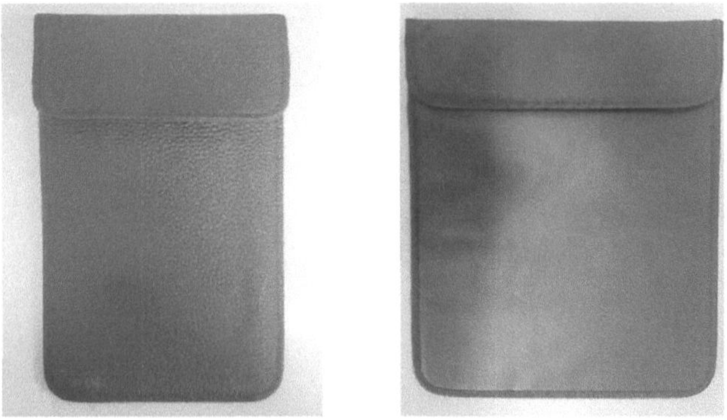
(a)
(b)

Figure 10.26 Mobile phone signal shielding bag (a) and Laptop radiation protection bag (b).

of mandatory disablement, preventing information leakage, telephone harassment, magnetic demagnetization, location tracking, and mobile phone radiation. It is suitable for centralized management of mobile communication devices in special locations such as various parties and government meetings, business negotiations, scientific and technological conferences, confidential offices, negotiation rooms, and military areas. Compared to traditional steel shielding cabinets, this mobile phone signal shielding box offers numerous advantages, such as flexibility in use, high performance, and personalized customization. It allows mobile phones to remain within the user's sight at all times, preventing incidents of mobile phone miscontrol and ensuring information security and economic security of the country, enterprises, and individuals. The product has passed expert review, been successfully approved for funding, and received nearly a hundred orders.

10.4.7 Outlook

Currently, the development and application of electromagnetic shielding paper-based functional materials face several major issues:

a) The research on electromagnetic shielding paper-based functional materials in China is primarily conducted by domestic universities, research institutes, and other institutions. Consequently, it is challenging to translate these research outcomes into practical applications, resulting in a low level of industrialization and market promotion.
b) While electromagnetic shielding paper-based functional materials are lightweight and thin like paper, they still suffer from disadvantages such as relatively low SE, narrow shielding frequency range, and intolerance to high temperatures. These limitations restrict their application in fields with higher requirements.
c) Due to the late start of research on electromagnetic shielding paper-based functional materials in China, foreign countries have occupied the technological high

ground in the field of novel electromagnetic shielding materials. This has, to a certain extent, compressed the space for independent research and development by domestic enterprises. Issues such as insufficient investment in research and development of electromagnetic shielding paper-based functional materials and other novel shielding materials, poor integration of industry, academia, and research, inadequate independent innovation capabilities, and severe homogeneity among low- and mid-range products persist.

With the rapid development of the global electromagnetic shielding, electromagnetic wave absorption, and heat dissipation management industries, market demand is constantly growing. Coupled with the advent of 5G (fifth-generation communication technology) and the artificial intelligence era, the global demand for electromagnetic shielding materials and related devices is expected to further increase. Future electromagnetic shielding materials are bound to evolve from their current single-functionality toward greener manufacturing processes, thinner and lighter materials, and integrated functionalities. As a novel electromagnetic shielding material, the electromagnetic shielding paper-based functional material aligns with the modern development trend of "thin, lightweight, broadband, and strong" shielding materials, holding vast market potential.

Domestic research institutions and enterprises should seize the opportunity presented by the vast markets of 5G and artificial intelligence, increase investment in upstream and downstream industrial chains, and break through core technologies in the field. Simultaneously, they should strengthen collaboration among domestic enterprises, research institutes, and universities in industry-academia-research cooperation. By combining market demands and leveraging the strengths of research institutes and universities in fundamental research, as well as the market sensitivity and ample funding of enterprises, they can achieve complementary advantages and strong collaborations. This will jointly promote the development of electromagnetic shielding materials in China, particularly electromagnetic shielding paper-based functional materials.

10.5 Paper-based Functional Materials for Decorative Base Paper

10.5.1 Overview

Decorative base paper is an industrial specialty paper made from high-quality wood pulp and inorganic fillers by special processing techniques. After printing and resin impregnation, it is mainly used for the lamination of impregnated film paper decorative boards and thermosetting resin decorative laminate boards, which can provide artificial boards for decoration, protection, reinforcement, and sealing. It can also effectively cover the inherent defects of artificial boards, giving them a beautiful appearance, while also improving their water resistance, heat resistance, weather resistance, wear resistance, and chemical resistance. In recent years, driven by the construction and furniture industries, China's artificial board industry has experienced rapid development, and the rapid growth of artificial boards has led to a strong demand for decorative base paper. Decorative base paper is now widely used in the furniture and interior decoration fields, and it is also an important laminating material for the building and home decoration industry, with significant implications for enhancing the added value of artificial boards and expanding their application areas.

The production of decorative paper backing originated in Germany, and it experienced rapid development in Europe in the 1950s. Although the history is not long, the production technology of decorative paper backing abroad has already become very sophisticated, often able to achieve the dual goals of reducing costs and improving the quality of decorative paper backing by using new materials and adjusting the amount of raw materials used. DuPont built the world's first chlorine-based titanium dioxide factory in 1959, and the chlorine-based process has higher technical difficulty, but the titanium dioxide produced has high opacity, good light stability, and high brightness, meeting the market demand for high-quality titanium dioxide. Applied to the production of decorative paper backing, it can significantly improve its whiteness and opacity. To this day, DuPont's R-794 is still the most commonly used titanium dioxide in the production of high-end decorative paper backing. At the same time, Dutt et al. also proposed using aluminum silicate and titanium dioxide together to reduce the use of titanium dioxide and improve the retention of the slurry, thereby achieving the goal of reducing costs. In addition to improving the performance of decorative paper backing from the raw materials used in papermaking, the research on surface treatment of decorative paper backing abroad has also gradually gained momentum in recent years. Soili et al. pointed out that using a single pigment or a combination of pigments and resin can be more beneficial for the surface coating of decorative paper after the coating, and can greatly improve the gloss and smoothness of the decorative paper. Wicher made a modified starch by adding water-soluble polymers and used it as a sizing agent to surface-size the decorative paper, finding that the paper's strength had improved. The surface structure of decorative paper has a great impact on its performance, especially smoothness, absorbency, and printability, so researching the surface coating structure of decorative paper for improving its printing performance and its application in artificial board decoration is of great significance.

The earliest decorative base paper in China was produced by Qianfeng Paper Factory in Shanghai in 1973, mainly for use by Yangzi Wood Factory to produce fireproof boards. At that time, the domestic market was mainly dominated by imported decorative base papers from Western Europe and the United States. However, in the mid-1990s, domestic production began to take off with the drive from the real estate and building decoration industries. In recent years, with the increasing demand for decorative base paper, more research has been conducted, and many scholars have continuously optimized the preparation of raw materials and production processes to improve the product quality of domestic decorative base paper. Zheng Huapei et al. solved the problems of poor durability and retention of domestic titanium dioxide in decorative base paper by coating titanium dioxide products with phosphate or phosphate salts and aluminum salts to prepare decorative base paper-specific titanium dioxide, with Zeta potential and retention rate and lightfastness comparable to that of DuPont R-794 decorative paper-specific titanium dioxide imported from abroad [69]. Hua Wang New Material Technology Co., Ltd. in Hangzhou met the requirements for high-speed impregnation of decorative base paper by adjusting the proportion of chemical auxiliaries in the production process of decorative base paper, reducing the permeability to 15 s/100 mL or less and increasing the impregnation speed to more than 45 m/min, thereby improving the production efficiency of the impregnation plant [70]. In addition, as people's demand for quality of life is increasing, various specialty decorative papers with unique properties have been developed, such as decorative papers with extremely high wear resistance achieved by adding abrasion-resistant materials to the pulp through processes such as internal addition, impregnation, and coating [71], or decorative papers with flame-retardant properties obtained by adding flame retardants to the paper [72], which greatly reduces fire hazards and losses caused by fires. The decorative paper industry has formed a complete industrial chain from raw paper to printing, impregnation, and surface treatment application in China, and has several industrial clusters in Zhejiang, Shandong, and Guangdong. With the rapid development of industries such as artificial boards, engineered wood flooring, furniture, and decoration, China's decorative paper industry is bound to be driven by these industries and enter a period of rapid development.

10.5.2 The Quality Requirements of Decorative Base Paper

Decorative base paper is the main raw material for printing decorative paper and decorative film paper, and its performance directly affects the quality of the final product. Therefore, in order to meet the needs of later reprocessing, there are corresponding technical requirements for the properties of decorative base paper, such as physical strength, surface properties, absorption, air permeability, etc.

Decorative base paper is different from other paper, after the printing and dipping process, and finally affixed to the surface of the wood-based board to play a decorative role, so it should avoid the phenomenon of exposing the bottom and prevent the color of the wood-based board itself from interfering with the surface color, which requires the decorative base paper to have a very high opacity to improve the covering of the matrix. Secondly, decorative base paper should also have a certain ability to absorb ink and adhesives. The base paper with good absorption has a strong affinity for ink and adhesives, and the ink and adhesives are easily diffused on its surface and inside, while the base paper with poor absorption will produce the phenomenon of uneven distribution of ink and adhesives, and a good evenness is also a necessary condition for uniform absorption of ink and adhesives. In addition, decorative base paper is inevitably strained during the rolling and printing process, and the printing and dipping process inevitably encounters water, so sufficient dry strength and wet strength are also particularly important. The smoothness of the decorative base paper determines the closeness of the contact between the base paper and the printing board, which has a great impact on the printing quality [73]. Smoothness can be improved by calendering operation, but it will interact with absorbency, and the relationship between the two needs to be balanced in production.

Better uniformity, smoothness, permeability, absorption, coverability, and tensile strength are all quality indicators that should be used for decorative base paper. Specific indicators can refer to the national standard GB/T 24989-2010. These key indicators complement each other and restrict each other. In the production process, appropriate preparation materials and production processes should be selected to lay the foundation for the production of qualified and exquisite decorative paper [74].

10.5.3 Functional Materials and Characteristics of Decorative Base Paper

In order to improve some characteristics of decorative base paper, reduce material loss, and meet postprocessing requirements, some materials with functional characteristics are usually added in the production process, such as inorganic fillers, retention agents, wet strength agents, and wear-resistant materials. These functional materials play a very important role in the quality of decorative base paper, production process, and production costs. Generally, they have the characteristics of high added value, special role, or obvious auxiliary role [75], and are essential materials for the production of decorative base paper.

10.5.3.1 Inorganic filler

1) Titanium dioxide

Titanium dioxide (TiO_2), also known as titanium dioxide, white powder, chemical stability, almost no reaction with other elements and compounds at room temperature. There are three crystalline forms of titanium dioxide in nature: Rutile, Anatase, and Brookite. Plate titanium type is an unstable crystal type with a plate shape and has no practical value in industry. Anatase type and rutile type have stable crystal type, which is an important white pigment and porcelain glaze, widely used in paint, plastic, rubber, ink, paper, ceramics, daily chemicals, medicine, food, and other industries. Compared with the anatase type, the rutile type is smaller and tighter because its unit lattice is composed of two titanium dioxide molecules, while anatase type is composed of four titanium dioxide molecules, which has greater stability and relative density, and also has a higher refractive index (Table 10.10) [76]. Refractive index is an important index of titanium dioxide used in decorative base paper, because of the optical property, when the refractive index of the pigment is equal to the refractive index of the surrounding medium, it is transparent, when the refractive index of the pigment is greater than the refractive index of the surrounding medium, it is opaque, the greater the gap between the two, the higher the opacity. Among the white pigments, rutile TiO_2 has the highest refractive index, and the difference of refractive index between it and the surrounding medium (pulp fiber and impregnated resin) is the largest, showing the highest hiding power, so it is the first choice for decorative base paper to improve the hiding power.

Secondly, whiteness and brightness are also major features affecting the application of titanium dioxide. Different particle sizes of titanium dioxide will have

Table 10.10 Refractive indices of common white pigments and resins.

White pigment	Refractive index	Resin	Refractive index
Anatase TiO_2	2.73	polystyrene	1.6
Rutile TiO_2	2.55	Acrylic acid	1.52 ~ 1.60
Kaolin	1.65	Melamine formaldehyde	1.52
Talc	1.65	Pulp fiber	1.57
Calcium carbonate	1.63		
Silica dioxide	1.54		

different scattering differences to the red light, green light, and blue light that the human eye is more sensitive to in visible light, so in the application, the human eye can distinguish different white shades. In general, the smaller the particle size of titanium dioxide, the stronger the scattering of blue light, showing the whiteness of the blue phase. The larger the particle size, the stronger the scattering of red and green light, showing the whiteness of the yellow phase. In addition, the color of titanium dioxide is also greatly affected by its production process. At present, there are two main production processes of titanium dioxide: sulfuric acid process and chlorination process. The sulfuric acid method removes impurities by different crystallization separation of solubility, while the chlorination method removes impurities by high-temperature rectification, and the degree of impurity removal is higher than that of the sulfuric acid method. Relatively speaking, titanium dioxide produced by the chlorination process has higher whiteness and brightness, finer and more uniform particle size, and more stable quality. Therefore, rutile TiO_2 produced by the chlorination process is used in high-grade decorative base paper at home and abroad to obtain stable color quality assurance [77].

Another important advantage of titanium dioxide is that it can absorb UV light, thereby reducing the damage of UV light to other components in decorative paper products and providing certain weather resistance. However, under UV irradiation and anaerobic conditions, titanium dioxide is easy to lose oxygen and grayish phenomenon, resulting in darkening of decorative paper products and reduced gloss. However, the coating treatment of the surface of titanium dioxide by phosphate-aluminum coating can not only effectively solve the phenomenon of gray change, but also improve the retention rate of titanium dioxide in decorative base paper [69, 78]. In the production process of decorative base paper, filler is the second largest component in addition to fiber raw materials. Titanium dioxide has become the most commonly used and ideal filler for decorative base paper with its high whiteness, good gloss, strong hiding power, and good weather resistance.

2) Kaolin

Kaolin ($Al_2O_3{\cdot}2SiO_2{\cdot}2H_2O$), also known as hydrated silicate China clay, white clay, is weathered by feldspar or mica, the chemical composition of $SiO_2$46%, $Al_2O_3$39%, H_2O13%. The crystal structure of kaolin is mostly hexagonal and laminated. Under normal conditions, the chemical properties are stable and not easy to oxidize and deteriorate, with high dispersion and good adsorption, and are widely used in papermaking, ceramics, rubber, and other industries. The appearance of kaolin is white powder, soft texture, whiteness is 80% to 86%, but it contains quartz and mica impurities, which must be purified before use. At present, there are two kinds of purification methods: wind separation method and water washing method, but the kaolin produced by water purification is more uniform, less impurities, and has higher whiteness and brightness, so it is more suitable as a filler to improve the whiteness and opacity of the paper. Although the covering power and optical properties of kaolin are not as good as titanium dioxide, kaolin is rich in reserves and cheap, which can partially replace titanium dioxide to reduce the production cost of decorative base paper. At present, with kaolin as the core and titanium dioxide as the coating layer, titanium dioxide and kaolin composite fillers, which are similar in whiteness and covering power to titanium dioxide, are prepared by mechanical method. Studies have shown that titanium dioxide coating kaolin is not a simple physical adsorption, but a chemical reaction between titanium dioxide and kaolin to form Si—O—Ti and al—O—Ti chemical bonds, so the coating effect is good, and the addition of titanium dioxide to decorative base paper can maintain good whiteness and opacity of the paper [79].

10.5.3.2 Retention Aid

The characteristics of high density and small particle size of titanium dioxide often lead to poor retention of the wet part, and its price is more expensive, which will significantly increase the production cost of decorative base paper. Therefore, it is of great significance to develop a retention system suitable for titanium dioxide.

Paper retention agents can be broadly divided into three categories: inorganic, modified natural products, and polymer. Inorganic retention agents are mainly aluminum sulfate (also known as alum), polyaluminum chloride, bentonite, colloidal silica, etc., which are used more in the early stage, and the retention effect is not obvious. The retention AIDS of modified natural products mainly include cationic starch, chitosan, guar gum, etc. The amount of such products is large, and the retention effect is not very obvious. The main polymer classes are polyacrylamide polymer, polyethylene imide, and polyamine, among which the most commonly used is polyacrylamide polymer. Polyacrylamide retention agents can be divided into cationic type (CPAM) and anionic type (APAM). Because CPAM itself carries cations, it can be directly combined with paper pulp and filler with anionic charge, and adsorb each other through the roles of “neutralization” and “bridging”, resulting in flocculation and retention effect. APAM has the same negative charge as pulp and filler, so it must be combined with aluminum sulfate to have a cationic surface to be effective when used, and it is generally only suitable for acidic papermaking systems, while the use of CPAM is not limited by pH value [80]. Therefore, CPAM is often used as a retention aid in the production of decorative base paper, but as a single retention agent, CPAM can easily produce large flocculants, which affects the evenness of decorative base paper and even leads to the reduction of paper strength [81].

In order to solve the problem of poor retention effect of inorganic and modified natural products, as well as the problem that high polymer is easy to produce too large flocs, people have begun to pay attention to the composition of the composite particle retention system of high polymer and micro and nano inorganic particles in recent years. At present, there are two kinds of particle retention systems commonly used: cationic starch-colloidal silica developed by Eka company in Sweden and cationic polyacrylamide-bentonite developed by Allied Colloids company in the United Kingdom. Both colloidal silica and bentonite particles can be well combined with the polymer retention AIDS in their respective systems to combine the fibers into smaller and more dense flocs, thus achieving a balance between the retention, filtration, and page forming effects of paper materials [82]. Because the mechanisms and effects of different particle retention systems are similar, paper workers often face difficulties in choosing. National Paper Chemical Engineering Technology Research Center Yao Xianping R & D team in order to try to achieve the localization of excellent paper chemicals, as well as to provide a reference for domestic paper workers to reasonably choose the particle retention system, based on the single product provided by domestic manufacturers, The retention and filtration effects of cationic starch-colloidal silica and cationic polyacrylamide-bentonite systems on paper performance were compared. The results show that the two systems can significantly improve the retention effect of paper material. Compared with the cationic polyacrylamide-bentonite system, a better retention effect can be obtained at a small amount, while the cationic starch-colloidal silica system needs to consume a larger amount to achieve a basically consistent retention effect. However, cationic starch-colloidal silica systems are more helpful in maintaining tensile strength [83]. Because the raw materials of decorative base paper are more expensive, it is more necessary to improve the retention of paper material, so the cationic polyacrylamide-bentonite system has been more widely used in the production of decorative base paper.

10.5.3.3 Wet Strength Agent

For decorative base paper, its application must be printed by ink and resin impregnation, and it needs to have higher wet strength than other papers, so it is necessary to find a strong wet strength agent to meet the processing requirements of printing and dipping. At present, the commonly used wet strength agents are urea formaldehyde resin (UF), melamine-formaldehyde (MF) resin, polyethylenimide (PEI), polyamide-epichlorohydrin resin (PAE), and so on. Among them, PEI was first used in paper production, but it has not been widely used due to the shortcomings of high production cost and difficult operation. UF and MF have been banned abroad in recent years due to the harm of free formaldehyde. However, PAE has relatively little influence on the absorptive properties of the finished paper, and its humidification strength performance is better than other wet strength agents, so it has become the most widely used wet strength agent in the production of decorative base paper [84].

As a water-soluble cationic thermosetting resin, under neutral and alkaline conditions, PAE can be adsorbed to the negatively charged pulp fiber to form a polymer network structure, that is, to form a dense water-resistant film, thereby improving the wet strength of paper, with high efficiency, nontoxic, and wide pH range of use characteristics. In addition, because PAE is positively charged, soluble in water, and has a large molecular weight, it also has a certain retention effect, which can improve the retention of fine fibers and fillers [85]. However, PAE is not easy to degrade after curing; it is difficult to reuse damaged paper, and the organic chlorine content is high, which is not conducive to environmental protection. In recent years, the adverse effects of traditional wet strength agents on the environment have increasingly attracted people's attention. Therefore, modifying PAE and sharing it with environmentally friendly additives to reduce its use in order to reduce the environmentally harmful organochlorine content in products and white water systems has become a current research focus. For example, by increasing the molecular weight and positive charge density of PAE, an efficient and environmentally friendly PAE is developed. The modified PAE has the characteristics of high solid content, high positive charge density and low organochlorine content, which can reduce the organochlorine content in the product and white water system, and is conducive to the retention and cross-linking of PAE, and can significantly enhance the wet strength of paper at a lower dosage, which is more than 35% lower than the amount of commercially available PAE. The production cost is greatly reduced [86].

Because of its unique properties, such as nanoscale size, large specific surface area, abundant surface hydroxyl group, excellent biocompatibility, mechanical enhancement, and barrier properties [87], nanocellulose has been widely considered by papermakers in the aspects of paper strengthening, retention, filtration, and coating. Some studies have found that combining nanocellulose and PAE into a binary system can further improve the dry strength and wet strength of paper, and the humidification strength of paper is especially significant [88–90]. The advantage of the binary system is that under the condition of using PAE first, nanocellulose is added to play the effect of its electronegative additives, through the opposite charge adsorption, more PAE is combined into the paper, and the polyelectrolyte composite molecular layer is formed on the fiber surface, which spreads and crosses each other, and increases the binding force between the fibers. Thus, higher paper strength requirements are met [91]. This also shows that under the premise of ensuring the same wet strength, the amount of PAE in the binary system of nanocellulose and PAE is significantly lower than that of adding PAE alone. At present, the research on papermaking wet strength agent has developed from a unitized single component to a diversified multicomponent, and the development and application of multicomponent system will become a new development trend of papermaking wet strength agent.

10.5.3.4 Wear-resistant Materials

Decorative base paper as furniture, decorative board and composite wood flooring and other architectural home decoration products of the finishing material, its wear resistance determines the service life of the final product, especially for the outer surface of composite wood flooring, surface wear resistance is particularly important, so usually in the production of decorative base paper to add some wear-resistant materials and impregnated MF resin to obtain a certain wear resistance. Al_2O_3, commonly known as corundum, has a Mohs hardness of up to 9, second only to diamond and a few special ceramic materials, ranking fourth. Its raw material sources are rich and cheap. In decorative base paper, the higher the amount of Al_2O_3 and the more uniform the dispersion, the better the wear resistance, but the high amount of Al_2O_3 will cause the appearance quality of the paper to decline, dispersion difficulties, and other problems [92]. At the same time, Al O will become transparent after impregnation with MF resin, which will not affect the beauty and use of decorative patterns, so Al_2O_3 has become the first choice to improve the wear resistance of decorative base paper.

In addition, MF resin can also give decorative base paper certain wear resistance. MF resin has stable chemical properties and a large rigid main chain structure, and can form a transparent film with high wear resistance, water resistance, and heat resistance after curing and cross-linking on decorative base paper. Coupled with the advantages of cost-effective, it has become the most commonly used impregnating resin for decorative base paper [93].

10.5.4 The Application of Functional Materials in Decorative Base Paper

10.5.4.1 Add in Pulp

Decorative base paper is usually made of bleached sulphite coniferous wood pulp or bleached sulfate hardleaf wood pulp as raw materials, adding titanium dioxide, kaolin, pigments, etc., and then adding certain additives to the obtained pulp to ensure that there is a large retention, dry and wet strength and absorption, and finally the decorative base paper is copied on the paper machine to meet the requirements. The production process is called in-pulp addition, which roughly includes four steps: pulp crushing, grinding, batching, and papermaking, as shown in Figure 10.27 [94].

1) Pulp crushing

The types and proportions of fibers used in the production of decorative base paper have important effects on the evenness, strength, smoothness, and absorption of

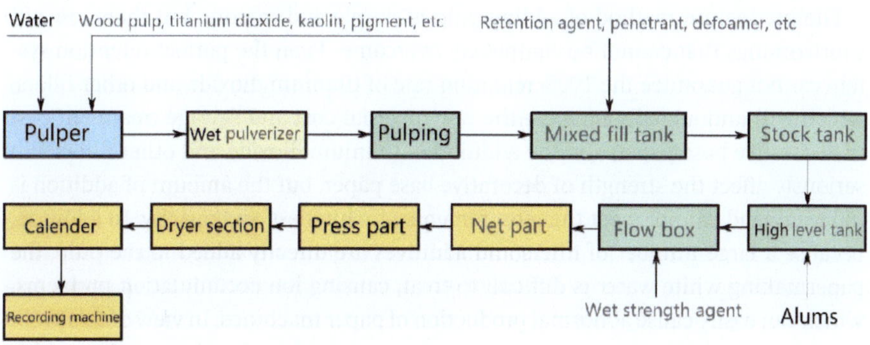

Figure 10.27 Pulp addition production process.

decorative base paper. The longer fiber of coniferous pulp can improve the strength of paper to a certain extent, but it will reduce the evenness and smoothness of paper. The fiber of broadleaf pulp is short and fine, and the paper made is good in evenness and smoothness, but the strength is poor. In terms of absorbency, broadleaf pulp is much higher than coniferous pulp. Therefore, 10% coniferous pulp and 60% broadleaf pulp are used in production to balance the properties of decorative base paper. In addition, a large amount of titanium dioxide must be added during the production process (addition amount of 35% to 45%) to ensure sufficient opacity [95], but in order to reduce costs, a certain amount of kaolin (5% to 15%) is usually added to reduce the amount of titanium dioxide [96]. Finally, through the pulp crushing equipment, wood pulp, titanium dioxide, kaolin, pigment, and water are prepared into a slurry with a concentration of 6% to 6.5%.

2) Grinding

In the grinding equipment, the pulp is subjected to friction and compression by a high-speed rotating plate. This action cuts and fibrillates the pulp fibers, achieving the desired finish while ensuring a more uniform and intimate mixture of the wood pulp, titanium dioxide, and pigments.

3) Ingredients

After grinding, the slurry is sent to the mixing and filling tank for further dilution, and various necessary additives are added to the slurry to make it evenly mixed with the slurry. In the slurry after grinding, 3% ~ 5% wet strength agent (PAE), 0.2% ~ 0.4% retention agent (cationic polyacrylamide-bentonite system), 0.1% ~ 0.3% alum, 0.05% ~ 0.15% were added to the slurry after grinding Penetrant (nonionic surfactant), 0.05% ~ 0.15% other additives (such as polyhydroxy-polyether defoamer), prepared into a concentration of 2% ~ 4% slurry.

4) Papermaking

The prepared stock is obtained by the paper machine mesh, press, drying, calender, and winder to meet the requirements of the decorative base paper.

The production method of adding pulp is simple and mature, but there are still shortcomings that cannot be completely overcome: Even the perfect retention system cannot guarantee the 100% retention rate of titanium dioxide and other fillers, which will undoubtedly increase the raw material cost and sewage treatment cost of decorative base paper, and the addition of titanium dioxide and other fillers will seriously affect the strength of decorative base paper, but the amount of addition is too small and cannot meet the requirements of whiteness and opacity. In addition, because a large number of fillers and additives are directly added to the pulp, the papermaking white water is difficult to treat, causing ion accumulation problems, which can easily cause abnormal production of paper machines. In view of this problem, in recent years, papermakers have put forward the way of surface coating, which realizes 100% retention of the filler and greatly saves the production cost.

10.5.4.2 Surface Coating

The surface coating production process is to apply titanium dioxide, kaolin, pigment, and other mixed coating liquid on the surface of decorative base paper, as shown in Figure 10.28. Among them, a certain amount of adhesive should be added to the preparation of the mixed coating solution to ensure sufficient viscosity. Because the decorative base paper has a high water absorption, the paper can be soaked in a uniform amount of adhesive, so the selection of adhesive in the coating solution should avoid the waterproof adhesive as far as possible. The results show that cationic starch (CS) and polyvinyl alcohol (PVA) have a good coating effect when used together. For the specific coating solution formula, see Table 10.11.

After the decorative base paper is coated on the surface, the coating can fill the surface pores of the paper, the paper fiber is uniformly coated, and a good, uniform, and tight coating structure can be formed, which can improve the printability. In addition, because titanium dioxide and other fillers are not directly added to the slurry, the surface coating makes the fillers concentrated on the surface of the paper, maximizing the hiding performance of titanium dioxide, reducing the amount of titanium dioxide, greatly saving the production cost, and significant economic benefits. The wet strength agent PAE is also coated, so that the PAE is not lost, and the amount of PAE is also reduced. Secondly, fillers and auxiliaries are not added in the papermaking process, which makes the ions and impurities in the papermaking and

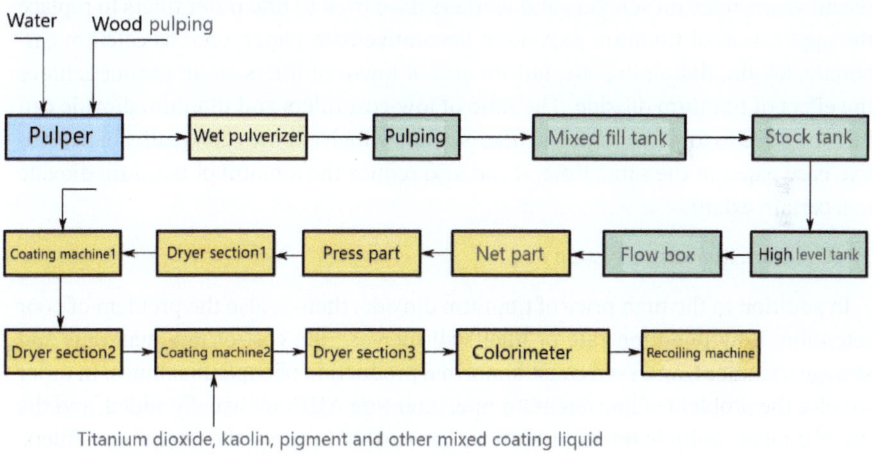

Figure 10.28 Surface coating production process.

Table 10.11 Decorative base paper mixed coating solution formula.

	Coated packing							
Material	Titanium dioxide	Kaolin	CS	PVA	Sodium polyacrylate	Ammonium hydroxide	Defoaming agent	Glycerinum
ratio	80	20	9	11	0.45	0.5	0.3	0.5

pressing process less, and the paper wastewater is easier to treat and reuse, reducing the environmental pressure and sewage treatment cost [97].

10.5.5 Outlook

Domestic decorative base paper is currently dominated by low-end products, and high-end decorative base paper still needs to rely on imports. The lower end of the market is highly competitive and less profitable. With the improvement of people's living standards, the demand for the high-end market continues to increase, and the development prospect is broad. Therefore, decorative base paper manufacturers should strictly control product quality and shift product positioning from low-end to medium-high-end. In addition, decorative base paper is an upstream product of wood-based panels, and its quality directly affects the development of the wood-based panel finishing industry. At present, there are still quality problems such as low whiteness, high production cost, and poor printability in domestic decorative base paper, and it is suggested that it can be improved through the following directions in the future:

1) Packing

Titanium dioxide is currently the best filler choice for the production of decorative base paper, but its price is expensive and will increase the production cost. In recent years, relevant scholars and workers have tried to find other fillers to replace the application of titanium dioxide in decorative base paper, such as calcium carbonate, kaolin, diatomite, etc., but the use of low-cost fillers alone cannot achieve the effect of titanium dioxide. The ratio of low-cost fillers and titanium dioxide can be controlled to form a composite filler system, which ensures the quality of decorative base paper at the same time. It can also reduce the amount of titanium dioxide to a certain extent.

2) Retention assistance system

In addition to the high price of titanium dioxide, there is also the problem of poor retention. Low retention rate of filler will increase the cost of raw materials and sewage treatment, and even cause abnormal production of paper machines. In order to solve the problem of low retention rate, retention AIDS are usually added, and the use of dual or multiple retention systems can further improve the retention of fillers. Therefore, the development of multiple retention systems will be the focus of future research to improve filler retention.

3) Surface coating

The surface structure of decorative base paper affects the printability of decorative base paper to a large extent. Studies have shown that the surface coating of decorative base paper with SiO_2 coating can effectively fill the surface pores of decorative base paper and improve its smoothness, wet and dry strength, and printability [98]. Therefore, the surface coating and surface treatment process should be further optimized on the basis of the existing research, which is of great significance to improve the decorative base paper in the decorative application and printing performance of wood-based board.

In the future, with further understanding of functional materials for decorative base paper and continuous optimization of the production process, in the near future, the product quality and technical level of domestic decorative base paper are expected to steadily improve, gradually replace imported products, and better meet the domestic market demand.

10.6 Thermal-sensitive Paper

10.6.1 Introduction

Thermal-sensitive paper, also known as thermal recording paper or thermal copy paper, is a kind of special-coating paper. By applying heat, the materials in the thermal functional layer of the thermal-sensitive paper undergo physical or chemical reactions to display images, texts, and barcodes. Different from traditional printers, thermal printers do not need ink, toner, thermal ribbons, and other consumables.

10.6.1.1 Development

In the 1930s, people used a hot pen as a heat source to record images and text on waxy thermal-sensitive paper, which was mainly used in electrocardiogram (ECG) recorders. In the 1950s, 3 M (Minnesota Mining Ink Character) launched "Thermo-fax", based on the reaction of thiocarbamide and heavy metal salts, can develop colors and record information with infrared heating. In the 1960s, NCR Company (National Cash Register Company) introduced a new thermal recording paper with leuco dye. Depending on thermal printing components miniaturization and thermal printers upgrading, leuco dye reactive type of thermal-sensitive paper has gradually replaced other types of thermal-sensitive paper and has become the most popular choice in the new era due to the advantages of simple production process, stable quality, and low cost.

10.6.1.2 Thermal Printing Mechanism

Based on the color development mechanism, there are two types of thermal printing – physical reaction type and chemical reaction type. The representative of the early physical reaction type is wax thermal-sensitive paper, which is composed of a paper substrate, coloring layer, and opaque wax layer. When the print head is heated and transfers heat energy to the paper surface, the opaque wax layer melts and becomes transparent, then the coloring layer shows and forms an image (Figure 10.29). Wax thermal-sensitive paper is rarely used because of its disadvantages, such as high printing temperature requirements and print heads that wear out easily. Nowadays, the leuco dye thermal-sensitive paper of chemical color development type, which is usually composed of a paper substrate and thermal functional layer, is the majority of thermal-sensitive paper on the market. During the thermal printing process, the color is developed by a chemical reaction between

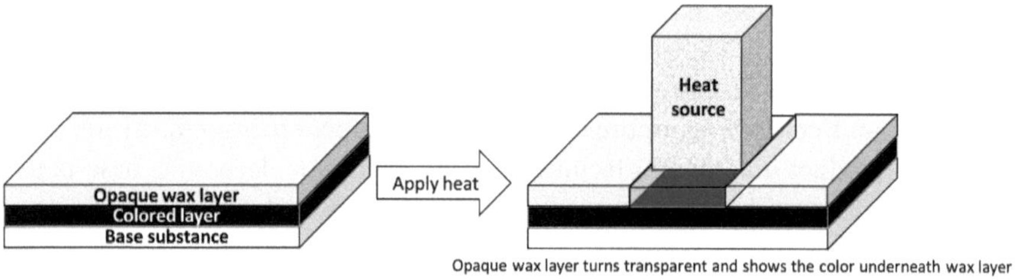

Figure 10.29 Schematic of the physical type of thermal-sensitive paper.

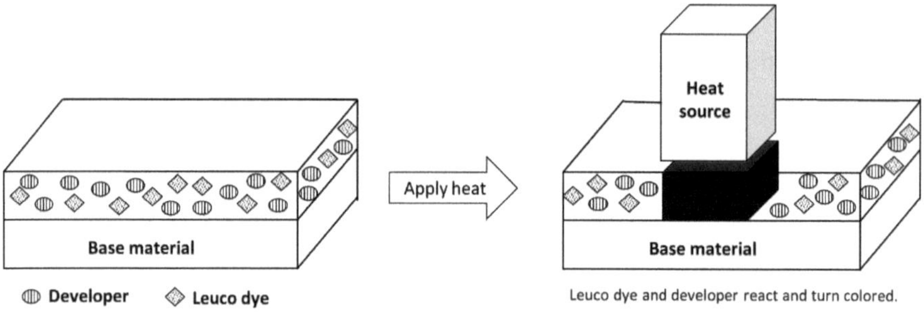

Figure 10.30 Schematic of the chemical type of thermal-sensitive paper.

the leuco dye and the color developer in the functional layer (Figure 10.30). For example, the color developer provides protons and makes the leuco dye protonated. After the leuco dye protonates, the lactone ring opens, leading to the colorless leuco dye becoming colored, and an image is formed on the paper.

10.6.1.3 Application

The popularization of information technology promotes the update and iteration of terminal output equipment, so that the use of thermal recording paper continues to expand in various fields.

1) **The Application of Cash Register**

As the retail industry keeps growing, the cash registers and terminal equipment applied availability rate continuously increases in malls, supermarkets, catering, entertainment, etc., which leads to the demand for thermal-sensitive paper growing rapidly. In order to adapt to the fast-paced lifestyle, the rapid response of printing equipment and the fast recording of printing paper put forward higher requirements. Thermal-sensitive paper, with the advantages of short printing time, high printing efficiency, and a single layer without additional consumables, is rapidly replacing the carbonless carbon paper, dominating the POS terminal printing field.

2) **The Application of Labels**

Direct-thermal label is mainly used in supermarkets, medical, logistics, and other scenes. In particular, the rise of cloud and big data has prompted a volume increase in network marketing. Meanwhile, it is also a challenge for the timeliness of logistics. The innovation of direct-thermal labels has enhanced the efficiency of the logistics and express delivery industry, and also reduced the manufacturing cost of express delivery forms. Direct-thermal labels are mainly used for information recording, and the bar code is required to be recognizable in different use environments. Therefore, the label is required to have excellent barcode clarity and durable image preservation.

3) **The Application of Tickets**

The iterative optimization of thermal-sensitive paper performance has expanded its application in the field of tickets, such as boarding passes, transportation tickets, various game tickets, performances, and gambling papers. Most of the information carried by thermal-sensitive paper has certain economic or collection values. Therefore, the thermal-sensitive paper used in these applications requires long-term image preservation and higher environmental tolerance.

4) **Manufacturer of Thermal-Sensitive Paper**

The expansion of the application field of thermal-sensitive paper has promoted the rapid growth of the demand for thermal-sensitive paper. At present, the main domestic thermal-sensitive paper manufacturers are Gold East Paper (Jiangsu) Industry Co., Ltd, Guangdong Guanhao High-Tech Co., Ltd, Henan Jianghe Paper Co., Ltd, Xianhe Co., Ltd, Chenming Group, and other enterprises. In addition, there are many small thermal-sensitive paper factories that have flourished in China. The international market of thermal-sensitive paper has also maintained a growing trend. Besides, Hansol Group, Koehler paper Group, OJI paper, Ricoh paper Group, and an increasing number of Chinese thermal-sensitive paper manufacturers are entering the global market to meet the worldwide demand for thermal-sensitive paper.

10.6.2 Quality and Requirements

Thermal-sensitive paper, as a medium for recording information, differs mainly from other recording paper in its color development method. Specifically, it relies on the thermal printer to apply energy to the paper's surface and print the desired text and images. Therefore, the print quality of thermal-sensitive paper is a key indicator of product quality. There are two indicators, static and dynamic sensitivity (Figure 10.31), of which, dynamic sensitivity better reflects the actual conditions during thermal printing.

Thermal printing has various application demands. For example, receipt printing requires a fast feeding speed to reduce customers' waiting time at checkout, while label printing demands high-quality barcode clarity to ensure it can be scanned easily. To meet the requirements, the Chinese National Standard also has two indicators to evaluate. First is image sensitivity, which measures the imaging rate (optical density, O.D.) at different energy levels—the energy level at which the O.D.

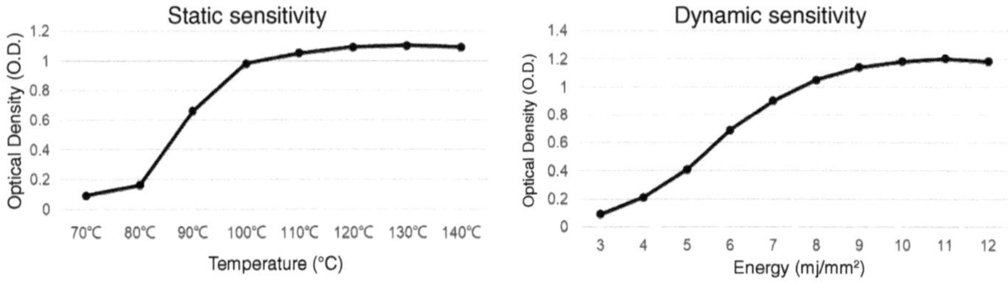

Figure 10.31 Static sensitivity (left), and dynamic sensitivity (right).

reaches 1.0. The other one is maximum optical density, which reflects the imaging ability (O.D.) when enough energy is applied.

Currently, most thermal-sensitive papers on the market are chemical reaction type, which has instability in its reaction product. The printed images on thermal-sensitive paper will fade over time, especially when it is exposed to intense light, higher temperature, high humidity, or contact with organic solvents. Therefore, image preservation under various conditions is also a quality criterion for thermal-sensitive paper. Short-term items like receipts and queue tickets, labels, lottery tickets, and transportation tickets may face more complex storage environments and require longer image preservation.

Beyond thermal printability, thermal paper—being a grade of cultural paper—is also subject to standard property requirements such as brightness, whiteness, smoothness, and stiffness.

Additionally, environmental safety in thermal-sensitive paper is also been considered. The latest Chinese National Standard, effective on April 1st, 2025, has demanded strict limits on Bisphenol A (BPA) content, <200 mg/kg for normal thermal-sensitive paper, and nondetectable for special thermal-sensitive paper.

The requirements of thermal-sensitive paper properties and environmental safety drive to strictly select materials and production process, and further enhance both the functionality and environmental safety of these daily used products.

10.6.3 Thermal Functional Materials

The most commonly used thermal-sensitive paper is a reaction type. According to its reaction mechanism and application requirements, a variety of functional materials are used.

10.6.3.1 Thermal-Sensitive Chemicals

The most important functional materials are thermal-sensitive chemicals, including developer, sensitizer, and leuco dye. As heat is applied, the developer protonates the leuco dye, causing a molecular rearrangement. This change switches the leuco dye from colorless to a colored form (Figure 10.32), making the image appear.

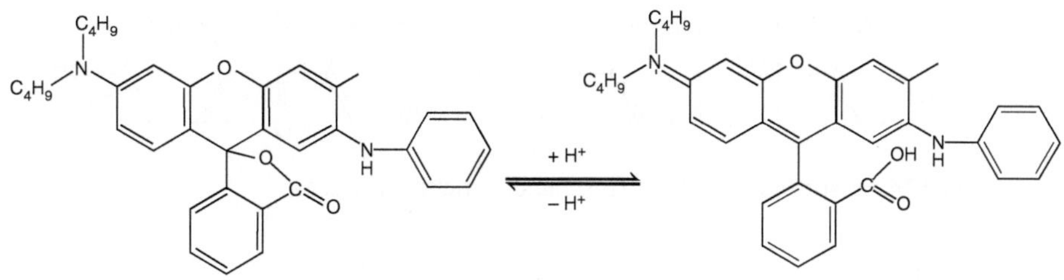

Figure 10.32 Leuco dye (ODB-2) development mechanism.

1) **Leuco Dye**

Leuco dye is an organic molecule, usually contains lactones or heterocyclic compounds, and is a major chemical that provides thermal imaging function. When leuco dye is protonated, its lactone ring opens and it switches from colorless (white or light color) to a specific color, such as black, blue, and red.

Lactone compounds are a class of molecules containing a lactone ring, typically featuring a carboxyl group (-COOR). Lactone compounds usually have a conjugated π system, which changes in length when the ring structure opens, allowing π electrons to move freely throughout the molecule and altering its light absorption properties. The lactone structure could stabilize the excited state through resonance, producing a specific absorption spectrum. The wavelength of the absorbed light determines the color of the compound.

The most used leuco dye in thermal-sensitive paper is ODB-2 (one dye blank 2, 2-Anilino-6-[dibutylamino]-3-methylfuoran). When ODB-2 is protonated, it switches from colorless (white) to colored form (black). There are other common black leuco dyes like ODB (one dye black, 7-Anilino-3-diethylamino-6-methyl fluoran) and S-205 (3-N-Isoamyl-N-ethylamino-6-methyl-7-anilinofluoran). Moreover, crystal violet lactone (CVL), 3,3-bis-4-Dimethylaminophenyl-6-dimethylaminophthalide) is a leuco-type to form a blue color (Figure 10.33).

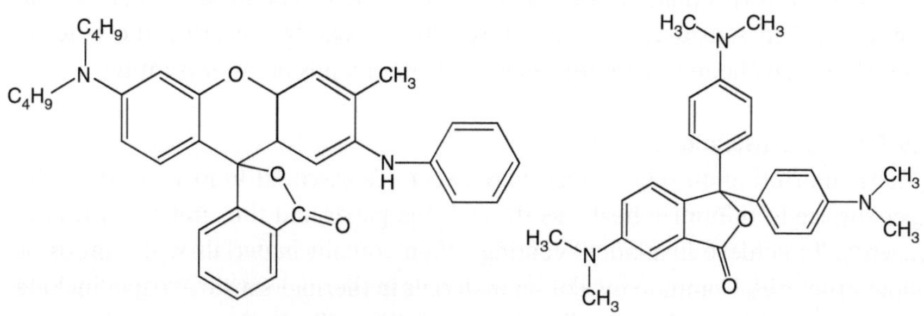

Figure 10.33 Leuco dye ODB-2 (left) and CVL (right) structure.

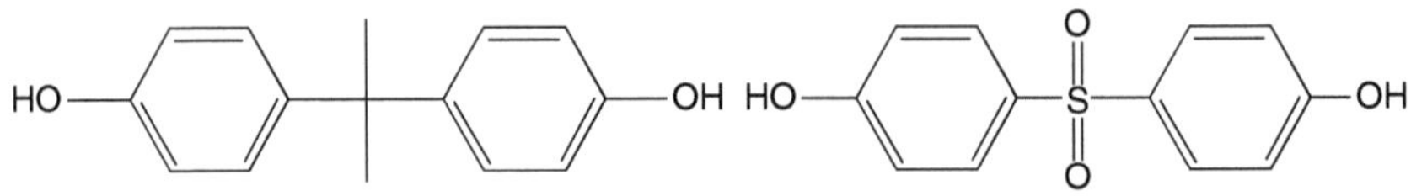

Figure 10.34 Bisphenol A (left) and Bisphenol S (right) structures.

2) **Developer**

Developer is a proton provider that can react with leuco dye in the molten state. Generally, developers are phenolic compounds. The most common developer is BPA (Bisphenol A, 4,4′-dihydroxy-2,2-diphenylpropane). Bisphenol A (BPA) has been widely used in thermal-sensitive paper due to its low cost. However, due to health concerns associated with BPA (potential threats to human health), some countries and regions, such as Japan and the European Union, have mandated the prohibition of BPA in thermal-sensitive paper. The Chinese national standard “GB/T 28210-2024 Thermal-sensitive paper” has also revised the BPA content restriction. As the restriction of BPA, Bisphenol S (BPS; 4,4′-dihydroxydiphenyl sulfone) has acted as a main alternative to BPA (Figure 10.34).

Apart from common bisphenol-based developers like BPA and BPS, there are other types of developers available. These include mono-phenolic developers such as BPS-MAE (4-[[4-[2-allyloxy] phenyl] sulfonyl] phenol), DBSP (2,4-diphenylsulfonylphenol), and D-8 (4-hydroxy-4′-isopropoxydiphenyl sulfone), and phenol-free developers like PF-201 (N-(p-toluenesulfonyl)-N′-(3-p-toluenesulfonyl-oxyphenyl) urea) and BTUM (4,4″-bis [p-toluenesulfonylurea]-diphenylmethane). Different types of developers can be selected based on different requirements.

3) **Sensitizer**

Sensitizers are chemicals that help developers and leuco dyes melt and react more quickly to generate color. Common sensitizers include BON (2-[phenylmethoxy]-naphthalene), DPE (1,2-diphenoxyethane), DPS (diphenyl sulfone), EGTE (1,2-bis [3-methylphenoxy] ethane), and stearamide. These chemicals enhance the color-developing sensitivity of thermal-sensitive paper by lowering the eutectic point of thermal chemicals, thereby meeting the demands of faster printing.

10.6.3.2 Heat Insulation Material

In thermal printing on thermal-sensitive paper, it’s essential to focus heat on the paper surface to minimize heat loss through the paper and thus improve printing efficiency. To achieve insulation, coatings often contain materials with porous or hollow structures. Common insulation materials in thermal-sensitive paper include inorganic fillers like calcined kaolin and organic fillers like hollow plastic pigments.

Insulation materials generally have porous or hollow structures and include substances like silica, diatomaceous earth, calcined kaolin, PCC (precipitated calcium carbonate), and ground calcium carbonate (GCC). Different porous materials offer varying insulation properties due to differences in pore structures.

Considering production costs and price differences, calcined kaolin is currently the most commonly used inorganic insulation material in thermal-sensitive paper. The primary organic insulation materials are hollow plastic pigments and foamed microsphere. However, due to the higher costs and application challenges of foamed microspheres, hollow plastic pigments are more widely used.

1) **Calcined Clay**

Calcined clay is a type of kaolin clay that has been heated at high temperatures. Kaolin is a natural mineral, forming a white, soft clay primarily composed of aluminosilicate minerals. During the calcination process, kaolin is heated to approximately 500 °C to 1100 °C, changing its physical and chemical properties. This process creates numerous micropores, transforming kaolin from its original plate-like or layered structure into an irregular and porous form.

This microporous structure and irregular particle shape give calcined clay enhanced insulating properties, which help reduce heat energy waste by preventing the direct penetration of heat through the base paper during thermal printing. In addition, the porous structure and high specific surface area of calcined clay give it a high oil absorption value, which helps to quickly absorb the melted material in the coating. This enhances the clarity of text and images during thermal printing.

2) **Hollow Plastic Pigment**

Hollow plastic pigments, also known as hollow spheres, are hollow particles made with plastic materials, commonly composed of styrene-acrylic compounds, as the shell material. The primary characteristic of these hollow spheres is their internal hollow structure. Since the air in the sphere has low thermal conductivity, these hollow spheres provide excellent heat insulation (Figure 10.35).

10.6.3.3 Other Functional Materials

1) **Lubricant**

The lubricants used in thermal-sensitive paper are typically lubricating materials like calcium stearate, zinc stearate, or paraffin. These materials have relatively low melting points, so when thermal-sensitive paper undergoes thermal printing,

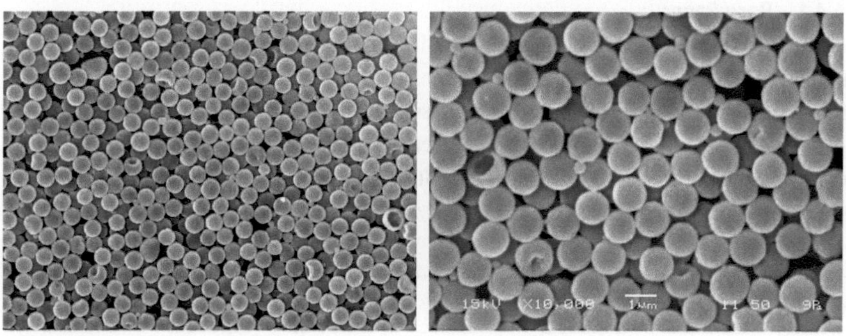

Figure 10.35 Hollow plastic pigment, from gold east paper.

the lubricant melts upon heating, providing early lubrication to the print head. This helps ensure smooth paper movement during thermal printing and reduces the risk of sticking.

2) **Protect Material**

For certain types of thermal-sensitive paper that require long-term image retention or are used in complex environments, such as lottery tickets, logistics labels, food labels, medical records, and securities, additional protective barriers are applied beyond the standard color-developing process. These barriers provide waterproofing, oil resistance, alcohol resistance, and other protections, ensuring that images and text remain readable for a certain period under varying conditions.

Such protective materials typically involve high polymers like polyvinyl alcohol or polyacrylic compounds, which, through a coating process, form a dense, mesh-like layer on the thermal coating surface, thereby safeguarding the printed images. To meet specific surface processing requirements, functional additives, like print performance enhancers, may also be incorporated into the polymer coating layer.

3) **Others**

In addition to the specialized functional materials commonly used in thermal-sensitive paper processes, other conventional coating components, such as adhesives, fillers, and other additives, are also selected and formulated based on the functional requirements of the coating. As a type of coated paper, the types and ratios of these materials are adjusted to meet the specific demands of thermal-sensitive paper applications.

10.6.4 Production Process

Thermal-sensitive paper is a type of multilayer coated paper. Through coating equipment, several layers (usually two to four layers) with different functional coatings are applied to the paper substrate to meet the functional requirements of thermal-sensitive paper (Figure 10.36).

10.6.4.1 Base Paper

Base paper is the foundation of thermal-sensitive paper, and it is a kind of lightweight coated (LWC) paper, which demands higher paper qualities. For instance, to ensure printing clarity, it requires good smoothness so that the coating can cover evenly. And, low wet-deformation could prevent curling after multilayer coating on one side. Moreover, sufficient paper strength reduce breaks in the paper during the coating process, etc.

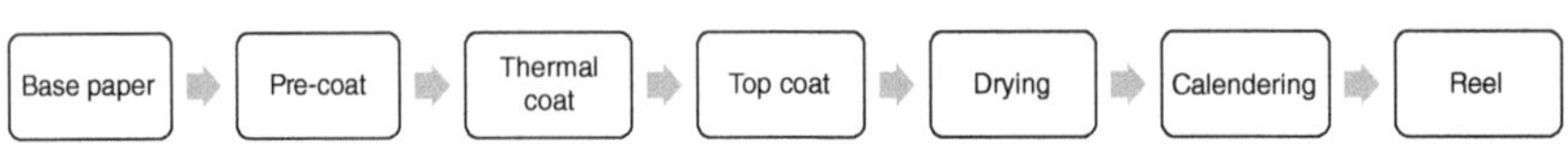

Figure 10.36 Process of thermal-sensitive paper production.

10.6.4.2 Precoat

By coating a layer of heat insulation coating on the surface of the base paper, it can not only provide the heat insulation function but better the smoothness of the paper surface, which helps heat remain on the thermal layer without heat loss through the paper. There are several ways to coat precoat, such as rod coater, blade coater, air knife coater, curtain coater, etc.

10.6.4.3 Thermal Coat

It is the main functional layer of thermal-sensitive paper, made by mixing various pretreated thermal materials in accordance with a certain proportion, and adding adhesive, filler, and other additives to prepare a thermal-sensitive coating, which is covered on the precoat. There are several ways to coat thermal coating as well, like rod coater, air knife coater, curtain coater, etc. Among them, curtain coating has become a more widely chosen coating method for coating the thermal layer in recent years, due to its advantages—no direct contact with the paper surface, no damage to the coating, and provides uniform distribution of the coating along the base paper shape, etc.

10.6.4.4 Top Coat Layer

For certain special thermal-sensitive paper with high requirements for image preservation, one or more layers of polymer-structured protective coating are applied on the thermal surface to extend the image preservation time.

10.6.4.5 Drying

Because of its temperature-sensitive characteristic, the temperature control of the drying process is very important. The common drying method of thermal-sensitive paper is hot air drying. After coating, the paper is dried through a number of dryers, in which the temperatures are gradually reduced. The paper surface temperature should be strictly controlled to ensure that the paper will not preimage.

10.6.4.6 Calendering

Calendering is also an important process for smoothing the paper surface. Higher smoothness can not only improve the thermal printing quality but also reduce the wear of the print head. However, it is necessary to strictly control the nip pressure to avoid preimage caused by frictional heating and affect the quality of the paper.

10.6.5 Prospect of Industry

After decades of development, the technology of thermal-sensitive paper has achieved high technical maturity and commercialization. The reaction type of thermal-sensitive paper is widely used in logistics, transportation, medical, and other fields. In the digital era, e-commerce logistics has driven demand for thermal-sensitive paper, but the trend of "paperless" has also impacted its growth. To ensure the sustainable development of the thermal-sensitive paper industry, it is essential to explore its application in other fields. On the one hand, the limitations of thermal-sensitive paper in practical applications, high preservation, high weather resistance, and multifunction are the future development trend of thermal-sensitive paper to expand it into new usage applications. On the other hand, with the innovations in new materials, it not only meets the usage needs but enhances the safety and environmental friendliness by breaking through the original imaging mechanism. It is also the future technology that the thermal industry focuses on.

References

1 Zhao Ping, S. and Xia, L.W. (2008). Additives in household paper. *Shanghai Paper & Paper* 39 (4): 51–56.
2 Konuma, A. Tissue paper containing liquid drug: EP1985755B1. 2011-2109-2121.
3 Bret, B., Leboeuf, J.F.Softening lotion composition, use thereof in papermaking, and resulting paper product: EP 0882155B2. 2005-2011-2009.
4 Wang, Xindong. A moisturizing paper towel and a preparation method thereof: CN107049834. 2017-2102-2116.
5 Zheng, T.Q. (2005). Correct coating method and applicable equipment of softener and cream. *Household paper* 7: 41–42.
6 Fujishima, A. and Hongda, K. (1972). Electrochemical photocatalysis of water at a semiconductor electrode. *Nature* 238: 53–58.
7 Shouxin, L. and Hong, L. (2006). *Fundamentals and Applications of Photocatalysis and Photocatalysis*. Beijing: Chemical Industry Press.
8 Tinglian, W. (2001). Hydrogen energy. *Energy Technology* 22 (3): 96–98.
9 Ye, J., Zou, Z., and Matsushita, A. (2003). A novel series of water splitting photocatalysts NiM2O6 (M = Nb, Ta) active under visiblelight. *International Journal of Hydrogen Energy* 28: 651.
10 Zou, Z., Ye, J., Sayama, K. et al. (2001). Direct splitting of water under visible light irradiation with an oxide semiconductor photocatalyst. *Nature* 414: 625.
11 Yong, Z., Yan-Ying, L., and Xiao-E, L. (1999). Nano titanium dioxide – – a new type of inorganic antibacterial agent. *Modern Chemical Industry* 19 (8): 46–48.
12 Qing, S.S. (1999). Application of photocatalytic reaction in environmental protection. *Environment and Development* 14 (3): 15–18.
13 Li, G., Zhang, L., and Fang, M. (2012). Facile fabrication of sodium titanate nanostructures using metatitanic acid ($TiO_2 \cdot H_2O$) and its adsorption property. *Journal of Nanomaterials* 875295.
14 Ying, L. and Liqin, W. (2011). Research progress on the application of nanomaterials in cultural relics protection. *Materials Review* 25 (Supplement): 34–37.
15 Lisong, Y. and Hui, S. (2000). Research progress and application of titanium dioxide photocatalysis. *Materials Review* 14 (12): 23–25.
16 Okamoto, K.J., Yamamoto, Y., and Tanaka, H. (1985). Heterogeneous photocatalytic decomposition of phenol over TiO_2 powder. *Bulletin of the Chemical Society of Japan* 58 (7): 2015–2022.

17 Linsebigler, A.L., Lu, G., and Yates, J.T. (1995). Photocatalysis on TiO_2 surfaces: principles, mechanism, and selected results. *Chemical Reviews* 95: 735–758.

18 Fengyu, S. (1998). The relationship between the size of titanium dioxide and photocatalytic activity. *Chinese Journal of Catalysis* 19 (3): 229–233.

19 Xinquan, X. (2012). Analysis of environmental protection problems in residential building decoration. *East China Science and Technology* 5: 131–132.

20 Wei, L. (2013). *Study on the Influence of Indoor Temperature and Noise on Work Efficiency*. Hangzhou: Zhejiang University of Science and Technology.

21 Wei, L. (2004). *Investigation and Analysis of Indoor Air Pollutant of Newly Renovated Residential - - Formaldehyde Pollution and Prevention Countermeasures*. Changchun: Jilin University.

22 Nian, L.S. (1999). Concern about indoor air pollution. *Labor Safety and Health* 9: 21.

23 Jingya, L. and Hong'en, L. (2010). Harm and Prevention of air pollution to human health after living room decoration. *Chinese Community Physicians: Medical Specialty* 12 (13): 220–220.

24 Feiyi, D. (2016). *Research on Development and Properties of Decorative Fabric for Air Purification by Negative Ions*. Xi 'an: Xi 'an Polytechnic University.

25 Hua, Q. and Haixia, D. (2007). Relationship between indoor air pollution and human health. *Environmental and Occupational Medicine* 24 (4): 426–430.

26 Fang, F., Zhan'en, Z., and Lijun, Z. (2001). Study on indoor pollution caused by building and decoration materials. *New Building Materials* 12: 39–41.

27 Ying, T. (2010). Study on indoor air pollution to human health and its prevention and control. *Anhui Journal of Preventive Medicine* 4: 304–306.

28 Zhangcai, H. and Zhiqian, Z. (2012). Source and control of indoor environmental pollutants. *Urban Construction Theory Research* 3.

29 Peng, T. and Yuehua., L. (2014). Study on the correlation between interior decoration materials and indoor environmental pollution. *Building Materials and Decoration* 21: 249–250.

30 Qiuna, Y. (2016). Study on the harm of indoor ambient air pollution to human body and its prevention and control. *Current Medicine* 22 (28): 11–12.

31 Fukahori, S., Ichiura, H., Kitaoka, T. et al. (2003). Capturing of bisphenol a photodecomposition intermediates by composite TiO_2-zeolite sheets. *Applied Catalysis B Environmental* 46 (3): 453–462.

32 Ichiura, H., Kitaoka, T., and Tanaka, H. (2003). Removal of indoor pollutants under UV irradiation by a composite TiO_2-zeolite sheet prepared using a papermaking technique. *Chemosphere* 50 (1): 79–83.

33 Ichiura, H., Kitaoka, T., and Tanaka, H. (2003). Photocatalytic oxidation of NOx using composite sheets containing TiO_2 and a metal compound. *Chemosphere* 51 (9): 855–860.

34 Ichiura, H., Okamura, N., Kitaoka, T. et al. (2001). Preparation of zeolite sheet using a papermaking technique Part II The strength of zeolite sheet and its hygroscopic characteristics. *Journal of Materials Science* 36 (20): 4921–4926.

35 Xiaoting, S. and Ya, G. (2016). Research status and application of aramid fiber. *Journal of Chengdu Textile College* 33 (3): 164–168.

36 Aimin, T., Chaofeng, J., and Xin, W. (2010). Effect of hot pressing process on strength properties of para-aramid paper. *Paper Science and Technology* 29 (6): 61–64.

37 Wensheng, W. and Shaohui, Y. (2009). Development status of fresh packaging at home and abroad. *Preservation and Processing* 9 (4): 44.

38 Xiaolin, Z. and Guoshun, Y. (2010). Review on application of Chinese herbal extracts in fruit preservation. *Hunan Agricultural Sciences* 11: 90–92.

39 Shouqing, Y. (2001). Study on the preservation of Wuxi peach at room temperature by AF atmosphere preservation paper. *Journal of Food and Biotechnology* 20 (6): 573–577.

40 Xianya, Z. (2003). New type of fruit preservation paper with high safety. *Hunan Packaging* 2: 45–45.

41 Bin, W., Linlin, C., Zhonghong, W. et al. (2012). Development and application of 1-methylcyclopropene preservative paper. *Food Science* 33 (24): 343–347.

42 Xiuyu, M., Yufeng, W., Jianqing, W. et al. (2017). Preparation of 1-MCP fresh-keeping paper and its application in nectarine preservation. *Chinese Journal of Packaging* 9 (1): 79–84.

43 Xisheng, S., Wenhui, W., Zhihua, W. et al. (2003). Effects of 1-MCP on postharvest physiology of apple. *Journal of Fruit Science* 20 (1): 12–17.

44 Lizheng, S., Gongnian, X., and Huifang, Z. (2010). Application of functional paper materials in fruit preservation. *Journal of Zhejiang University of Science and Technology* 22 (6): 507–511.

45 Yu, Hao. A kind of nano fresh-keeping carton: CN204489574U. 2015-2107-2122.

46 Wenyi, B., Chen, X., Fei, S. et al. (2015). Preparation of cellulose/chitosan blended transparent film and its antibacterial barrier properties. *Journal of Macromolecules* 1: 49–56.

47 Ščetar, M., Kurek, M., and Galić, K. (2010). Trends in fruit and vegetable packaging – a review. *Croatian Journal of Food Technology, Biotechnology and Nutrition* 5 (3/4): 69–86.

48 Jianqing, W. and Bin, L. (2008). Study on the effect of sulfur dioxide sustained-release preservative on cherry preservation. *Packaging Engineering* 29 (11): 11–12.

49 Yonghua, Z., Xinguo, S., Yunbo, Y. et al. (2000). Effect of SO_2 on refrigeration effect of loquat. *Journal of Nanjing Agricultural University* 23 (2): 89–92.

50 Shuli, Y., Da, W., and Yike, Z. (2015). Development of cinnamon essential oil microcapsule antibacterial paper and its preservation effect on Cherry Tomatoes. *Packaging Engineering* 36 (13): 47–51.

51 Shukla, K. Fenugreek impregnated material for the preservation of perishable substances: US6372220 B1. 2002-2004-2016.

52 Junying, L. and Xiyuan, G. (2016). Research progress on detection of key pathogenic microorganisms of fruit spoilage. *Journal of Food Safety and Quality Inspection* 9: 3510–3515.

53 Song, Jianmin, Wang, Dehai, Huang, Xiangjun, et al. Fruit and vegetable storage active fresh-keeping carton and its preparation method: CN109695179A. 2019-2104-2130.

54 Luo, Laikang, Xu, Guang, Guo, Fengyuan. Fruit and vegetable fresh-keeping agent, fruit and vegetable fresh-keeping paper and their preparation method: CN102349571A. 2012-2102-2115.
55 Min, L., Guangxue, C., Zhaohui, Y. et al. (2017). Research progress of fruit and vegetable preservation materials. *Chinese Journal of Packaging* 9 (5): 86–94.
56 Tuerdi, Shen Jizeng, Jin, Tao, et al. Laminated grape fresh-keeping paper: CN1279020. 2001-2001-2010.
57 Wu, Xuemin, Guo, Xinyu, Wang, Yin. A new type of automatic release fresh-keeping medicine with fresh-keeping function of fruits, vegetables and flowers: CN109527074A.2019-2103-2129.
58 Lim, L.K., Lee, K.E. Packaging systems for the control of relative humidity of freshfruits, vegetables and flowers with simultaneous regulation of carbon dioxide and oxygen: WO2008076075A1. 2008-2106-2126.
59 Tian, S. and Xiaoming, Z. (2018). Research progress of electromagnetic shielding materials. *Journal of Textile Science and Engineering* 35 (2): 118–122.
60 Yingjian, L., Ming, Y., and Yi, G. (2019). Shielding mechanism and current status analysis of electromagnetic shielding materials. *Value Engineering* 1: 159–162.
61 Wenjun, Z. (2017). *Preparation and Properties of Carbon Fiber Paper Based Shielding Materials*. Xi 'an: Shaanxi University of Science and Technology.
62 Lin, L. and Dong, Z. (2015). Research progress of electromagnetic shielding materials. *Functional Materials* 46 (3): 3016–3022.
63 Yanhui, D. (2009). *Development and properties of high performance carbon fiber shielding paper*. Xi 'an: Shaanxi University of Science and Technology.
64 Luong, N.D., Pahimanolis, N., Hippi, U. et al. (2011). Graphene/cellulose nanocomposite paper with high electrical and mechanical performances. *Journal of Materials Chemistry* 21 (36): 13991–13998.
65 Lee, T.W., Lee, S.E., and Jeong, Y.G. (2016). Highly effective electromagnetic interference shielding materials based on silver nanowire/cellulose papers. *ACS Applied Materials & Interfaces* 8 (20): 13123–13132.
66 Linxin, Z. (2008). Development and properties of carbon fiber conductive shielding paper. In: Xi 'an: Shaanxi University of Science and Technology.
67 Song, W.L., Fan, L.Z., Cao, M.S. et al. (2014). Facile fabrication of ultrathin graphene papers for effective electromagnetic shielding. *Journal of Materials Chemistry C* 2 (25): 5057–5064.
68 Cao, W.T., Chen, F.F., Zhu, Y.J. et al. (2018). Binary strengthening and toughening of MXene/Cellulose nanofiber composite paper with nacre-inspired structure and superior electromagnetic interference shielding properties. *ACS Nano* 12 (5): 4583–4593.
69 Huaping, Z.H.E.N.G. and Jun, L.I.U. (2016). The preparation method of titanium dioxide for decorative paper coated with phosphorus-aluminum compound[J]. *Modern Industrial Economy and Informationization* 119 (11): 59–61. 66.
70 Fan, Huiming, Ge, Lifang. Production process of high speed impregnated printing decorative base paper: CN201110047678.1. 2011-2102-2128.

71 Changsheng, J. (2006). Research on wear resistance of laminate wood floor and domestic wear resistant paper. *Forest Products Industry* 33 (5): 25–29.

72 Sheng'en, W., Cuiying, L., and Xiongzhen, W. (2010). Application of flame retardants in papermaking. *Hangzhou Chemical Industry* 40 (2): 23–26.

73 Shulan, S. and Fuwang, H. (2015). *Analysis and Detection of Pulp and Paper Making*, 194–196. Beijing: China Light Industry Press.

74 Janfeng, X., Ling, L., and Jiahao, Y. (2015). Preparation and characterization of self-emulsified styrene acrylate surface treatment agent for decorative base paper. *Scientia Silvae Sinicae* 51 (3): 116–123.

75 Beihai, H. and Meiyun, Z. (2014). *Principle and Engineering of Papermaking*, 62–64. Beijing: China Light Industry Press.

76 Zaili, Z., Chunliang, L., and Xin, C. (2017). Application of titanium dioxide in thin sheet special printing paper products. *Materials Manufacturing and Application* 3: 21–24.

77 Zhirong, L. (2015). Effect of titanium dioxide on optical properties of decorative paper. *Paper Chemicals* 46 (5): 37–40.

78 Chengyi, C., Yugang, Y., and Aihua, S. (2012). Study on photofastness of TiO_2/aluminum phosphate composite pigment. *Ningbo Chemical Industry* 4: 23–30.

79 Ran, Y., Lanfeng, H., Zhong, L. et al. (2017). Preparation and application of titanium dioxide and kaolin composite fillers. *China Paper Making* 36 (11): 37–43.

80 Ke, W. and Shiju, Y. (2005). Application and development of filter aid and particle retention system for papermaking. *Hubei Paper Making* 1: 15–18.

81 Wendel, K., Schwerzel, T., Hirsch, G. Aqueous polymer dispersions: US 5358998. 1994-1910-25.

82 Albinsson, C.J., Swerin, A., and Ödberg, L. (1995). Formation and retention during twin-wire blade forming of a fine paper stock. *TAPPI Journal* 78 (4): 121–128.

83 Junchao, W., Nannan, C., Lijun, W. et al. (2018). Comparison of application effects of two representative particle retention and filtration systems. *Hangzhou Chemical Industry* 48 (4): 44–47.

84 Ruiheng, L. and Shiyu, F. (2007). Influencing factors of decorative base paper quality and production process control. *Shanghai Paper Making* 38 (5): 18–21.

85 Yuanyuan, B., Yanhao, L., Chunli, Y. et al. (2016). Research progress of environmental protection wet strength agents for papermaking. *Chinese Journal of Paper Making* 31 (4): 49–54.

86 Cui, M., Wen, L., Fanjin, M. et al. (2019). Development and application of high-efficiency and environmentally friendly PAE wet-strength agent new products. *Paper Chemicals* 31 (6): 19–22.

87 Alemder, A. and Sain, M. (2008). Isolation and characterization of nanofibers from agricultural residues – wheat straw and soy hulls. *Bioresource Technology* 99 (6): 1664–1671.

88 Wang, A.J., Wang, L.J., Jiang, J.M. et al. (2019). Reinforcing paper strength by dual treatment of a cationic water-soluble polymer and cellulose nanofibril. *Paper and Biomaterials* 4: 34–39.

89 Ahola, S., Österberg, M., and Laine, J. (2008). Cellulose nanofibrils – adsorption with poly (amideamine) epichlorohydrin studied by QCM-D and application as a paper strength additive. *Cellulose* 15 (2): 303–314.

90 Yan, Y., Shuyuan, H., and Chunli, Y. (2013). Enhancement effect of PAE/NCC binary system on paper. *Chinese Journal of Paper Making* 28 (2): 8–11.

91 Wagberg, L., Forsberg, S., Johansson, A. et al. (2002). Engineering of fibre surface properties by application of the polyelectrolyte multilayer concept. Part I: modification of paper strength. *Journal of Pulp & Paper Science* 28 (7): 222–228.

92 Jun, L. (2006). Analysis on production technology of surface wear resistant paper. *Tianjin Paper* 2: 8–12.

93 Rongxing, W. (2018). *Study on Modification of Melamine Formaldehyde Resin by Tri (2-hydroxyethyl) Isocyanurate for Impregnated Film Paper*. Beijing: Chinese Academy of Forestry.

94 Zhijun, H. (2019). *Practical Technical Course of Special Paper*, 130–131. Beijing: China Light Industry Press.

95 Andreas, K., Primoz, P., Sergej, M. et al. (2010). On the performance of a melamine-urea-formaldehyde resin for decorative paper coatings. *European Journal of Wood and Wood Products* 68: 63–75.

96 Ge Lifang. Production technology of decorative base paper with low ink consumption: CN201110047677. 2011-2102-2128.

97 Keben, T., Bao, Z., Xia, L. et al. (2013). Development of new Decorative base paper production process. *East China Paper* 44 (3): 32–34.

98 Zhengjian, Z., Qilian, Z., Mingzhi, Z. et al. (2018). Effect of coating on surface properties and printability of wood-based board decorative base paper and its mechanism. *Scientia Forestry Sinica* 54 (1): 111–120.

Index

a

Abaca fiber 135–136
ABO/RhD blood type diagnostic paper 266
activated carbon fiber 109, 113, 140–141
active modified atmosphere 316
active packaging materials 317
adhesive composite 237
aerogel thermal insulation materials 215
air filtration
  diffusion effect 143–144
  electrostatic effect 145
  gravity effect 144–145
  inertia effect 142–143
  interception effect 142
air pollution degradation 303–305
alumina fiber (AF) 112–113
amino silicone oil 290–291
animal fibers
  chitin fiber 141–142
  silk fibers 141
  wool fibers 141
anthocyanins 318, 324
antibacterial preservation paper 312–314, 317–326
aramid fiber (AF) 100–102, 110, 115, 116, 125–127, 307, 308
aramid fibrid 117–118, 124, 125
aramid nanofibers (ANF) 118–119
aramid paper substrates 307
aromatic polyamide (aramid) fiber 224
asbestos fibers 140, 208
ASSURED criteria 250, 251
ASSURED standards 265
automotive filtration 6, 201, 240–241
AVEBE 11

b

bactericidal and bacteriostatic 306
basalt fiber (BF) 110–111
base paper 51, 58, 149, 160, 360
battery separator paper 91–92, 243–244
bedside testing 249
Benders Paper Cup Company 219
bioactive paper 249–253, 257
Brownian motion 143, 144, 227, 231

c

cake filtration 146, 230–231
calcined clay 359
calendering 344, 361
Canada's CelluForce company 70
carbon fibers (CF) 1, 9, 99, 109, 115, 140–141, 336
carbon nanotubes 204–205, 334, 336
carboxylated nanocellulose 294
carboxymethylation methods 294
cash register, application of 354
Cassie-Baxter model 167, 168
cationic etherified nanocellulose 293
cationic quaternary ammonium salt softeners 289
cationic starch 3, 13, 43–48
cationization 23, 37–39, 69
cellulose-inorganic complex 185–186
cellulose nanocrystals (CNC) 69, 71–72

Paper-Based Functional Materials: Preparation and Applications, First Edition.
Xianping Yao, Genrong Chen, Zhen Yao, and Liping Zheng.

cellulose nanofibrils (CNF) 69, 72–73
ceramic fiber 112–113, 211, 215, 304
Changchun Dacheng Industrial Group Company LTD. 12
chemical analysis filter paper 155
  chromatography paper 160–161
  classification of 156
    qualitative filter paper 156–158
    quantitative filter paper 156–161
  diffusion-type and dispersion-type filter papers 159–160
  pH test functional material 160
  Shuangquan brand 155–156
chemical etching method 256–257
chemical modification method 176, 177, 184
  to improve both water and oil absorption 187
  to improve oil absorption 186
    esterification treatment 186–187
    graft modification 187
    lye treatment 186
  to improve water absorption 184
    cellulose 185–186
    cellulose-inorganic complex 185–186
    etherification and cross-linking 184
    graft copolymerization 184–185
chemical reaction type 353, 356
chemiluminescence method 258–259
China Starch Industry Association 12
chitin fiber 141–142
chitosan fibers 142
coating combined treatment technology 328–329
coffee ring phenomenon 266, 268
colorimetric analysis 260
combustion process 204–206
commercial products 95, 247
conduction band 300
conductive fabric shielding materials 332, 333
continuous fiber-reinforced thermoplastic composite (CFRTP) products 9
continuous flow and pipeline-based green manufacturing new process 20–24
corundum 349
cotton fiber 135, 137–139, 156, 157, 185
coupling agent modification 186
COVID-19 pandemic 245
cream moisturizing paper 283
cream paper
  definition of 283
  Gravure roller coating method 295–296
  high-precision spraying method 296
  human safety 284
  moisturizing 287
  odorlessness 284, 288
  online spraying method 295
  rotary spraying method 295
  safety testing of 285
  softness 286–287
crystal phase defects 302
cutting method 252, 255

d

decorative base paper
  ingredients 350
  kaolin 345–346
  overview 341–343
  paper making 350
  pulp crushing 349–350
  quality requirements of 343–344
  retention aid 346–347
  surface coating production 351–352
  titanium dioxide 344–345
  wear-resistant materials 348–349
  wet strength agent 347–348
deep-bed filtration 230, 231
depth filtration 146, 231
developer, chemicals 358
differential fibers 115–119
diffusion deposition 227
diffusion effect 143–144, 160
digital transfer printing technology 39
direct casting method 216–217
direct interception 227

direct-thermal label 354, 355
drawing method 107, 252, 254
drying 18, 19, 21, 42, 53, 56, 361
dry process technology 19, 20
DuPont Company 100, 102, 112

e

Ecosynthetix 35
edible coating materials 316–317
electrical insulation paper 125–127, 213
electrochemical method 257–259
electrochemiluminescence (ECL) method 259
electromagnetic compatibility 330, 331, 333
electromagnetic shielding paper
  applications of 338–340
  conductive fabric shielding materials 333
  development of 333–334
  filled composite shielding materials 333
  forming techniques and functional characteristics of 336–338
  intrinsically conductive polymer shielding materials 333
  overview 330–331
  preparation of 336
  shielding mechanism of 331–332
  surface conductive shielding materials 333
electrospinning 153, 154, 169, 173, 238, 308
electrospun nanofibers 239
electrostatic deposition 227
electrostatic effect 145
energy-storing paper 9
engine filtration paper
  applications 147
  development of 147–148
  materials 148–151, 153, 154
    areas of application 149
    processing methods for filter manufacturers 149
    production processes 149
environmentally friendly paper coating adhesive 35
  modification of starch through cationization 37, 39
  paper surface chemicals 33
  research and development 35
  starch-based coating adhesives 33
  ultra-fine grinding starch particles (UGSP) 34
  viscosity reduction process 36
  viscosity stability of starch 36–37
enzyme-converted starch 15, 50
enzyme-linked immunosorbent assay (ELISA) 260, 261
esterification 36–37, 186–187
esterification treatment 186–187
ethylene removal preservation materials 314–315
expanded graphite-based flame retardants 204
expert group evaluation method 286

f

face masks 244–246
fatty acid derivative softener 289–290
fiber dispersion 114–116, 119–121, 233
filled composite shielding materials 332, 333
filter paper 212
filter paper-based functional materials
  applications of 147
  long fibers 135–142
filtration 142
  air 142–145
  liquid 145–147
filtration materials 139, 141, 148, 153, 154, 223, 237–241
filtration paper 147–154, 223, 226
First International Conference on Nanocellulose Materials 70
flame retardants for paper-based functional materials
  application of 210
    electrical insulation paper 213

flame retardants for paper-based functional materials (*contd.*)
- filter paper 212–213
- honeycomb paper 211–212
- packaging materials 210–211
- prospects 213–214
- wallpaper 210
- classification of 201
  - inorganic flame retardants 202–204
  - nano flame retardants 204–205
  - organic flame retardants 202
- combustion process 205–206
- preparation of 207
  - impregnation 209–210
  - internal addition 208–209
  - surface coating 209
- thermal decomposition of fibers 206–207

flexible supercapacitors 308
flexographic printing 252, 255
fluorescence method 257–259
foam forming technology 8
food additives 260, 263
forest resources
- China's paper production 16
- energy conservation and emission reduction 17–18
- wood fiber raw materials 16–17

forward osmosis membrane technology 238
Fourdrinier paper machine 56, 108
fresh-keeping cardboard 329
fruit and vegetable preservation paper
- active packaging materials 317
- antibacterial preservation paper materials 317–326
- definition of 309–310
- development of 310–314
- edible coating materials 316–317
- ethylene 314–315
- modified atmosphere packaging materials 315–316

fuel filter paper 181

g

gas filtration and separation
- filtration process 227–228
- mechanism 227

gas-phase flame retardation mechanism 206
glass fiber (GF) 106–108, 111, 112, 140, 145, 226
graft copolymerization 184–185, 187
graft modification 187
graphene 205
graphene materials 335
gravitational deposition 227
gravity effect 144–145
Gravure roller coating method 295–296
Guidelines for National Priority Development in High-tech Industrialization 13

h

halogen-based flame retardants 201, 202, 207
handfeel value 286
Handle-O-meter method 286
Hangzhou Papermate Science & Technology Co., Ltd. 12, 13, 17, 18
heat bonding composite 237–238
heat exchange flame retardation 206, 207
heat insulation material 358–359
heat-seal tea bag filter paper
- materials 162–163
- process equipment and flow 163

heat transfer printing paper 39–43
high-efficiency air filters (HEPA) 241, 243
high-performance fibers 99
- classification of 100
- inorganic fiber 99
- organic fibers 99

high-performance inorganic fiber
- basalt fiber 110–112
- carbon fiber 108–110
- ceramic fiber 112–113
- glass fiber 106–108

high-performance organic fibers 99
  aromatic polyamide fibers 100–102
  polyimide fiber 102–104
  poly-p-phenylenebenzobisoxazole (PBO) fiber 104–105
  ultra-high molecular weight polyethylene (UHMWPE) fiber 105–106
high-precision spraying method 296
High-Speed Flying Knife Mixer 21
high-temperature jet cooking technology 59–62
hollow plastic pigment 358, 359
hollow spheres 218, 359
honeycomb paper 211–212
household paper, applications in 92–94
hyaluronan 294
hyaluronic acid 294–295
hydrated silicate China clay 345
hydrophilic and lipophilic paper-based functional materials
  application
    facial oil blotting paper 189–190
    humidity regulation cling paper 188–189
    industrial wiping paper 187–188
    kitchen absorbent oil-absorbing paper 188
    medical absorbent paper 189
  preparation
    chemical modification method 184–187
    physical modification method 182–183
    surface coating method 183–184
hydrophilic and oleophobic paper-based functional materials
  application
    grease-proof paper 190–191
    greaseproof paper for food packaging 192–193
    oil repellent 191–192
    oil-water separation membrane 193
    prospects 194–195
  preparation 190
hydrophobic and lipophilic paper-based functional materials
  application 180
    fuel filter paper 181
    oil-water separation 180–181
  preparation 179–180
hydrophobic and oleophobic paper-based functional materials
  application 177
    intelligent response to functional materials 178, 179
    microfluidic device 178
    oil-water separation 177, 178
    self-cleaning 178
  preparation 171–173
    chemical modification method 176–177
    nanoparticle deposition method 176
    physicochemical modification method 173–174
    surface coating method 175–176
hydroxyalkyl starch 3, 13, 15

i

immunoblotting 260
immunodiffusion 260
immunofluorescence 260
impregnation method 151, 173, 178, 179, 209–210, 219
inclined screen formers 233
inertia effect 142–143
inertial deposition 227
infiltration combined treatment technology 327, 328
inkjet printing method 252, 256
innovative crosslinking formulation design 42–43
innovative production processes and equipment 26
innovative synthetic process technology 42
inorganic fiber
  asbestos fibers 140
  carbon and active carbon fibers 140–141

inorganic flame retardants
  expanded graphite-based 204
  metal hydroxide flame retardants 202–203
  nitrogen-based 204
  phosphorus-based 203–204
inorganic minerals 4
inorganic phosphorus-based flame retardants 203–204
inorganic thermal insulation materials 215
in-pulp addition 349
instant testing 249
interception effect 142, 143
intrinsically conductive polymer shielding materials 332, 333
ion accumulation 350

j
Japan's Nippon Paper 70, 303

k
kaolin 345, 346
kaolin clay 15, 359
Kevlar® 100, 101
Kozo bark fiber 137

l
labels, application of 354–355
Lab on a Chip (LOC) 248
lateral flow assays (LFAs) 247
Leuco dye 353, 356–358
liquid filtration
  principle of 145–146
  types
    cake filtration 146
    depth filtration 146–147
liquid filtration and separation
  filtration types 228–230
  fundamental of filtration 230–231
long fibers for filter paper 135
  from non-plants 138–142
  from plants 135–138
lubricants 58, 359
lye treatment 186
Lyocell fiber 140

m
Manila hemp fibers 136
mechanical bonding composite method 238
melt-blown nonwoven fabrics 244
membrane filtration 145, 229, 243
metal hydroxide flame retardants 202–203
metal thermal insulation materials 215
microfluidic device 176, 178, 249
Microfluidic Paper-based Analytical Devices (μPADs) 248
microporous filtration 229
Miniaturized Total Analysis Systems (μ-TAS) 248
mixed crystal effect 302
modified atmosphere packaging materials 315–316
molecular structure design 25, 27
mulberry fiber 136–137

n
nanocellulose (NC) 5, 69
  applications of 74
  characteristics of 73
  classification of 71
    bacterial nanocellulose 73
    cellulose nanocrystals 71–72
    cellulose nanofibrils 72
  development of pretreatment technologies 69
  domestic technology companies 71
  main characterization methods 80–82
  microfibrillated cellulose (MFC) 73–74
  modification techniques 79–80
  paper barrier 87–91
  paper filling 86–87
  in paper making 82–83
  paper reinforcement 83–85
  preparation 74
    biological methods 79
    chemical methods 78–79
    mechanical methods 74–78
  problems and prospects 94–95

sulfuric acid hydrolysis, CNC based 70
nanocellulose derivatives 292, 293
nano flame retardants
carbon nanotubes 204–205
graphene 205
nano silicon dioxide 205
nanoindentation 307
nano silica dioxide 205
nanotechnology 176, 194, 215, 334
nano TiO_2 photocatalytic material
air pollution degradation 303–305
aramid fiber 307–308
bactericidal and bacteriostatic 306
crystal phase effect 302
crystal size 301
flexible supercapacitors 308
mixed crystal effect 302
principle of 300–301
structure of 299–300
superhydrophobic 306–307
National Key Technological Renovation Project 18
National Starch and Chemical Company (NSCC) 11
natural biomass materials 339
natural cellulose 69, 74, 77, 79, 83, 184, 201
Neisseria meningitidis 261
nitrogen-based flame retardants 201, 204, 206
Nomex fibers 224–225
Nomex® 100, 101
non-heat-seal tea bag filter paper
materials 161–162
process equipment and workflow 162
non-plant fiber long fibers
inorganic fibers 140–141
synthetic fibers 138–140
non-wood fiber papermaking 24–26
nonwoven fabric composite technology
adhesive composite 237
heat bonding composite 237–238
mechanical bonding composite method 238–240
nonwoven fabrics 141, 181, 232–234, 239, 244, 302, 326
N-type semiconductors 297

o

odorlessness 284, 288
base paper 284
cream impact 284, 286
water quality 284
oil-water separation 171, 177–178, 180–181, 190, 193–195, 223
Oji Paper, Japan 70
online spraying method 295
organic flame retardants
halogen-based 202
organic phosphorus-based 202
organic phosphorus-based flame retardants 202–204
organic thermal insulation materials 214, 215
organometallic complexes 298
original fibrous cellulose 69
oxidized starch 3, 13, 15, 27–31, 50, 58, 162, 328

p

paper-based analytical detection technology
application cases of 264
blood typing paper chip and paper-based structure 265–268
composite substrate for 272–273
heavy metal ions in water based on optical transmittance density 268
high-throughput detection of antibiotics in milk 268–271
chemical modification method
chemical etching method 256–257
inkjet printing method 256
paper modification method 257
photolithography method 255–256
plasma treatment method 256
detection methods of 257
chemiluminescence method 258–259

paper-based analytical detection technology (*contd.*)
colorimetric method 260
electrochemical method 258
electrochemiluminescence method 259
fluorescence method 259
development of 247–251
in environmental monitoring 262–263
fabrication methods for 251–253
in food safety control 263
bacteria 264
food additives 263
pesticide residues 263–264
future prospects of 273, 274
in medical diagnosis 260
analysis of cells 262
analysis of nucleic acids 261
analysis of proteins 260–261
colorimetric method 262
for microfluidic detection chips and traditional materials 250
physical barrier/deposition method
cutting method 255
drawing method 254
flexographic printing method 255
physical inkjet etching method 254
screen printing method 255
wax patterning method 253–254
paper-based functional materials 124, 125
aramid fiber based electrical insulation paper 125–127
aramid/mica insulation paper 127–129
definition of 1–2
development trends 8
energy-storing paper 9
foam forming technology 8
new degradable paper mulch 9
new products development 8–10
polystyrene nanocomposite coatings 9
Sharklet microstructure 9
single-use plastics 10
stretchable paper technology 8
economic and social value 6–8
high-temperature resistant polyimide fiber 130–131
inorganic minerals 4
natural polymers 3
para-aramid honeycomb core material 129–130
petroleum-based organic polymers 4–5
properties and applications of 5–6
raw materials of 2
paper-based microfluidic detection chip technology 249
paper-based strip tests 248
paper forming 121–123, 130, 153, 253, 255, 304
papermaking 350
China's 13
European 12
in Japan 12
starch-based 13
papermaking chemicals 1, 2, 6, 7, 25, 27, 33, 44, 48, 50, 64, 82, 88, 94, 287, 304, 335
papermaking combined preparation technology 327–328
papermaking fillers online surface modification 62–64
papermaking method 9, 217
papermaking technology 8, 48, 82, 103, 119, 183, 304, 307, 327
paper modification method 257
paper packaging materials 191, 210, 211
paper retention agents 346
paper strengthen 123–124, 348
para-aramid fibers 101, 119, 307
para-aramid pulp (PPTA-pulp) 115–117
passive modified atmosphere 316
pathogenic bacteria 258, 264, 306
pesticide residues 263–264
petroleum-based organic polymers 4–5
phosphate ester starch 13, 15, 34, 47
photocatalytic materials
energy 297–298

environmental governance 298–299
overview 296–297
sterilization and antibacterial effects 299
photodegradation 102, 308
photolithography method 252, 255–256
physical inkjet etching method 254
physical modification method 182–183
physical reaction type 353
Plant-based long fibers
Abaca fiber 135–136
cotton fiber 137–138
Kozo bark fiber 137
mulberry fiber 136–137
softwood fiber 138
plasma treatment method 252, 256
point-of-care testing (POCT) products 249, 251
polyacrylonitrile (PAN) fiber 224, 226
polyester fiber 106, 115, 139, 183, 187, 188, 226, 238
polyester (PET) fiber 226
polyimide (PI) fiber 102, 103, 225
polymer softener
mechanism of action 291
particle size impact 291–292
polyphenylene sulfide (PPS) fiber 225
poly-p-phenylenebenzobisoxazole (PBO) fiber 104–105
polypropylene (PP) fiber 138, 139, 225
polystyrene nanocomposite coatings 9
polytetrafluoroethylene (PTFE) fiber 105, 223–224
polyvinyl chloride (PVC) 191, 210, 328
pre-coat layer 361
protect material 360
pulp crushing 349–350

r
recycling waste paper 32
retention aid 3, 48, 62, 63, 208, 217, 346–347, 352
retention assistance system 352
retention rate analysis 65
rotary screen former 233, 234
rotary spraying method 295
Rotoformer cylinder machine 122

s
sandwich combined design technology 329–330
Schelkunoff formula 332
screen filtration 229
screen printing method 252, 255
self-cleaning 4, 170, 177, 178, 306, 307
semiconductor band theory 300
semi-dry process technology 18, 20
sensitizers, chemicals 358
Shuangquan brand filter paper 155, 156
silicon carbide fiber (SCF) 112, 113
silicone-based softener 290
silicone quaternary ammonium salt 290
silk fibers 141
sodium hyaluronate 294
softener SME-4 289, 290
soft moisturizing paper 283
softwood fiber 138
solar energy 297, 298, 309
sol-gel method 113, 173, 175, 180, 308
solid-liquid filtration 228, 229
solvent method 19
spectrophotometry 29, 307
spunbond fiber web 244
starch-based functional products
alternative wood fiber raw materials 16–17
application of 13–15
application technologies of 46, 47
China's paper industry 65–67
coating application technology 57–58
comprehensive economic and technical evaluation 64–65
green technology applications and equipment 59–64
surface sizing technology 49–53
wet-end 47–49
continuous flow and pipeline-based green manufacturing new process 20–24

starch-based functional products (*contd.*)
  dry process 19–20
  energy conservation and emission reduction 17–18
  in maintaining forest resources 15–16
  for papermaking 13
    environmentally friendly paper coating adhesive 33–39
    for heat transfer printing paper 39–43
    high-retention surface sizing starch 27–31
    interference-resistant specialized enhancer for recycled paper 32–33
    non-wood fiber 24–27
  preparation methods of 18
  semi-dry process technology 20
  starch and 11–13
  wet processing technology 18–19
starch modification 11
steady-state filtration 228
storage-oxidation catalysts 304
strawberries
  characteristics of 318, 322
  enzyme activity characteristics of 324
  nutritional components of 322–324
  surface microorganisms of 324–326
stretchable paper technology 8
suction filtration self-assembly process 335
sulfuric acid hydrolysis 69, 70, 79
superhydrophobic 4, 167, 169, 171–181, 306–307
surface coating method 173, 175–176, 183–184, 209, 217–218
surface coating production 351–352
surface conductive shielding materials 332, 333
surface modification 62–64, 77, 79, 101, 214, 251
surface sizing 27
  application of 31–32
  environmentally friendly properties of 30
  improvement of printing effects 31
  and oxidized starch 27–31
  during recycled paper repulping 29
  starch sizing amount on 29
surface sizing technology 50
  coating application technology 57–58
  drawback of additives 50
  film sizing 52
  functional products 49–50
  Gate roll surface sizing 52
  interlayer application technology 53, 54
  selection of agents 50
  selection of equipment 50–52
  shared chemical technology 52–53
  size press application 51
  spray technology 55–57
surface wettability of paper-based materials
  hydrophilic and lipophilic interface 169–170
  hydrophilic and oleophobic interfaces 170–171
  hydrophobic and lipophilic interface 169
  hydrophobic and oleophobic interface 169
  surface of the object 167–168
synthetic fibers
  Lyocell fiber 140
  polyester fiber 139
  polypropylene fiber 139
  viscose fiber 139
  Vylon fiber 138–139

t

tea bag filter paper
  preparation process of
    heat-seal tea bag filter paper 162–163
    non-heat-seal tea bag filter paper 161–162
    paper-based functional materials 164
    technical parameters 163–164

principle of 161
Teijin Company 102
TEMPO catalytic oxidation method 70
TEMPO oxidation 70, 77, 294
thermal coat 360, 361
thermal copy paper 353
thermal functional materials 356–360
thermal insulation paper based functional materials
applications of 218
in daily life 219
in food packaging 219
insulation thermal paper used in power sector 218–219
classification of 214
aerogel thermal insulation materials 215
inorganic thermal insulation materials 215
metal thermal insulation materials 215
organic thermal insulation materials 214
preparation of 216
direct casting method 216–217
papermaking method with mixed fibers 217
surface coating method 217–218
wet end addition 217
thermal transfer principle 215–216
thermal printing 358
mechanism 353–354
thermal recording paper 353, 354
thermal-sensitive paper
cash register, application of 354
chemicals 356–358
development 353
functional materials 356
labels, application of 354–355
manufacturer of 355
quality and requirements 355–356
thermal printing mechanism 353–354
tickets, application of 355
3D paper-based microfluidic chip 249
tickets, application of 355
tissue softness analyzer (TSA) method 286
titanium dioxide (TiO_2) 219
decorative base paper 344–349
packing 352
tobacco sheet 91
top coat layer 361
transmission line theory approach 331

u

ultra-high molecular weight polyethylene (UHMWPE) fiber 99, 105–106
umami of papermaking 13
unsteady-state filtration 228

v

vacuum drying functional starch production 18
valence band 300
Van der Waals deposition 227
viscose fiber 138–140
viscosity reduction techniques 36, 37
Vylon fiber 138–139

w

water filtration 15, 87, 94, 114, 243
wax patterning method 253–254
wear-resistant materials 344, 348–349
Wenzel model 167–169
wet end addition method 217
wet-end application technology 47–49
wet-laid forming technology 113
differential fibers 115–119
fiber dispersion 119–121
high-performance fibers 114, 119
paper forming 121–123
paper strengthen 123–124
wet-laid nonwoven fabric
fiber bonding and reinforcement technology 234–236
formation equipment of 233–234
fundamental of 232–233
production process 232
types of fibers 232

wet-laid paper machine 336
wet process technology 18
wet strength agent 188, 284, 347–348, 350, 351
wool fibers 141
World Health Organization (WHO) 250, 298

y

Yantai Tayho Advanced Materials Co., Ltd. 101, 115
Young's model 167

z

Zylon® 104